international
AIR POWER
REVIEW

AIRtime Publishing
United States of America • United Kingdom

international AIR POWER REVIEW

Published quarterly by AIRtime Publishing Inc.
US office: 120 East Avenue, Norwalk, CT 06851
UK office: CAB International Centre, Nosworthy Way,
Wallingford, Oxfordshire, OX10 8DE

© 2002 AIRtime Publishing Inc.
B-24 cutaway © Aerospace Publishing Ltd
Rafale cutaway © Mike Badrocke/Aviagraphica
Photos and other illustrations are the copyright of their respective owners

Softbound Edition ISSN 1473-9917 / ISBN 1-880588-38-2
Hardcover Deluxe Casebound Edition ISBN 1-880588-39-0

Publisher
Mel Williams

Editor
David Donald
e-mail: airpower@btinternet.com

Assistant Editors
John Heathcott, Daniel J. March

Sub Editor
Karen Leverington

US Desk
Tom Kaminski

Russia/CIS Desk
Piotr Butowski, Zaur Eylanbekov
e-mail: zaur@airtimepublishing.com

Europe and Rest of World Desk
John Fricker, Jon Lake

Correspondents
Argentina: Jorge Felix Nuñez Padin
Australia: Nigel Pittaway
Belgium: Dirk Lamarque
Brazil: Claudio Lucchesi
Bulgaria: Alexander Mladenov
Canada: Jeff Rankin-Lowe
France: Henri-Pierre Grolleau
India: Pushpindar Singh
Israel: Shlomo Aloni
Italy: Luigino Caliaro
Japan: Yoshitomo Aoki
Netherlands: Tieme Festner
Romania: Danut Vlad
Spain: Salvador Mafé Huertas
USA: Rick Burgess, Brad Elward, Peter Mersky, Bill Sweetman

Artists
Mike Badrocke, Chris Davey, Zaur Eylanbekov, Keith Fretwell,
Mark Rolfe, Andrei Salnikov, John Weal, Iain Wyllie

Designer
Zaur Eylanbekov

Controller
Linda DeAngelis

Origination by Chroma Graphics, Singapore
Printed in Singapore by KHL Printing

International Air Power Review is published quarterly in two editions (Softbound and Deluxe Casebound) and is available by subscription or as single volumes. Please see details opposite.

Acknowledgments
We wish to thank the following for their kind help with the preparation of this issue:

Martin Bowman
Paul E. Eden
Del Holyland, Martin-Baker
Phil Jarrett
Tom Newdick
Stan Piet
Clive Richards, AHB
Michael Stroud

The author of the Rafale article would like to thank Yves Robins, Jean Camus, Yves Kerhervé, Philippe Rebourg, and Pierre Delestrade for their kind help, especially in organising the familiarisation flight.

The author of the 320 Squadron feature would like to thank LTZ G. Walraven and LTZ v/d Haak of the MARVO, LTZ2 A. Lelijveld of VIPCO Valkenburg, KLTZ F. Noom (former commander of 320 Sqn), the Marine AudioVisuele Dienst Sectie Fotografie, Institute for Maritime History, Royal Netherlands Navy and especially Commandeur (retd.) H. van der Kop.

Correction: The five cabin photos that appeared on page 21 of IAPR Vol. 2 should have been credited to Marco P.J. Borst. We regret the oversight.

The editors welcome photographs for possible publication but can accept no responsibility for loss or damage to unsolicited material.

Subscriptions & Back Volumes

Readers in the USA, Canada, Central/South America and the rest of the world (except UK and Europe) please write to:
AIRtime Publishing, P.O. Box 5074, Westport, CT 06881, USA
Tel (203) 838-7979 • Fax (203) 838-7344
Toll free 1 800 359-3003
e-mail: airpower@airtimepublishing.com

Readers in the UK & Europe please write to:
AIRtime Publishing, RAFBFE, P.O. Box 1940,
RAF Fairford, Gloucestershire GL7 4NA, England
Tel +44 (0)1285 713456 • Fax +44 (0)1285 713999

One-year subscription rates (4 quarterly volumes), inclusive of shipping & handling/postage and packing:
Softbound Edition
USA $59.95, UK £48, Europe £56/EUR 89, Canada Cdn $93,
Rest of World US $79 (surface) or US $99 (air)
Deluxe Casebound Edition
USA $79.95, UK £68, Europe £76/EUR 123, Canada Cdn $122,
Rest of World US $99 (surface) or US $119 (air)

Two-year subscription rates (8 quarterly volumes), inclusive of shipping & handling/postage and packing:
Softbound Edition
USA $112, UK £92, Europe £105/EUR 169, Canada Cdn $174,
Rest of World US $148 (surface) or US $188 (air)
Deluxe Casebound Edition
USA $149, UK £130, Europe £146/EUR 236, Canada Cdn $229,
Rest of World US $187 (surface) or US $227 (air)

Single-volume/Back Volume Rates by Mail:
Softbound Edition
US $16, UK £12, Europe £12/EUR 19, Cdn $25 (plus s&h/p&p)
Deluxe Casebound Edition
US $20, UK £17, Europe £17/EUR 27, Cdn $31 (plus s&h/p&p)

All prices are subject to change without notice.
Canadian residents please add GST. Connecticut residents please add sales tax.

Shipping and handling (postage and packing) rates for back volume/non-subscription orders are as follows:

	USA	UK	Europe	Canada	ROW (surface)	ROW (air)
1 item	$4.50	£3	£4/EUR 6.40	Cdn $7.50	US $8	US $16
2 items	$6.50	£5	£6/EUR 9.60	Cdn $11	US $12	US $27
3 items	$8.50	£7	£8/EUR 12.80	Cdn $14	US $16	US $36
4 items	$10	£9	£10/EUR 16	Cdn $16.50	US $19	US $46
5 items	$11.50	£11	£12/EUR 19.20	Cdn $19	US $23	US $52
6 or more	$13	£12	£13/EUR 20.80	Cdn $21.50	US $25	US $59

Volume Four
Spring 2002

CONTENTS

MAJOR FEATURES PLANNED FOR VOLUME FIVE
Focus Aircraft: Lockheed Martin F-22 Raptor, **Warplane Classic:** Saab J 35 Draken, **Technical Briefing:** Kamov Ka-50
'Hokum' family, **Air Power Analysis:** India, **Air Combat:** Gunships in Vietnam: Part 2, **Variant File:** Dornier Do 217

PROGRAMME UPDATE

Eurofighter Typhoon

Following an extensive lay-up, Eurofighter EF2000 prototype DA4, one of the two development aircraft based at BAE Systems' Warton facility in England, recently re-entered the flight test programme. During the lay-up the aircraft's avionics and power-generation systems were upgraded and it completed ground trials associated with the first phase of the defensive aids sub-system (DASS) at the company's electronic warfare test facility at Warton.

The aircraft's first flight back in the programme was used to verify a variety of systems including engines, radar, AMRAAM missile integration, and ground proximity warning system (GPWS). Subsequent flights will involve flight trials to support the fighter's initial clearance into service next year. Initial tests will concentrate on the weapon system integration between the AMRAAM and ASRAAM missiles, radar and GPWS. Subsequent flights will include assessment of the DASS, and live AMRAAM flight trials. The AMRAAM flight trials will lead up to the first fully guided firing of an AMRAAM against an airborne target, at the Benbecula deep-sea range off northwest Scotland early in 2002.

In Germany, the Luftwaffe is reassessing original plans for only 40 of its planned 180 Eurofighters to be equipped for ground-attack

With the UK's Chief of the Air Staff aboard, DA 4 roars into the air from its Warton base in December 2001 following an extensive refit. After initial system trials, flight assessment and analysis of the DASS, including on-range flight trials, will be carried out at RAF Spadeadam in Cumbria, the RAF's electronic warfare test range.

as well as air defence, from later procurement, with a view to doubling multi-role EF 2000 totals. From its DM25 billion (US$11.5 billion) Eurofighter programme, Germany is currently due to receive 28 single-seat and 16 two-seat aircraft from the first production batch (Tranche 1), 58/10 in Tranche 2, and 61/7 in Tranche 3. Software for main air-to-surface roles is planned for introduction in Tranche 2, between 2005-2010, and further expanded in Tranche 3 for full

swing-role capability, from 2011-2014.

As the Luftwaffe's first nominated Eurofighter unit, JG 73 will replace its MDC F-4F Phantom

PROJECT DEVELOPMENT

Japan

New transport/patrol aircraft
The Japanese Defense Agency has selected Kawasaki Heavy Industries to lead the development of a new transport aircraft for the Air Self-Defence Force (JASDF) and a new maritime patrol plane for the Maritime Self-Defence Force (JMSDF). The agency expects to purchase approximately 40 aircraft for the JASDF, to replace 31 Kawasaki C-1 transports, and 80 aircraft for the JMSDF, replacing the 88 Kawasaki/Lockheed P/EP-3Cs currently in service.

Funding will come from the respective US$1.2 billion C-X and $1.6 billion P-X funding allocations, and the two types are expected to employ a mainly common airframe. The first to fly in prototype form is scheduled to be the P-X in 2006, powered by four possibly indigenous medium-thrust turbofans. The C-X, to follow from 2007, will have two larger imported turbofans, to lift a 26-tonne payload.

South Korea

T-50 rolled-out
First flight of the Korean Aircraft Industries (KAI)/Lockheed Martin T-50 Golden Eagle transonic digital fly-by-wire advanced trainer is expected in mid-2002, following its formal roll-out at Sachon on 31 October 2001, over three months ahead of schedule.

Developed in conjunction with Lockheed Martin from a US$2 billion programme, the T-50 closely resembles an 80 per cent scale F-16, with underwing lateral intakes for its 78.7kN (17,700 lb) F404-GE-402 afterburning turbofan. The 94 aircraft planned for RoKAF procurement will include some A-50 light combat versions. The first production aircraft will follow six development prototypes by late 2005.

United States

EW Super Hornet variant
Boeing recently completed an initial flight demonstration of an F/A-18F that had been configured to serve as an Airborne Electronic Attack (AEA) concept aircraft. The flight tests measured noise and vibration data and assessed the Super Hornet's flying qualities while equipped with three AN/ALQ-99 jamming pods and two external fuel tanks. A modified Super Hornet is one of the platforms under consideration in a Department of Defense analysis of alternatives to replace the Navy's EA-6B Prowler electronic warfare aircraft.

Boeing tankers?
The USAF could soon become a customer for the Boeing 767 tanker transport aircraft and the US Congress is currently reviewing a proposal that would provide the service with 100 aircraft under a 10-year lease arrangement at a cost of approximately $22 billion.

Osprey progress
A National Aeronautics and Space Administration (NASA) panel has recommended that the USMC resume flight-testing of the Bell Boeing MV-22B Osprey tilt-rotor aircraft. The Osprey fleet was grounded in December 2000, in the wake of the second of two fatal crashes.

The programme has undergone numerous reviews since then and a number of systems have been redesigned and software has been

Following the roll-out ceremony on 31 October 2001, the T-50 prototype is due to make its maiden flight in late spring 2002. Three further full-scale development (FSD) aircraft will join the test progamme, before the first production example is due to fly in late 2005.

and MiG-29 air defence squadrons at Laage from 2003, followed by similar re-equipment of JG 74's two F-4F squadrons at Neuburg. Tornado IDS replacement will not begin in the JBG 31 ground-attack wing at Nörvenich until 2007-09, but some withdrawals will start this year with planned disbandment of JBG 34 at Memmingen.

Airbus A400M

Airbus and its partners expected to finalise orders for 196 A400M military transport aircraft by mid-November 2001. The orders include 73 aircraft for Germany, 50 for France, 27 for Spain, 25 for the United Kingdom, 10 for Turkey, 8 for Belgium and one for Luxembourg. Italy had earlier announced plans to purchase 16 aircraft, however the possible withdrawal of Italy has raised concerns regarding the viability of the $16 billion project. Italy's Defence Minister recently announced that its air force did not need the 16 aircraft it had planned to buy. The programme's minimum requirement is for 200 aircraft, however an Italian withdrawal would leave Airbus Military with just 196 orders.

Boeing Sikorsky RAH-66

Lockheed Martin's Missile and Fire Control division has delivered an engineering and manufacturing development (EMD) version of its Night Vision Pilotage System (NVPS) for the US Army's RAH-66 Comanche armed reconnaissance helicopter to Boeing, which is the mission equip-

ment package integrator. The NVPS completed safety-of-flight testing in Orlando before shipment to Boeing's System Integration Laboratory (SIL) in Philadelphia, where integration testing is being conducted. Once the SIL testing is complete, the equipment will be returned to Lockheed Martin where it will be equipped with instrumentation in preparation for installation on the RAH-66A for subsequent flight tests.

The US Army is proposing to reschedule the development programme, allocating additional funds from the initial production budget into the engineering and manufacturing development (EMD) phase to address a number of system

shortfalls. In addition the number of EMD airframes may be reduced to 11 to keep costs down. To compensate for the lower production budget it is proposed that annual full-rate production be boosted from 62 to 96, reducing the overall production run by some eight years with savings in the region of US$3 billion. These savings would allow the US Army to purchase 1,213 Comanches as planned, without the programme going significantly over budget.

Showing the differing tail arrangements, RAH-66 Prototypes 1 and 2 are continuing to test avionics and mission systems as part of the EMD phase.

rewritten. Significant design changes have been made to the hydraulic system, the engine nacelle and flight control software, which were blamed for the 11 December 2000 crash that killed the four crew members.

The aircraft had been assigned to Marine Tiltrotor training Squadron VMMT-204 at MCAS New River, North Carolina, where eight Ospreys are currently grounded. Two additional USMC Ospreys are located at NAS Patuxent, River, Maryland and the USAF has two prototypes of the CV-22B Special Operations variant at Edwards AFB, California. Although the aircraft has remained grounded, production has continued and a number of new aircraft are also stored at Bell Helicopter's facility in Amarillo, Texas. The Marines hope to resume flight testing in the spring of 2002 once the aircraft have been modified, and have estimated that it will take one to two years for the aircraft to become operational.

American Merlins
AgustaWestland and Lockheed Martin have announced an agreement that will allow the latter to market the EH-101 helicopter to the US military. Lockheed Martin, which developed the systems that equip the Royal Navy's Merlin

HM.Mk 1 anti-submarine helicopters, will initially attempt to sell the Merlin to the USAF as a combat search and rescue (CSAR) aircraft.

MH-60S progress
HC-3 at NAS North Island, California commenced training with the MH-60S in August 2001. Following this, HC-5 at Andersen AFB, Guam, and HC-6 at Naval Station Norfolk, Virginia, will be the first fleet units to operate the MH-60S.

In preparation for its initial deployment, the MH-60S entered operational evaluation (OPEVAL) testing during November 2001. The tests, which will include operations from a variety of ships and facilities, will determine the operational effectiveness and suitability of the helicopter in the combat support role, and are scheduled to be completed in January 2002.

HC-11 at NAS North Island will be the last combat support (HC) squadron to operate the H-46 Sea Knight and is scheduled to receive its first MH-60S in March 2003, meaning that the Sea Knight could be phased out of the navy inventory by 2004. The MH-60S will assume the combat search and rescue (CSAR) role now assigned to the HH-60H in 2006.

Once the two new Seahawk models are in service, it appears

likely that each deployed carrier will have one MH-60S squadron assigned to provide SAR/CSAR, vertical replenishment (VERTREP) and airborne mine countermeasures (AMCM) for the entire carrier battle group (CVBG). Additionally,

an MH-60R squadron will operate from the carrier in the anti-submarine warfare role and light airborne multi-purpose system (LAMPS) missions from escort vessels. By 2009 older HH-60Hs will replace UH-1Ns and UH-3Hs, used in the

Initial flight trials of Boeing's Airborne Electronic Attack concept, based around the F/A-18F Super Hornet airframe, have commenced. The aircraft is seen here configured with a single AN/ALQ-99 jamming pod on each outboard wing pylon and a further AN/ALQ-99 on the underfuselage station.

Above: F-22 Raptor 4006 is seen on a test flight over the Californian desert in late 2001. Note the apparently different nozzle arrangement at the end of each jetpipe. The arrival in January 2002 of Raptor 4007 will relieve some of the pressure to meet the dedicated initial operational Test and Evaluation phase, scheduled for April 2003.

(USAASOTD) at Fort Bragg, North Carolina, conducted the tests and the aircraft were flown from Pope AFB, which adjoins Fort Bragg. Demonstrations included the deployment of up to 92 paratroops at a time using doors located on both sides of the aft fuselage, and the rapid ground evacuation of the aircraft by up to 128 troops.

At the conclusion of testing, the aircraft returned to Lockheed Martin's Marietta, Georgia, facility in preparation for delivery to the Rhode Island Air National Guard's 143rd Airlift Wing. The first two arrived in Providence, Rhode Island on 2 December 2001.

Raptor developments
On 21 September 2001 the F-22 Combined Test Team met another

milestone when Raptor 4005 launched an AIM-120C AMRAAM within lethal range of its target, scoring a 'hit' against the unmanned target aircraft. The demonstration was the first live test of the avionics suite's ability to track and destroy a target. The demonstration took place at approximately 40,000 ft (12192 m) over the Pacific Missile Test Range.

Boosting the Combined Test Force's programme, Raptor 4007 arrived at Edwards AFB on 5 January 2002 from Lockheed Martin's Marietta plant. The aircraft will be used as a second missile test aircraft and is equipped with integrated sensor fusion capabilities that encompass electronic warfare, radar and communications capabilities.

Above: Developed by Composites Technology Research Malaysia and Excelnet, with support from BAE Systems, the Eagle Aerial Reconnaissance Vehicle (ARV) was unveiled in October 2001. Able to be flown either manned or unmanned, the aircraft can be fitted with a range of surveillance sensors.

Right: The first of 100 Bell UH-1Ys for the USMC made its maiden flight on 20 December 2001.

SAR role at Naval Air Stations.

The MH-60S will also replace the MH-53E Sea Dragon in the airborne mine countermeasures (AMCM) role, which will result in a merger of the Combat Support (HC) and Mine Countermeasures (HM) communities. The last MH-53Es will be retired in 2010. The Atlantic and Pacific Naval Air Forces will each eventually operate individual fleet

readiness squadrons (FRS) for the new models.

C-130J-30 testing
The USAF, US Army and Lockheed Martin recently completed a joint test programme that validated the C-130J-30's ability to conduct paratroop airdrop operations. The US Army Airborne and Special Operations Test Directorate

UPGRADES AND MODIFICATIONS

Australia

AP-3C deliveries begin
Redelivery last October of the first two of 17 RAAF Lockheed AP-3C Orion maritime patrol aircraft to be upgraded by Raytheon with new missions system equipment was reportedly over three years behind the originally-planned schedule. As Phase II of the RAAF's Project Air 5276 programme, the Orion avionics upgrade was delayed by missions and flight systems software integration problems, which have helped to push the originally quoted January 1995 $A747.5 million (now US$377.2 million) costs to $A880 million (US$440 million).

After a prototype AP-3C upgrade conversion by the former E-Systems in Greenville, Texas,

production modifications are being undertaken by Raytheon Systems Australia at Avalon, Victoria. Three more upgraded Orions will be redelivered by April 2002, for Phase II programme completion by early 2004. Phase IIB of Project Air 5276, now in hand, involves procurement of three ex-USN P-3B Orions from AMARC storage for RAAF training and transport use.

Canada

New aerial tanker plans
Plans were revealed in late 2001 by the Canadian Armed Forces to regain a strategic air refuelling capability, lost in 1997 following the withdrawal of its two Boeing CC-137 (707) tanker/transports. Rectification is planned from

conversion of two of the CAF's five Airbus CC-150 Polaris with air refuelling equipment. Modification, in co-operation with the German air force and industry, is proposed to bring the aircraft to A310 MRTT (multi-role tanker/transport) standard with underwing Flight Refuelling Mk 32B hose-and-drogue pods. The Canadian CC-150s are planned for redelivery in 2003-04, for a target programme cost below $C100 million ($63 million)

Egypt

Apache upgrade contract
Long-discussed plans for upgrade and remanufacture of 35 EAF AH-64A attack helicopters to next-generation AH-64D Apache standards, announced in September 2000, finally resulted in US$400 million FMS contract signature on 3 December 2001. Boeing was

authorised to begin procuring long-lead items in March 2001 for the US$241.9 million Egyptian upgrade order, for redeliveries from 2003. The AH-64D's associated Lockheed Martin/Northrop Grumman AN/APG-78 Longbow fire-control radar is not included in the upgrade package.

Germany

A310 MRTT conversions
The first phase of a two-stage programme to convert four of seven GAF Airbus A310-304s as tanker/transports was completed in December 2001 by the Airbus A310 MRTT Air Force Consortium.

It involved initial modification by DASA's Elbe Flugzeugwerke subsidiary in Dresden, with a 3.58 x 2.57 m (11.74 x 8.43 ft) forward-fuselage cargo door and reinforced floors. Three similar conversions

Above: Increasing the current range of F-16 in-flight refuelling (IFR) options is the newly revealed IAI Lahav probe modification, seen here on IDF/AF F-16 Barak 397. The installation offers the F-16 the flexibility to refuel from boom- or drogue-equipped tankers.

Below: The IDF/AF has commenced operations with CH-53s armed with up to eight laser-guided IAI MBT Nimrod missiles, believed to have a range in the region of 11 nm (20 km). This example is seen firing a Nimrod during initial trials in the 1980s.

then followed by Lufthansa Technik at Hamburg/Fuhlsbuttel.

Second-stage work is now beginning in Hamburg to install tanker equipment, comprising a Flight Refuelling Mk 32B hose-and-drogue pod under each wing. Provision is also made for an additional or optional rear-fuselage fly-by-wire refuelling boom, and a rear-cabin operator's position.

India

Il-38 upgrade launched

Work on a US$205 million programme to upgrade the mission systems avionics and operational capabilities of five Ilyushin Il-38 'May' maritime patrol/ASW aircraft was started last year by the original manufacturers in Moscow. Ilyushin Aviation received a US$75 million sub-contract from the Indian Defence Ministry to undertake structural alterations to the aircraft, operated by Indian Naval Air Squadron 315 from Dabolim, near Goa, since their delivery in 1976.

Initial modifications will include additional weapons attachment points, plus torpedo launching and air-refuelling capabilities. India's Defence Research & Development Organisation (DRDO) is also scheduled to supply new mission systems costing some US$30 million. Enhanced electronic warfare and intelligence (EW/ELINT) systems will be supplemented by new radar, acoustic, imaging infra-red, MAD, digital communications, navigation and avionics equipment, from another US$100 million contract.

MiG-27 modernisation

Upgrades of some 40 of 165 MiG-27L Bahadur ground-attack

Following a decision to retain the MiG-29 as its primary fighter, Bulgaria's defence minister has announced that a tender will be issued covering an upgrade for the air force's fleet of 21 MiG-29 fighters. Seen at Graf Ignatievo AB in late 2001, this is one of only three MiG-29s currently airworthy, the others grounded by a lack of parts.

fighters licence-built by HAL between 1984-1997 are planned from US$130 million funding in India's 1997-2002 military procurement budget. In Russia, the Irkutsk Aircraft Building Industrial Association (IAPO) is bidding to participate with HAL in installing new Indian, French and Israeli avionics in the upgrade package, which may also include an air-refuelling system.

Malaysia

Tanker conversions

Lockheed Martin recently began a tanker conversion programme involving two Royal Malaysian Air Force C-130Hs and will equip the transports with Flight Refuelling Limited equipment and an 1,800-US gal (6814-litre) fuel tank developed by Aero Union Corp. Two other Malaysian C-130Hs will be modified to the stretched C-130H-30 configuration by the Malaysian company Airod Sendirian Berhad.

Poland

'Hind' upgrade plans

Further progress has been reported with the Zl828 million (US$202.7 million) programme for missions systems upgrades of the Polish Army Air Force's 30 or so Mil Mi-24D 'Hind-D' and 14 Mi-24V

'Hind-E' attack helicopters. Apart from meeting NATO standards, new digital avionics will confer day/night attack capabilities with eastern or western weapons.

Polish 'Hind' update contenders include teams led by BAE Systems, using Hellfire/Brimstone, Rafael NT-D Dandy or NT-S Spike series anti-tank missiles (ATMs); IAI's Tamam Division, also with NT-D/S or similar ATMs; and Rostvertol/Mil's extant Mi-35M, with AT-6/9 Shturm or Ataka missiles. Also competing are Thales and SAGEM, using Euromissile's HOT-3 ATMs; and South Africa's Advanced Technologies & Engineering Mk III upgrade package, with Denel Kentron Ingwe or Mokopa ATMs.

Russia

Su-27/30 upgrade progress

Further progress with Russian air force Sukhoi Su-27/30 single- and two-seat combat aircraft avionics upgrades included the start of service flight-tests last year of two two-seat prototypes. These comprised the Su-30KN (s/n 302), which first flew in March 1999, as the Su-30MKR, and the Su-27UBM (s/n 20), delivered from IAPO on 6 March 2001. The upgraded twin-stick Su-30KN differs from the Su-27UBM only by its more extensive operational avionics and

equipment. These include a port-side retractable air refuelling probe, and digital electronic control for its Saturn Lyulka AL-31F turbofans.

Limited avionics upgrades, costing about US$1 million per aircraft, were planned for up to 20 VVS Su-27UBMs and Su-30KNs by late 2001. From about 650 Su-27 series built by KnAAPO and IAPO for the VVS, 70-80 per year will be upgraded from 2002. Initial changes comprise a single 12.7 x 12.7 cm (5 x 5 in) MFI-55 liquid-crystal cockpit MFD, new head-up display and up-front control panel, plus GPS, for new ground-attack capabilities. Operations will then be possible with Kh-31A/P (AS-17 'Krypton'), Kh-29T (AS-14 'Kedge'), and other precision-guided munitions.

Additionally, work has reportedly also started by Federal Research & Production Centre (FRPC) Salyut to uprate the Su-27/30's Lyulka AL-31F turbofan from its current 122.58 kN (27,557 lb) thrust to 150 kN (33,730 lb). Bench running started last year and will reach 147 kN (33,069 lb) by late 2002. Some Russian industry reports claim that work on further AL-31 thrust-vectoring development from complex swivelling nozzles in the Su-30 series is tapering off, since similar agility is being achieved from advances in fly-by-wire software.

NASA's testbed Hornets

Smoke generators are used to show the paths of wing tip vortices behind the two NASA F/A-18s currently employed in the Autonomous Formation Flight programme. Mapping the vortex pattern allows onboard software to place the trailing aircraft in the optimum position, reducing fuel burn by 15 per cent.

Left: This modified F/A-18A is used by NASA Dryden as the testbed for the Active Aeroelastic Wing project, researching the advantages of using flexible wings for roll control at transonic and supersonic speeds.

Saudi Arabia

Upgrades for AWACS fleet

Boeing has begun upgrading the Royal Saudi Air Force's (RSAF) fleet of five E-3A AWACS aircraft with new mission computers, hardware and software, as part of a contract worth approximately US$60 million. The upgrade will bring the RSAF aircraft to the same configuration as the USAF E-3B/C fleet. The aircraft are being modified at Boeing's Seattle, Washington, facility and the initial two Sentries were retrofitted during 2001 and the remainder will be completed in 2002. Saudi Arabia purchased five E-3As and eight KE-3A tankers in 1981 under Peace Sentinel and the first aircraft were delivered in June 1986.

Spain

Seahawk upgrades

Upgrade of the avionics and weapons of the first of six Spanish Naval Aviation (AAAE) Sikorsky S-70B-1 Seahawks, from a $77.4 million mid-2001 FMS contract, was completed by Lockheed Martin Systems Integration (LMSI) in late 2001. Due for completion by 2004, this also includes procurement of six new similar Seahawks, to double the AAAE's ASW/ASuW helicopter inventory.

The new equipment fit includes new nav/com systems, a forward-looking infra-red (FLIR) sensor, plus provision for AGM-114B/K/L Hellfire attack missiles, for additional air-to-surface roles.

Sweden

Hercules tanker

The Swedish Air Force recently signed a contract with Lockheed Martin to convert of one of its C-130E transports to aerial tanker configuration. The aircraft will be equipped with Flight Refueling Limited (FRL) equipment and an 1,800-US gal (6814-litre) fuel tank developed by Aero Union Corp. The conversion will take place in late 2002, and the modified aircraft will be returned to service in 2003.

Switzerland

Hornet enhancements sought

Having taken delivery of 26 MDC SF/A-18Cs and eight two-seat SF/A-18Ds from a SwFr3.5 billion (US$2 billion) contract, the Swiss air force (SFF) has become the first non-US operator of these Hornet versions to instigate avionics upgrades. The SFF has earmarked SwFr220 million (US$128.7 million) for SF/A-18 avionics improvements, through a two-phase programme, planned for 2004-06 completion.

Initial upgrades will include an open architecture mission computer, tactical aircraft moving map capability (TAMMC), an active interrogator/transponder IFF, and NATO-standard Link 16 or multi-function information distribution system (MIDS) data-link terminals. Next-generation high off-boresight close-combat AAMs, in conjunction with a helmet-mounted sight/display system, are planned for second-phase installation by 2006.

Operational availability of the 33 Swiss Hornets has recently dropped below 50 per cent because of inadequate spares provisioning. This is now being rectified, and 12 more surplus F/A-18C/D Hornets were being sought for continued SFF operation through 2010.

Thailand

Helicopter upgrade plans

Further modernisation and expansion of Thai army (RTA) helicopter capabilities will result from structural, engine and rotor-blade overhaul and repairs of 19 Bell UH-1Hs, from over 80 ex-US Army Iroquois and six Bell AH-1F attack helicopters acquired between 1970-90. The Hueys will receive new avionics in their US$22 million upgrade by Israel Aircraft Industries, and will be supplemented by about 30 more surplus UH-1Hs from US Excess Defence Articles and Foreign Military Sales programmes, for similar refurbishment.

Three Sikorsky UH-60 Black Hawks costing some US$30 million were also ordered in 2001 by the RTA, for delivery in 2002 to reinforce Thailand's narcotic interdiction programmes.

United Kingdom

100th Tornado GR. Mk 4

BAE Systems recently delivered the 100th Tornado GR.Mk 4 mid-life update (MLU) strike aircraft to the Royal Air Force. The GR.Mk 4 programme provides the earlier strike, trainer and reconnaissance variants with structural improvements and an avionics system enabling the latest weapons and systems to be progressively introduced. Systems include a thermal imaging and laser designator system (TIALD), which allows the aircraft to deliver laser-guided bombs, a forward-looking infra-red (FLIR) system and night vision goggle (NVG)-compatible cockpit displays. The aircraft is also equipped with a laser inertial navigation system and an integrated global positioning system, and ground proximity warning system. The GR.Mk 4 programme will be completed in May 2003.

United States

C-5 update progresses

The USAF's Aeronautical Systems Center at Wright-Patterson AFB, Ohio, has awarded Lockheed Martin a $1.1 billion contract associated with the C-5 Reliability Enhancement and Reengining Program (RERP). The RERP will provide the Galaxy fleet with new General Electric CF6-80C2 turbofan engines and systems, along with minor structural enhancements to ensure the C-5 fleet will remain operationally viable through at least 2040. The contractor had previously been awarded a contract to develop an avionics modernisation program (AMP) for the Galaxy fleet.

ICAP III Prowler Flies

The first Northrop Grumman EA-6B equipped with the Increased Capability III system made a successful 1-hour, 45-minute maiden flight on 16 November 2001 at Northrop

The cluster bombs contained within wind corrected munitions dispensers (WCMDs) have been used for the first time in combat during Operation Enduring Freedom missions against Taliban positions in Afghanistan. Here an Eglin AFB F-16C drops a WCMD during the test programme.

Grumman's facility in St Augustine, Florida. The aircraft is one of two prototypes being modified as part of the $200 million ICAP III development programme. The new variant is equipped with sophisticated software that enables the system to change the jamming frequency as quickly as modern radars change their transmission frequency to avoid jamming.

Other improvements include an integrated communications jamming system, a provision for the Navy's Link 16 data link, and new displays and controls. The system will be installed in all current fleet EA-6B Prowlers and it will achieve initial operational capability in 2005.

USMC TAV-8B upgrade
Naval Aviation Depot Cherry Point recently finished upgrading the first of 17 TAV-8B two-seat operational trainers for VMAT-203. Under the two-phase program, known as TAV-8B Upgrade Program (TUP), each of the two-seat trainers will be completely rewired and equipped with capability improvements.

The improvements include night vision equipment, more powerful F406-RR-408 Pegasus engines, structural enhancements and the latest mission software, and will provide the trainers with the same capabilities as the fleet AV-8Bs. In recent years the single-seat AV-8Bs have been equipped with night attack and radar capabilities, however none of these upgrades was installed in the trainers. The new systems will ensure that VMAT-203, the fleet readiness squadron, is able to provide student pilots with a training tool more representative of the current fleet aircraft.

Tomcats drop JDAMs
The Naval Weapons Test Squadron (NWTS), at Naval Station Ventura County/Point Mugu recently tested the interface between the F-14D and the GBU-31 Joint Direct Attack Munition (JDAM) when it conducted captive carriage tests of the weapon.

The F-14D is capable of carrying up to four JDAMs, which can be delivered, simultaneously, to four different targets from distances outside the range of some air defence systems. VF-143 and VF-11, both assigned to Carrier Air Wing Seven (CVW-7), became the first F-14B units to drop live JDAMs during exercises at NAF El Centro, California, earlier this year.

Initial F-16 CFT trials
Lockheed Martin recently completed Phase I flight testing of new conformal fuel tanks (CFTs) designed for the F-16 fighter. The

tests utilised aerodynamic 'shapes' installed on F-16C serial 87-0353, operated by the 40th Flight Test Squadron at Eglin AFB, Florida. A total of 24 test flights and 65 flight test hours was flown between March and August 2001, demonstrating loads, flutter, stability and control.

According to the manufacturer the flying qualities of a CFT-equipped F-16 differ little from that of a standard aircraft. In fact, with CFTs installed the aircraft retains its full 9-g capability and flight envelope. A set of two CFTs installed on the upper fuselage surface provides the fighter with 440 US gal (1666 litres) of additional fuel and offers extended mission range, time on station or time engaged in combat. In addition, the CFTs increase the F-16's payload flexibility by eliminating the need for external under-wing tanks.

The flight test F-16C was subsequently returned to the contractor's Fort Worth, Texas, facility and was fitted with a functional set of CFTs. Ground systems testing of fuel transfer began in September 2001. Functional flight testing began in late 2001, conducted at both Fort Worth and Eglin.

Blackhawk upgrade progress
Sikorsky has begun evaluating the first three US Army UH-60 Blackhawks that will be upgraded to the new UH-60M configuration under the Blackhawk Upgrade programme. The service plans to equip 1,200 helicopters with a strengthened fuselage, new composite spar wide-chord blades, more powerful engines and new digital cockpit displays. The composite spar wide-chord blade will provide 500 lb (227 kg) more lift than the current UH-60L blade and the General Electric T700-GE-701D engine, which is currently under development, will provide another 400-500 lb (181-227 kg) of additional payload.

Once modified, the aircraft will be capable of carrying a greater payload with lower maintenance costs than the current fleet, and will gain an additional 20 years of service life. In addition to rebuilding the three earlier aircraft, Sikorsky is building one new UH-60M under the terms of a $219.7 million research, development, test and evaluation contract. The UH-60M will replace the UH-60L as the standard new production configuration, beginning in 2007.

The maiden flight of the first modified Blackhawk is scheduled for 2003 and low-rate initial production of UH-60M aircraft will begin in 2004.

Bulgarian AF at Cooperative Key 2001

Above: Bulgaria's 35 Sukhoi Su-25K 'Frogfoot' ground-attack aircraft will require NATO-compatible upgrades if they are to remain in service long-term.

Above: Deploying to Graf Ignatievo for the Partnership for Peace (PfP) Cooperative Key exercise was this Mil Mi-17, normally based at 24 Airbase Krumovo.

Above: The Bulgarian air force's fleet of Su-22M4 'Fitter-Ks' has recently undergone major overhaul and had new avionics installed by TEREM's Plovdiv factory.

Above: Some 85 MiG-21s of various marks were present at Graf Ignatievo, although most were in long-term storage. Only around 12 MiG-21bis are currently operational.

Above: The AB-206 is the newest aircraft type in the Bulgarian air force inventory. It was used during the exercise for communications and liaison.

PROCUREMENT AND DELIVERIES

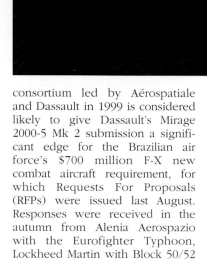

The first two AP-3Cs have entered RAAF service following their redelivery in autumn 2001. Three more will be returned to service by May 2002.

Algeria

Beech 1900 deliveries

Deliveries started in mid-2001 of the first of 12 Beech 1900D twin-turbo-prop light transports ordered by the Algerian air force a year earlier. Six of the new aircraft are being operated on general transport, communications and utility roles, but the remaining six are equipped with multiple sensors and avionics for electronic surveillance and combat support missions. In addition to Raytheon/Hughes integrated synthetic aperture radar (HISAR) in a ventral fairing, Algeria's ELINT Beech 1900s incorporate Wescam Type 16 forward-looking infra-red (FLIR), SATCOM data-link, and BAE Systems Sky Guardian radar detection and ESM equipment, in support of ground operations against Islamic fundamentalist insurgents.

Austria

F-16s to replace Drakens?

The Defense Security Co-operation Agency has notified the US Congress that Austria may purchase 30 F-16C fighters in a contract worth an estimated US$1.74 billion. In addition to the fighters, Austria has requested three spare engines, 13 various AMRAAM missiles and 15 Sidewinder practice or training missiles. The aircraft will replace Saab Drakens currently in service.

Bahrain

Avro RJ85 order

BAE Systems has delivered a four-turbofan Avro RJ85 multi-role transport ordered by the Bahrain Defence Force from a US$25 million contract. Its operation is expected by the Bahrain Amiri air force's VIP unit from Shaikh Isa Air Base, alongside the BAAF's single Boeing 727 and 747, and Gulfstream G.II/G.III.

Brazil

F-X RFP submissions sought

A 20% shareholding in EMBRAER acquired by a French industry consortium led by Aérospatiale and Dassault in 1999 is considered likely to give Dassault's Mirage 2000-5 Mk 2 submission a significant edge for the Brazilian air force's $700 million F-X new combat aircraft requirement, for which Requests For Proposals (RFPs) were issued last August. Responses were received in the autumn from Alenia Aerospazio with the Eurofighter Typhoon, Lockheed Martin with Block 50/52 F-16C/Ds, the SAAB/BAE Gripen, and Rosoboronexport with the MiG-29SMT and Sukhoi Su-27/30.

Canada

Tutor replacement

Canada's Department of National Defence (DND) has begun looking for a replacement aircraft for the Air Force's 'Snowbirds' Air Demonstration Squadron. The DND will look at a number of options including retaining the unit's CT-114 Tutors beyond 2006, using the CF-188 Hornet and acquiring new jet or turboprop trainers. It is also probable that the team will be reduced in size from the current nine aircraft display to as few as four aircraft, and the DND has requested prices for four, six and nine aircraft as well a reserve aircraft.

With the exception of 15 aircraft operated by the 'Snowbirds', the CT-114 is not operational with the air force. As far as new aircraft are concerned, the turboprop Raytheon T-6A Texan II is already in service with the air force as the Harvard II and would be the least expensive option. Also operational with the air force is the BAE Systems Hawk trainer, which is used by the Royal Air Force's 'Red Arrows' display team. Although no display teams currently fly the Texan/Harvard, several teams use other turboprop trainers.

Chile

Citations delivered

The Fuerza Aérea de Chile or Chilean Air Force took delivery of three Cessna Citation CJ1 business jets in late 2001. They will be based at El Tepual Air Base in Puerto Montt.

Helicopter re-equipment

Both the Chilean air force (FACh) and Army Aviation (CAEC) are planning to modernise their helicopter fleets from acquisition of new equipment. In this respect, the FACh is already leading the way with long-term procurement of the Bell Textron 412EP, from deliveries of four between mid-2000/2001, towards total requirements for up to 16, to replace its eight Bell UH-1Hs. The Bell 412 is also competing with the Eurocopter AS 532 Cougar, Mil Mi-171 and Sikorsky UH-60 for an exactly similar CAEC requirement, accompanying its planned disposal of three of its eight Aérospatiale SA 330F Pumas, several of its dozen Aérospatiale SA 315B Lamas, and all 13 of its Enstrom 280FX Shark training helicopters.

Czech Republic

Gripen wins new contract

As expected, following withdrawal last May of competing bids, the SAAB-BAE Gripen International was selected on 10 December to provide the Czech air force's MiG-21 replacement. From an original requirement for 36 new fighters, Czech procurement is now planned of 24 Gripens, through a reduced CKr 50 billion ($1.34 billion) contract. Its conditions include Czech language documentation, 150 per cent industrial offsets, and deliveries for initial operational capability by late 2005.

Mi-35s to join helicopter fleet

The Czech Republic's Defence Minister has announced the purchase of six Mil Mi-35 combat helicopters from Russia and the modification of Mi-24 helicopters already in Czech service.

Denmark

EH101 contract signed

Selection of the tri-RTM322-turboshaft EH101 by Denmark in the Nordic Helicopter Programme last September was followed on 7 December by a DKr3 billion (US$363 million) contract with EH Industries for 14 of these multi-role helicopters. Signed by Maj Gen Fynbo, head of RDAF Materiel Command, the contract will be

Two of the ROKAF's 10 Raytheon Hawker 800XPs are seen here. The aircraft are used for Sigint/Elint surveillance duties, with mission equipment installed by Goodyear. There are at least two configurations, as displayed here, with differing ventral fairings and tail antennas.

completed by AgustaWestland at its Yeovil factory

Challenger delivered

Bombardier delivered two Challenger 604s to the Royal Danish Air Force at Værløse Air Base in the autumn of 2001. The first aircraft, serial C-168, arrived on 13 November 2001 and the second, serial C-172, was delivered nearly a month ahead of schedule on 21 November 2001. The RDAF took delivery of its first Challenger 604 for use in a multi-role capability in 1999 and exercised options for two additional aircraft in 2000. All three aircraft will be assigned to 721 Squadron at Værløse, and used in a multi-role capability, supporting maritime surveillance missions such as fisheries inspection and anti-pollution; as well as search and rescue operations. With a maximum range of 4,077 nautical miles (7550 km) the aircraft will also provide long-range VIP transportation capability and medical evacuation. The Challengers were constructed at Bombardier's facility in Montreal, Quebec, however the mission equipment and interior were installed at the company's Tucson, Arizona facility.

France

Rafale deliveries

By 2008 the French air force (AdA) will have taken delivery of 57 Dassault Rafale multi-role fighters, and naval aviation of 19 Rafale Ms. This is about a dozen fewer than earlier expected, from planned Dassault production of about one aircraft per month, and AdA will not receive its first definitive Rafale F3s until 2008. Only seven Rafales had been delivered by late 2001, and service entry dates of 2002 for Aéronavale and 2005 with AdA are six or seven years behind original schedules.

All Rafales will be equipped to launch MBDA SCALP/EG cruise missiles, but funding has been

US Army in Korea

Above: The first AH-64D Apache Longbow helicopters destined for the 2nd Infantry Division in South Korea arrived onbaoard MV Green Wave during October 2001, with the final 12 or 24 aircraft flying in to their new base at Page Army Air Field, Camp Page in Chunchon, Korea, from Pusan at the end of the month. The aircraft are assigned to 1-2nd Aviation (Attack), which is the fourth active-duty US Army unit to be equipped with the Longbow version.

Right: The US Army's 4-160th Special Operations Aviation Regiment (SOAR) is now operational in the theatre equipped with MH-47E Chinooks. The unit is based at Camp Walker, within the joint US/RoK Taegu Air Base.

excluded for Aéronavale's long planned second aircraft-carrier in the 2003-08 defence programme. Continued development of MBDA's ASMP-A supersonic nuclear missile will allow its use by AdA Mirage 2000Ds in 2007, and Rafales a year later.

Greece

RJ-145 Erieye deliveries begin

Deliveries started in September 2001 to the Hellenic air force (HAF) of the first two of four EMBRAER RJ-145 twin-turbofan transports equipped with the 450-km (243-nm)-range Ericsson Erieye dorsal planar array radar and associated NATO-interoperable equipment for AEW&C roles. Ericsson received a US$600 million HAF order in late 1999 for its Erieye systems, for which training started

in Greece on two loaned Swedish air force twin-turboprop SAAB S 100B Argus AEW&C aircraft in mid-2001. These are operated from Elefsis air base by the HAF's newly-formed 380th AEW&C Squadron, which will also receive the RJ-145s.

Hungary

Gripen lease signed

Final agreement was signed in Budapest on December 20 between the Swedish Defence Materiel Administration (FMV) and Hungary's Department of Defence on leasing 14 Swedish air force Gripens for Hungarian use. Delivery of 12 single-seat JAS 39As and two twin-seat JAS 39Bs is due between late 2004-June 2005, for 10 years' lease, with eventual purchase options, after their upgrade to current Batch 3 JAS 39C/D NATO-

interoperable standards.

Modifications, including Have Quick radios, data-link communications and an air-refuelling system, will be funded from Hungary's estimated $500 million contract. Operational and tactical training, mainly in Sweden, is included for 15 pilots, 32 maintenance technicians, two flight simulator technicians and five fighter controllers. Armament is excluded in the lease, apart from the single-seat Gripen's internal 27-mm Mauser BK 27 cannon.

Iran

More Mi-17s ordered

Following a bilateral agreement last October on military and technical co-operation, a new $150 million Iranian Defence Ministry contract was announced in November for up to 30 more Mil Mi-171Sh transport helicopters. This was the third placed with the Ulan Ude Aircraft Plant since 1998, with overall totals of 56. Of these, two batches of five and 21 were delivered from late 1999 to late 2001. Some deliveries from the third contract, which includes options in the second order, will be armed attack versions of the Mil Mi-171Sh.

Recce Viggens from Flygvapnet's SWAFRAP (Swedish Air Force Rapid-reaction unit) undertook their first major overseas NATO PfP deployment during Exercise Cooperative Key in Bulgaria. SWAFRAP is allocated six AJSF 37 Viggens from 1. Spaningsflygdivision, F21 Norrbottens Flygflotiilj at Luleå-Kallax.

Israel

Gulfstream Vs ordered

Israel's Ministry of Defence has placed an order for three Gulfstream V business jets with Gulfstream Aerospace. The aircraft will be modified for use as special electronic mission aircraft (SEMA) at Gulfstream's Savannah, Georgia, production facility, however the mission equipment associated with their surveillance role will be installed locally following delivery to Israel. The contact, which is worth $206 million, includes a $32 million option for a 10-year logistic support programme.

Japan

Coast Guard Gulfstreams

The Japanese Coast Guard has ordered two Gulfstream V special mission aircraft from Gulfstream Aerospace in a deal worth approximately US$100 million, which includes engineering support for modification and integration of the mission systems. The aircraft will provide Japan with an all-new ocean surveillance and rescue capability, and will be equipped with a high performance airborne surveillance radar, a forward-looking infra-red system. Thales in France is developing the mission equipment.

Fokker Services in the Netherlands will modify the aircraft at its facility at Woensdrecht Air Base and will arrive at the end of 2002. Delivery of the two modified aircraft is scheduled for the second quarter of 2004.

Jordan

New primary trainer order

Signature of a new contract was announced last November with Slingsby Aviation for 16 T67M260 Firefly two-seat composite primary trainers for the Royal Jordanian air force. When delivered from late 2002 with spare parts, ground support equipment and technical services, the Fireflies will presumably replace 15 RJAF BAe Bulldog Mk 125/As operated since 1974.

Kuwait

AH-64D interest renewed

A 1998 FMS Letter of Offer & Acceptance (LOA) for 16 Boeing AH-64D Apache attack helicopters, which was allowed to lapse, was due for renewal by Kuwait, according to Washington reports late last year. Kuwait's new $640 million AH-64D package was expected to include Boeing/Lockheed Martin AGM-114 Hellfire fire-and-forget ATMs, but not Longbow millimetric-wave fire-control radar. KAF procurement interest was also reported to include Boeing AWACS aircraft, plus four Lockheed Martin C-130J tactical transports and two KC-130J tanker/transports.

Malaysia

Super Hornet in the future?

Malaysia has apparently begun discussions with Boeing covering the purchase of F/A-18E/F Super Hornets. According to reports, if Malaysia does purchase the advanced version it will exchange its eight F/A-18Ds as part of the package.

Namibia

New ASNDF equipment

Delivery began last year to the Air Squadron of the Namibia Defence Force of the four NAMC K-8 Karakoram basic/advanced jet-trainers, with Honeywell TFE731-2A-2A turbofans and provision for armament, ordered in late 2000.

Earlier ASNDF equipment deliveries from China have reportedly included two Harbin Y-12-II twin-turboprop light transports, which arrived in 1997. Further ASNDF expansion was expected in mid-2001 from reported leases from Libya of two each Mil Mi-8 'Hip' transport and Mi-24 'Hind' attack helicopters.

Oman

F-16 FMS contract proposed

In a major change from previous British combat aircraft procurement policies, the Royal Air Force of Oman (RAFO) was awaiting Congressional approval late last year for a $1.12 billion US equipment package through Foreign Military Sales auspices. The RAFO's Jaguar International ground-attack aircraft, which are still undergoing upgrades to RAF Jaguar GR.Mk 3 standards, will be supplemented by 12 Block 50 Lockheed Martin F-16C/D multi-role combat aircraft with a wide range of US weapons, plus support equipment, services and training.

Poland

C-295 transports ordered

Following earlier selection of the C-295M twin-turboprop tactical transport from additional Antonov bids for the An-32 and Lockheed Martin Alenia's C-27J, a US$211 million Polish air force (WLOP) contract was signed last year with EADS/CASA for eight aircraft. Delivery is scheduled between 2003-05, from the C-295 contract, containing 'advantageous' offsets for PZL Warsaw Okecie, in which EADS/CASA has acquired a 41 per cent shareholding, and Polish industries.

Portugal

EH101 selection

Recent initial FAP orders for 10 RTM322-powered NH Industries NH-90 transport helicopters were accompanied last December by selection of the tri-RTM322-turboshaft EH101, to meet Portuguese naval requirements for 12 new medium-lift helicopters for search and rescue, combat SAR and fishery protection roles. Chosen after competitive evaluation against the Eurocopter Cougar Mk 2+ and the Sikorsky S-92, the EH101s will be produced by AgustaWestland at Yeovil, from a 350 million euro ($315 million) contract.

Following delivery between 2004-06, two of Portugal's EH101s, acquired with 50 per cent EU funding, will be allocated to fishery protection. Some will also be Azores-based, where their long-range and three-engined reliability are deemed important. EH101 funding is being made available, despite 30 per cent cuts over the previous year in the 2002 Portuguese defence budget, to around Esc252 billion ($1.132 billion).

Deliveries of the IAR 330 Socat armed Puma continue to the Romanian air force's 61st Helicopter Group. The nose turret houses a FLIR while the gun is a trainable GIAT THL20 20-mm cannon.

Singapore

Apaches ordered
The Republic of Singapore has formally agreed to purchase 12 additional AH-64D Apache attack helicopters from Boeing. Singapore had previously purchased eight aircraft and held options for the latest purchase. Boeing delivered the 200th AH-64D to the US Army on 23 August 2001 at its Mesa, Arizona, facility and has delivered 45 Apache Longbows to international customers.

Slovenia

New support aircraft orders
Recent Slovenian military aircraft procurement has included two AS 532AL Cougar transport helicopters from a Euro 27 million (US$24.3 million) army contract with Eurocopter last summer. Deliveries to the 15th Air Brigade are due to begin in early 2003 A Dassault Falcon 900EX long-range VIP transport costing US$30-35 million was also ordered in October 2001.

Spain

C-295 delivered
EADS CASA delivered the first of nine C-295 military transports to the Spanish Air Force's Ala 35 (wing) at Getafe Air Base on 15 November 2001. The wing has operated the CN-235 since 1986. The C-295 is equipped with a cargo hold that is 9.8 ft (3 m) longer than the CN-235, providing a 50 per cent increase in load capability over similar ranges. Powered by two 2,645-hp (1972-kW) Pratt & Whitney Canada PW127G engines the C-295 is capable of carrying up to 73 troops, five standard containers or 27 stretchers. Ala 35's C-295s will assume some of the humanitarian and peacekeeping missions currently assigned to the larger C-130 Hercules.

Sweden

Nordic NH-90 requirements
As participants in the Nordic Helicopter programme, the Swedish Armed Forces NH-90 helicopter commitment has emerged as 18 Hkp 14s for transport and SAR roles, costing SKr6 billion ($564 million), with options on seven more. Norway has contracted for 14 NH-90s for ASW and coast guard missions, with options for another 10 SAR versions. Finland's 20 NH-90s will be employed on tactical troop transport and SAR duties, and all will be powered by Rolls-Royce Turbomeca RTM322 turboshaft engines.

United States

USMC Citation Encores
The US Marine Corps has accepted the first two of three UC-35D

Right: 93-054 from the 157th Fighter Squadron, 20th TFW, RoKAF is one of 72 F-16C/D Block 52s wholly licence-built in Korea by a KAI consortium (led by Samsung Aerospace), out of a total order for 120. The 157th FS is based at Seosan AB alongside its sister unit – the 121th FS.

Below: Until South Korea receives its new T-50 jet trainers later in the decade, the RoKAF's Training Command will continue to lease 30 Northrop T-38 Talons to fulfil its advanced training needs. This example is operated by the 189th TTS based at Yeacheon.

Citation Encore operational support aircraft from Cessna and assigned them to MCAS Miramar, California, and MCAS Futenma, Japan. The aircraft join two similar UC-35C Citation Ultras that are already in USMC service. One additional UC-35D will be delivered to the Marine Corps Reserve at NAF Washington/Andrews AFB, Md. The two UC-35C Citation Ultras are assigned to the Marine Corps Reserve at NAS New Orleans Joint Reserve Base, Louisiana.

Texan II full production
On 3 December 2001 the USAF gave its approval for the Raytheon T-6A Texan II to enter full-rate production, clearing the way for the award of a follow-on contract, valued at $1.4 billion if all of its options are accepted. The USAF and US Navy have already placed orders for 168 T-6As and 49 have been delivered to the USAF at Randolph AFB, Texas, and Moody AFB, Georgia.

New ARL-M system ordered
Northrop Grumman has been awarded a US$8.9 million modification to an existing contract to integrate an Airborne Reconnaissance Low-Multifunction (ARL-M) system into a de Havilland Canada DHC-7. The contractor had previously been authorised to procure the airframe to be modified. Once complete, the RC-7B will replace the O-5A Airborne Reconnaissance Low-Imagery (ARL-I) system aircraft that crashed in Colombia in July 1999.

Command post contract
The US Army has selected Raytheon to design and manufacture the Army Airborne Command and Control System (A2C2S) under a $110 million contract covering the development, low-rate initial production and the first year of full-rate production. A2C2S will be installed in UH-60L Blackhawk helicopters, turning the aircraft into airborne command posts. As part of the four-year contract, the company will deliver 34 systems.

Hawkeye 2000 delivered
The first production Hawkeye 2000 (A-179) flew for the first time at Northrop Grumman's St Augustine, Florida, facility in mid-October and was publicly unveiled on 23 October 2001.

C-130J-30 delivered
Lockheed Martin delivered the first two C-130J-30s to the Rhode Island Air National Guard's 143rd Airlift Wing in Providence, Rhode Island on 2 December 2001. The aircraft are the first stretched C-130Js for the USAF and the first -30Js to be equipped with the computer-controlled enhanced cargo handling system (ECHS) that allows precise airdrop event sequencing and quick conversion from cargo floor tie-downs to rollers for palletised cargo. The wing will receive a third aircraft by the end of 2001 and will eventually operate a mixed fleet of standard C-130Js alongside the stretched models.

AIR ARM REVIEW

Canada

Oldest Hornet retired
On 19 October 2001, CF-188B serial 188901 was delivered to the Canada Aviation Museum in Ottawa, Ontario, Canada. The aircraft, which last served with 410 'Cougar' Tactical Fighter Operational Training Squadron at CFB Cold Lake, Alberta, was the first Hornet delivered to the Canadian Armed Forces on 24 October 1982. After being secured for display by personnel from 3 Wing at CFB Bagotville, Quebec, the aircraft was placed on display at the museum, which is located at Rockcliffe Airport in Ottawa.

Czech Republic

Excess aircraft for sale
The Czech Republic has announced plans to sell its fleet of 15 Sukhoi Su-25K ground attack aircraft along with three Mil Mi-24D attack helicopters, six Mil Mi-17 and 17 Mi-2 helicopters, as well as Tu-154 and L-410 transport planes. The Vzdusne Sily Armady Ceske Republiky (Air Forces of the Army of the Czech Republic) is currently receiving new L-159 attack aircraft and searching for a new fighter aircraft. The JAS 39 Gripen has been selected to fulfil the CzAF's fighter requirement.

United Kingdom

Navy leaves Prestwick
RNAS Prestwick lost its last operational Fleet Air Arm aircraft when 819 Naval Air Squadron was decommissioned on 1 November 2001. The squadron had flown the Sea King HAS.6 from the station, which is also known as HMS *Gannet*. A detachment of Sea King HU.5s from 771 NAS at RNAS Culdrose will continue to operate from the station's Search and Rescue facility.

United States

Texan II training at Moody
The first class of student pilots has began flight training in the Raytheon T-6A Texan II with the 3d Flying Training Squadron (FTS) at Moody AFB, Georgia on 20 November 2001. The 13 USAF and two US Navy students, who are assigned Joint Specialized Undergraduate Pilot Training (JSUPT) Class 02-01, will complete the six-month course in April 2002. Subsequent classes will begin at three-week intervals and the 3rd FTS will train about 250 students annually.

US Coast Guard changes
The last HU-25A Falcon departed Air Station Borinquen at Rafael Hernandez Airport in Puerto Rico on 27 September 2001. The air station had only recently begun operations with the Falcon but the US Coast Guard is currently reallocating its assets and will operate the type from just four sites. Meanwhile, the service has begun taking delivery of updated HU-25C+ and HU-25D models from Northrop Grumman's California Microwave division in Maryland. During October 2001 Air Station Miami took delivery of a 'new' VC-4A from the Aircraft Repair and Supply Center at Elizabeth City, North Carolina. The Gulfstream I had previously served with the National Air and Space Administration (NASA).

USMC Sabreliner retired
The last CT-39G Sabreliner in service with the US Marine Corps was turned over to the US Navy on 14 September 2001 at NAS Pensacola, Florida. Known as the 'Scalded Dog', the aircraft had been assigned to MCAS Futenma, Japan's Headquarters and Headquarters Squadron (H&HS) as an operational support aircraft.

The Sabreliner flew its first mission in support of the 1st Marine Air Wing on 17 January 1999 and accumulated more than 2,600 flight hours by 6 September 2001 when its last OSA mission ended. The Sabreliner departed from Futenma on 10 September 2001 and made stops in Japan, Alaska and Montana before arriving at Pensacola, where it will see further service as a trainer.

Futenma received a Cessna UC-35D on 18 October 2001. The aircraft will serve in the operational support role and is assigned to the Air Station's Headquarters & Headquarters Squadron.

OSACOM to restructure
The US Army has announced plans to restructure its Operational Support Airlift Command (OSACOM) over the next three years. Besides reducing the size of the fleet it will be consolidated at a smaller number of facilities throughout the country. As part of the restructuring the fleet will see a reduction of 30 airframes during fiscal year 2002. This will result in the retirement of 19 C-12s in February 2002 followed by all 11 C-26Bs in September.

The remaining aircraft will be consolidated at 15 hub locations around the country between 2003 and 2004. The location of those sites has not been determined. Although the aircraft are in constant demand there have been continual shortfalls in funding, including $10 million in fiscal 2001 and $13.4 million in unfunded requirements for fiscal 2002.

Harriers return to sea
US Marine Corps Marine Expeditionary Units (MEU) are once again deploying with AV-8B Harriers assigned to their Aviation Combat Elements (ACE). As a result of the recent groundings of the Harrier fleet due to problems with their Rolls-Royce F402 Pegasus engines, no aircraft have been assigned to the two most recent MEU deployments in the Atlantic and Pacific, nor have any been assigned to the Unit Deployment Program (UDP) in Japan. With modification to engines complete the Harrier was once again cleared for deployment and VMA-311 deployed to MCAS Iwakuni, Japan, in support of the 31st MEU. Harriers subsequently deployed to the Western

Priority has been granted in Malaysia's latest defence budget for a new medium-lift transport helicopter to replace some 30 RMAF S-61A-4s currently in service. This example was present at Langkawi in October 2001.

Four Royal Australian Air Force F/A-18 Hornets deployed from RAAF Williamtown on 9 November 2001 as part of the country's contribution to the war on terrorism. The RAAF is also contributing two P-3Cs and a Boeing 707 tanker.

Pacific/Persian Gulf with the 15th in August and to the Mediterranean Sea with the 26th MEU in September.

Former Air Force One retired
The Secretary of the Air Force has announced that C-137C 72-7000, one of two Boeing 707s that served as Air Force One, will be placed on static display at the Ronald Reagan Presidential Library in Simi Valley, California. The aircraft made its final flight as a Presidential transport on 29 August 2001, when it transported President Bush from San Antonio, Texas to Waco.

The library will build a hangar to properly house and display the aircraft, which served President Reagan throughout his term in office and flew 445 missions as Air Force One during its nearly 29 years of service. The aircraft made its final flight on 8 September 2001 when it flew from Andrews AFB,

Maryland to San Bernadino International Airport, California.

DC-130 delivered to NWTS
On 24 May 2001 the Naval Weapons Test Squadron (NWTS) at Point Mugu, California, received the first of six C-130 transports that will eventually replace the unit's NP-3Ds. Originally designated as a DC-130H serial 65-0979 last served the USAF's 46th Test Wing at Duke Field, Florida, as an NC-130H but will likely revert to its previous designation in US Navy service. In addition to launching sub-scale aerial targets and providing airborne logistics support the aircraft will replace the NP-3Ds in a number of roles.

Raptors for Langley
The results of a recently completed environmental impact statement indicate that Langley AFB, Virginia is a suitable location

The 418th Flight Test Squadron flew its last EC-18B Advanced Range Instrumentation Aircraft (ARIA) mission on 24 August 2001, and both aircraft are now at Lake Charles, Louisiana, for conversion into E-8C JSTARS battlefield reconnaissance aircraft.

for 72 F-22A Raptor fighters. The base is already home to 66 F-15C/Ds operated by the 1st Fighter Wing and the USAF's preferred site for the initial operational wing of Raptors. Air Combat Command hopes to begin basing the F-22A at Langley in 2004.

USMC Hornets realigned
The 'Checkerboards' of VMFA-312

at MCAS Beaufort, South Carolina have transferred their F/A-18Cs to the 'Crusaders' of VMFA-122 and begun transitioning to the F/A-18A+.

The squadron has also been transferred from CVW-3 to CVW-1. VMFA-115, now equipped with the F/A-18+, and VMFA-122 will rotate operational deployment cycles with CVW-3.

OPERATIONS AND DEPLOYMENTS

United States

Lead Mobility Wings reduced
Air Mobility Command has announced that effective with Aerospace Expeditionary Force Cycle 3, which begins in March 2002, it will realign and reduce its Lead Mobility Wing (LMW) mission from five to two wings. Under the new structure the 60th Air Mobility Wing (AMW) at Travis AFB, California, and the 305th AMW, McGuire AFB, New Jersey, will share the LMW duties.

The LMW is responsible for providing mobility leadership for the AEF, and an initial response team (IRT) for humanitarian relief operations, disaster response and contingencies. The IRT deploys rapidly to assess local conditions, prepares the airfield to serve as a reception base for humanitarian relief, and assist in the bed-down of follow-on forces. The IRT also

works with an AMC Tanker Airlift Control Element (TALCE) to facilitate the flow of aid. Travis and McGuire were chosen as LMWs in part to link them respectively with the 615th Air Mobility Operations Group (AMOG) at Travis and the 621st AMOG at McGuire to enhance their crisis-response capabilities. The AMOGs are tasked with mobile command and control and airfield operations, and are capable of rapidly deploying to austere locations.

Beginning with the AEF 3 cycle the LMW rotation will be increased from 90 to 120 days and each will be aligned with one of the USAF's two combat-response Aerospace Expeditionary Wings. The 60th AMW will respond alongside Air Combat Command's 366th Wing, stationed at Mountain Home AFB, Idaho and the 305th AMW will be paired with the 4th FW at Seymour Johnson AFB, North Carolina.

Above; Joint JASDF, USAF and USMC Exercise Cope North 02-1 was successfully concluded on 20 November 2001. Based at Kadena AB, 44th and 67th FS F-15C/Ds take on fuel from a 909th ARS KC-135R during a defensive CAP sortie.

Below left: USS Nimitz (CVN 68) resumed operations in the autumn of 2001 following major modernisation, including the installation of a new radar antenna mast. Two E-2Cs approach the carrier prior to deployment aboard.

Below: USS John C. Stennis (CVN 74) departed San Diego on 12 November bound for the war in Afghanistan. The carrier was in theatre in time to join in the last phase of strikes against Taliban positions as this VF-211 'Checkmates' F-14A demonstrates leaving for a dawn raid in late December 2002.

Enduring Freedom

Fighting the War against Terrorism

Within hours of the terrorist atrocities committed in the US, fingers were pointing at Osama bin Laden, his al-Qaeda network and his Taliban hosts. A month after the attacks, the United States, supported by the UK, launched Operation Enduring Freedom with the aims of smashing the terrorist network, capturing or killing bin Laden, and removing the Taliban from power in Afghanistan.

Airpower employed in Operation Enduring Freedom centred on the USAF's heavy bombers, such as this B-1B (top) launching from Diego Garcia, and carrierborne attack aircraft, like this VFA-195 F/A-18 Hornet looming from the steam of Kitty Hawk's catapult.

Week 1 – 7 October/13 October

During the night of 7 October US and UK forces began the aerial assault on Afghanistan. The opening attacks followed predictable lines, with command and control facilities, air defence sites and airfields throughout the country bearing the brunt of the strikes. Airfield targets included Herat, Shindand, Shebergan, Mazar-e-Sharif, Kunduz, Jalalabad and several in the Kabul region. Four terrorist training camps were also hit, initiating the campaign against the al-Qaeda terrorist network. The scale of these first-night attacks was surprisingly limited: around 40 missions were flown by manned aircraft, including two B-2 missions and around 13 by B-1s and B-52s operating from Diego Garcia. The aircraft-carriers *Enterprise* and *Carl Vinson* launched around 25 strike aircraft, mostly F-14 Tomcats. US Navy surface ships and Royal Navy submarines launched 50 TLAM (Tomahawk) missiles between them.

Along with offensive action, the US also embarked on its first humanitarian aid mission, flying two C-17s from Ramstein in Germany to drop the first of over 2 million daily ration packs to refugees. The first drop was made along the Pakistan border, where refugees had

44 hours from Missouri – B-2 Spirit missions

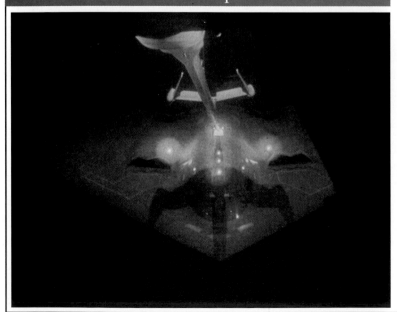

On the first three nights of the war the 509th Bomb Wing dispatched two B-2s to bomb key targets in Afghanistan. The missions were launched from Whiteman AFB, in Missouri, terminating in a one-hour, engines-running refuelling turnround and crew-slip at Diego Garcia, whereafter the B-2 returned to Whiteman non-stop. These gruelling efforts all topped 40 hours flying time (the longest was 44 hours) for the outbound/bombing leg, and over 30 hours for the return leg to Whiteman. It is believed that the B-2s employed GAM-113 GPS-guided 5,000-lb (2268-kg) penetration bombs against underground bunker targets, as well as the more conventional 2,000-lb (907-kg) JDAMs.

B-52 – pounding the Taliban

From the first day of the conflict, B-52s operated daily over Afghanistan. Ten B-52Hs were stationed at Diego Garcia, from where the 28th Air Expeditionary Wing launched five missions on each day. B-52s employed a variety of weapons, including 2,000-lb (907-kg) JDAMs, free-fall Mk 82s for area attacks and WCMDs (Wind-Corrected Munition Dispensers) with BLU-97 sub-munitions. In addition, B-52s were routinely used for psywar duties, using M129 leaflet bombs. In the latter part of the war B-52s were launched on long-endurance missions with no preassigned targets. They loitered over Afghanistan waiting for emerging targets to be identified by reconnaissance assets. Attack co-ordinates were relayed to the circling bomber within minutes.

Left: Armourers load 500-lb (227-kg) Mk 82s on board a B-52 at Diego Garcia. These weapons were used widely against Taliban front lines, especially to the north of Kabul and around Mazar-e-Sharif.

A B-52H is seen during a post-strike refuelling (above), while below four aircraft prepare to launch from Diego Garcia on 7 October for their first bombing mission of the war.

Diego Garcia, in the British Indian Ocean Territories, was home to the 28th AEW which supported operations by 10 B-52Hs, eight B-1Bs and KC-10A tankers. The base also played host to four F/A-18A Hornets from the RAAF's No. 77 Squadron which provided local air defence as part of Australia's contribution to the anti-terrorist effort. Missions from Diego Garcia routinely lasted 12-15 hours, including time on station over Afghanistan. The heavy bomber was an ideal weapon for use in Afghanistan, given the complete lack of any air threat. The ability to target new locations in minutes, combined with the proven accuracy of GPS-guided weapons, allowed B-52s and B-1s to operate in a close support role. Despite flying less than 10 percent of the combat sorties, the 28th AEW bombers dropped over 70 percent of the total ordnance expended in Afghanistan.

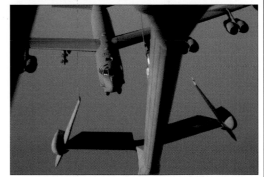

gathered to escape the impending hostilities. The fact that these missions could be flown on the first night of the war highlighted the small threat posed by the Taliban's air defences.

However, it was vital to ensuing operations that those defences were completely defeated, and the next three days' raids continued to hit air defence sites and airfields, albeit at a modest pace. On 8 October (Day 2) only 13 targets were hit by five-eight land-based bombers and 10-15 carrierborne aircraft, while even fewer sorties were mounted on 9 October. Tomahawks were also launched on 8 October (15) and 10 October (three). Humanitarian flights continued with two sorties daily, attention turning to the refugees in the north along the Turkmenistan border on Day 4. From Day

3 RAF tankers flying from Seeb in Oman routinely supported US Navy aircraft.

On 11 October (Day 5) only seven targets were hit and no humanitarian flights were mounted. On the next day, 12 October, no sorties were flown as US Central Command took stock of the situation so far.

Operations resumed on Saturday 13 October with renewed vigour. With the air defences

effectively nullified (although US forces would occasionally have to re-target SAM sites over the coming weeks), attentions turned fully to Taliban military installations and al-Qaeda facilities. Raids against Taliban military targets, such as motor pools and military training facilities, had begun on the third night of the campaign, and would occupy the US forces for some time. On 13 October the humanitarian aid effort was

Using a K-loader, 42 TRIADS (Tri-Wall Aerial Delivery System) boxes are pushed into a C-17 at Ramstein (above), containing a total of 17,200 HDRs. The boxes split open to dispense the ration packs. Tragically, at least one civilian was believed killed by the food drops. Shown right are two C-17s during the long transit to Afghanistan.

Humanitarian aid

Food drops to refugees by C-17 began on the first night of the war, and lasted until late December, when road and river traffic had been re-established. As well as HDR rations, C-17s also dropped wheat, dates and blankets in the latter stages.

The basis of the food drops was the HDR (Humanitarian Daily Ration), providing a nutritionally-balanced 2,200-calorie meal.

For food drops from medium altitude, the C-17 had to depressurise, requiring the use of oxygen masks. The C-17's unique roller system allowed automatic drops.

FRONT

BACK

Left: Leaflets were inserted in with the HDRs to leave no doubt as to the purpose for which they were intended. The food drops were were viewed as a vital part of the US public relations effort, both in Afghanistan and in the international community.

Right: A C-17 lands at Ramstein on 8 October after the first food-drop mission.

doubled to four C-17 sorties, but Central Command suffered its first public relations setback when a JDAM dropped by an F/A-18 accidentally landed in a residential area near Kabul airport instead of on its intended target – a military helicopter.

Week 2 – 14 October/20 October

The second week of Enduring Freedom continued with attacks aimed primarily against Taliban military infrastructure, with ongoing raids on al-Qaeda facilities. On the ground there was little in the way of action, the opposition Northern Alliance waiting in their positions for US raids to weaken the Taliban front lines before opening the ground war.

Week 2 was notable for the first appearances of the AC-130 (15 October) and F-15E (17 October), the latter believed to be operating from Kuwait. On 14 October B-52s dropped leaflets for the first time, scattering 480,000 over four areas of the country, while the psychological warfare effort was aided by the introduction of EC-130E(RR) aircraft flying Commando Solo radio broadcast missions.

From 16 October – Day 10 – operations picked up dramatically in intensity, with some 85 carrierborne attack aircraft being launched. On the following day the newly-arrived *Theodore Roosevelt* launched its first strikes. The carrier had sailed to the northern Arabian Sea to relieve the *Enterprise*, but for a few days all three carriers launched strike aircraft. Subsequently, *Kitty Hawk* arrived with a partial air wing and launched F/A-18s on combat missions.

During the raids of 16 October, an F-14D crew targeted a warehouse believed to be housing Taliban vehicles, but which actually held food and other relief stores for the ICRC (International Committee of the Red Cross). The following day, Taliban forces seized two warehouses which were in use by the UN World Food Programme. On a positive note, on 18 October the C-17 humanitarian aid effort passed 500,000 ration packs dropped.

In the preceding days there had been considerable speculation over the roles played by US and UK special forces, but on 19 October the US staged a major special forces assault on an airfield and barracks complex near Kandahar.

Troops, including Rangers, were inserted by parachute from MC-130s and helicopters on a mission to gather intelligence about the whereabouts of the Taliban/al-Qaeda leadership and to destroy command and control facilities. The same day the US incurred its first casualties when two Rangers were killed in the crash of a UH-60 (also reported as an MH-60K) close to the Pakistan border. The Black Hawk was recovered the next day by a Marine TRAP team flying CH-53Es from *Peleliu*.

The week ended on a sour note for US Central Command with two further bombs going astray, this time hitting a residential area in northwest Kabul after release by a Tomcat.

Week 3 – 21 October/27 October

During the third week of Enduring Freedom US forces turned up the pressure on Taliban front lines, although the campaign against military infrastructure and al-Qaeda continued unabated. Among the latter class of targets were cave complexes near the Taliban's southern stronghold of Kandahar. Missions against the front lines were conducted in the two areas

B-1 Lancer at war

Eight B-1Bs were deployed to Diego Garcia from Ellsworth AFB, South Dakota, and Mountain Home AFB, Idaho, as the lead element of the 28th Air Expeditionary Wing. From the Indian Ocean base they launched, on average, four sorties a day over Afghanistan. Following the initial campaign against fixed, pre-assigned targets, B-1s began to roam over designated 'kill-box' areas, waiting for emerging targets to be identified by reconnaissance and ground-based intelligence sources. The ability to re-target the JDAM inflight was of inestimable value in this role, and the 2,000-lb (907-kg) GBU-31 became the B-1's most useful weapon. Free-fall Mk 82s and M129 leaflet bombs were also dropped. In the latter stages of the campaign it was mooted that the B-1s should move to a base in Oman, probably Thumrait, from where they could maintain 24-hour coverage over Afghanistan, each aircraft flying a 12-hour shift on-station. This had not occurred by the time of the fall of the Taliban in mid-December.

Under the USAF's Expeditionary Air Forces concept, the 28th Bomb Wing at Ellsworth was the nominated lead wing for the period 1 September to 30 November 2001 (AEF 8). The heavy bombers at Diego Garcia were the most visible evidence of the USAF's enormous contribution to Enduring Freedom.

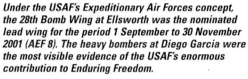

Above: With its wings spread for take-off, a B-1B lumbers down Diego Garcia's taxiway, closely followed by a KC-10A tanker. On 12 December a B-1 was lost in a non-combat related accident. Thankfully, all four crew ejected safely and were plucked from the Indian Ocean by a US Navy destroyer.

Right: Armourers on Diego Garcia worked extremely hard to keep the bombers supplied. Nine heavy bombers were launched on most days, and considering that a B-1 can lift 84 Mk 82s, the bomb-loading exercise took on massive proportions. The strap-on guidance system distinguishes the JDAM family.

The B-1B can carry 24 of the 2,000-lb (907-kg) GBU-31 JDAM (Joint Direct Attack Munition). The GPS guidance provides accuracies in the region of 10 ft (3 m). The standard weapon (below) is based on the Mk 84 warhead, but a penetrating version is also available (below right), employing the BLU-109 warhead for use against caves and bunkers. Enduring Freedom is believed to have provided the combat debut for this version.

Above: GBU-31 JDAMs sit on trolleys, waiting to be loaded on to B-1s for the next Enduring Freedom mission. As well as warhead options, the GBU-31 can be configured to explode above ground, upon impact or in penetrating mode, presenting a range of options. Widespread use of the weapon in Afghanistan stretched US stocks.

F-14 – 'Bombcat' and Fast FAC

F-14s flew from three carriers: *Enterprise* (VF-14 and VF-41 with F-14As), *Carl Vinson* (VF-213 with F-14Ds) and *Roosevelt* (VF-102 with F-14Bs). The Tomcat's LANTIRN capability, second crew member and good endurance made it ideal for the Fast FAC role, loitering over the battlefield to designate targets for the shorter-legged Hornets, as well as delivering its own ordnance.

Above: The twin undernose sensor fairings identify this Tomcat as an F-14D – from VF-213 aboard Carl Vinson.

Above: CVW-8 aircraft from Enterprise, *including a VF-41 F-14A, refuel from a KC-10 on 5 October, two days before the campaign began. The Tomcat is carrying a full live air defence load.*

As well as LGBs, F-14s also used free-fall weapons such as these 1,000-lb (454-kg) Mk 83s being loaded on to a VF-102 'Diamondbacks' aircraft (above left). Contrary to earlier reports, VF-102's aircraft were not JDAM-capable. Shown below left is a VF-213 F-14D returning to Carl Vinson.

Above: A VF-102 F-14B prepares for a night launch from Roosevelt *on 16 November. Tomcat crews were in the thick of the action from the first day. On at least one occasion an F-14 crew, with all other ordnance expended, descended to strafe ground targets from low level with the M61 Vulcan cannon.*

where the Northern Alliance was strongest: around the northern town of Mazar-e-Sharif and on the Shomali Plain to the north of Kabul.

As well as pre-planned target sets, US forces also operated over engagement zones, circling while waiting for what became known as 'emerging targets'. The sight of B-52s lazily orbiting overhead, pulling contrails, was to become the enduring image of the air war. Flexible mid-air targeting became a vital part of the air campaign, with target locations being fed rapidly to attack aircraft through datalinks. Intelligence was provided by ground sources or by UAVs such as the RQ-1 Predator.

Although daily mission rates remained high – typically 60-80 aircraft – aircraft often returned to their carriers with weapons still aboard as no targets had 'emerged' during their time over the engagement zone. On the other hand, some crews resorted to using guns in strafing attacks as they had expended all their bombs.

There were, inevitably, further targeting mistakes. On 21 October a malfunctioning laser-guided weapon landed outside an old people's home in Herat, while a second ICRC warehouse complex in Kabul and nearby residential area were bombed by accident on 25 October.

Week 4 – 28 October/3 November

A high rate of operations was maintained throughout the week, with 55-90 combat sorties being mounted daily. Missions were divided between pre-planned attacks against Taliban and al-Qaeda targets (including cave and tunnel complexes), and engagement zone

Psychological warfare

As part of the overall psywar effort, the EC-130E(RR) Rivet Rider aircraft from the 193rd SOS, Pennsylvania ANG, worked hard on Commando Solo radio broadcast missions.

From early in the campaign US Central Command conducted a concerted psywar effort against Taliban forces, as well as targeting displaced refugees and the Afghan population at large to ensure that broad public opinion in Afghanistan remained 'on-side and on-message'. EC-130 Commando Solo aircraft flew regular radio broadcasting missions, which became especially important after official Afghan radio stations had been knocked out. Leaflets were dropped to advise of the Commando Solo frequencies. Psywar leaflets were also widely dispensed over Taliban troop concentrations, using M129 leaflet bombs dropped by B-1s and B-52s. These leaflets urged Taliban fighters to switch sides, and warned of the consequences should they choose to fight on. Two anti-Taliban examples are shown here: on the left is a leaflet showing what was reported to be the car of the Taliban leader (Mullah Omar) and an obvious message, while the leaflet at right implores the reader to "Drive out the foreign terrorists" on the front and asks "Is this the future you want for your women and children?" on the back.

FRONT

دهشت افگنان
جنبی را بیرون کنید

پردي دار
اچنکي وشمو

BACK

آیا برای آینده زنان و اطفال خود
این نوع زندگی را میخواهید؟

آیا داسی ژوندون دخیل
ښخو او ماشومانو لپاره
غواړئ؟

Hornet swarm

By virtue of the numbers available on the four carriers used in 'OEF', the Hornet flew the greatest number of sorties, although this entailed a sizeable refuelling effort. Hornets were used to drop a variety of precision-guided weapons, mostly GBU-12 and GBU-16 LGBs (often designated by F-14s or by sources on the ground) and 1,000-lb (454-kg) GBU-32 JDAMs.

Above: Despite leaving some of its air wing behind in Japan (including the F-14As of VF-154), Kitty Hawk joined in the fray with regular Hornet attacks. Here a VFA-195 aircraft prepares to launch with four GBU-12s.

Above left: This self-portrait shows a VFA-94 pilot.

VFA-82 'Marauders' operated from Roosevelt. Here two of the squadron's aircraft are seen launching (above), and recovering with unexpended ordnance (below).

Above: A VFA-195 F/A-18C returns to Kitty Hawk, still carrying a GBU-12. On 'emerging target' missions it was not uncommon for ordnance to be brought back.

A VFA-97 Hornet taxis on Carl Vinson with a GBU-16 (above), while below is a VMFA-251 aircraft about to launch from Roosevelt with two GBU-12s.

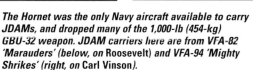

The Hornet was the only Navy aircraft available to carry JDAMs, and dropped many of the 1,000-lb (454-kg) GBU-32 weapon. JDAM carriers here are from VFA-82 'Marauders' (below, on Roosevelt) and VFA-94 'Mighty Shrikes' (right, on Carl Vinson).

Above is VFA-82's 'CAG-bird' returning to Roosevelt, while below is VFA-94's JDAM-armed 'CAG-bird' refuelling from a KC-135.

Below: Returning to its carrier on 30 October, this Hornet carries a single AIM-7 Sparrow medium-range AAM under the wing and AIM-9 Sidewinders on the wingtip launch rails.

Above and above right: The Boeing E-3 fleet flew from Thumrait in Oman, and involved aircraft from both the USAF's 552nd Wing and the RAF's No. 8 Squadron.

AWACS cover

In the opening days the AWACS force was used in its traditional air defence role, but as the threat of air opposition evaporated, the flying radar stations were primarily used to direct the flow of strike aircraft into and out of Afghanistan, and to assist refuelling rendezvous.

While most AWACS operations over Afghanistan itself were handled by the E-3 Sentry, Navy Hawkeyes were used from their orbits over Pakistan to direct strike aircraft through assigned corridors. The E-2 crew took control of aircraft as they left the carrier, guided them through post-launch refuelling and the ingress route, and then handed them off to the Sentry. As strike aircraft left Afghanistan the reverse procedure was undertaken. These Hawkeyes are from VAW-124 'Bear Aces' in Enterprise *(above, on first night of war), and VAW-123 'Screwtops' in* Roosevelt *(left and far left).*

sorties against front-line areas. The pressure was increased considerably in the Mazar-e-Sharif and Shomali regions, with B-52s carpet-bombing Taliban troop positions.

During the night of 30 October the millionth HDR ration pack was dropped to refugees. The feat had been accomplished in 61 sorties covering over 400,000 air miles. The C-17 effort had stabilised at two missions daily. The psychological warfare effort also continued, with both leaflet-dropping and Commando Solo broadcasts. The two efforts were often co-ordinated,

with leaflet drops in areas where food had been delivered.

On 2 November a US military helicopter (believed to be a USAF MH-53M Pave Low) was lost in Afghanistan, presumably on a special forces mission. The downed aircraft was subsequently bombed by VF-102 Tomcats with laser-guided bombs to prevent sensitive material from falling into the wrong hands. An RQ-1B Predator UAV was also lost, believed to be the victim of pitot icing.

On 3 November the USMC's Harriers joined

the fray, flying their first mission from USS *Peleliu*. USMC F/A-18s had already been in action as part of CVW-1 aboard *Roosevelt* (VMFA-251).

Week 5 – 4 November/10 November

As fighting between Northern Alliance troops and Taliban/al-Qaeda forces intensified around the strategically important town of Mazar-e-Sharif, so Central Command stepped up its bombing of troops in the region. On

Battlefield surveillance

Finding al-Qaeda/Taliban forces and their hideouts was one of the bigger challenges faced in Enduring Freedom. Much of the work was performed by drones, including the RQ-4 Global Hawk, flying from al-Dhafra in the UAE, also base to French Mirage IVs and U-2Ss. The Boeing E-8C J-STARS (below and below left) was deployed forward to the Gulf region.

Above: Royal Air Force assets played a significant role in the surveillance effort, using two Canberra PR.Mk 9s and a single Nimrod R.Mk 1 flying from bases in Oman.

Marine TRAP mission

Following the downing of a US Army UH-60 (or MH-60) near the Pakistan border the previous day, the 15th MEU(SOC) launched a TRAP mission on 20 October to recover the helicopter. CH-53Es, refuelling from KC-130s, flew to the crash site, where Army engineers quickly stripped the UH-60. The CH-53Es took the UH-60 away, the airframe carried beneath one of the helicopters, landing at a base just inside Pakistan for refuelling. While at this base, the US helicopters came under fire from someone on the ground. After returning fire with the CH-53's 0.50-in calibre machine-guns and personal weapons, the Marines regrouped and took off quickly to clear the area, leaving the UH-60 airframe behind. However, they had not refuelled, and had to move to another base, where they refuelled from other helicopters in the recovery package. All made it safely back to *Peleliu*, but then had to return for the Black Hawk. This time, the Marines went in with US Navy air cover, and AH-1 escort. EOD specialists were carried to ensure that the UH-60 had not been booby-trapped. The lift was successful and the UH-60 recovered.

With a CH-53E in the background, Marines rehearse setting up a defensive perimeter on the deck of Peleliu, in preparation for a TRAP mission.

Right: On 26 November this RAF Hercules arrived at Bagram to fly out Northern Alliance delegates to the conference in Bonn, aimed at creating a provisional Afghan government. An SBS team had earlier been inserted into Bagram to secure the airfield.

4 November MC-130H Combat Talon IIs dropped two BLU-82 15,000-lb (6804-kg) 'daisy-cutter' bombs on Taliban forces in an attempt to break the deadlock on the front lines. The use of this weapon had a significant psychological effect, in addition to its considerable destructive power against 'soft' targets over a wide area.

Attacks were also mounted in the Herat region, which was also under pressure from Northern Alliance troops, while the campaign against al-Qaeda tunnels and caves continued.

On 4 November a civil-registered DHC-6 Twin Otter was filmed landing at an airstrip at Sherkat, to the north of Kabul. The aircraft, believed to have been operating from Tajikistan, is thought to have brought in special forces, perhaps a surveillance team to assess the possibility of using the airfield for further operations. Later in the day at least two US-operated Mi-17s brought in further personnel to the strip.

Much of the success of the US bombing campaign was attributable to the large number of US and UK special forces deployed into

Afghanistan to reconnoitre targets and to provide laser designation. Special forces were also instrumental in advising Afghan opposition groups, and were occasionally spotted by camera crews during joint operations.

On 5 November reports surfaced of US gunship helicopters being used in attacks against a specific building in downtown Kabul.

At the time of writing no details had been released of this attack, but the gunship was clearly an ideal weapon to use against an important target in a heavily populated area.

At the end of the week Taliban resistance finally crumbled in Mazar-e-Sharif and Herat, both cities falling to Northern Alliance forces with US/UK special force assistance. The

USAF transports – supplying the front line

With its forces spread throughout the Middle East region and on the far-flung island of Diego Garcia, the US required a massive logistic effort by Air Mobility Command to keep its forces supplied and flying. C-5 Galaxies (left and right) were at the forefront of operations, assisted by C-17s (below) and C-141s. US bases in Europe and the Mediterranean, such as Morón in Spain and Incirlik in Turkey, were very busy as AMC transports staged through on their way to the forward deployment bases.

S-3Bs were vital to the carrierborne strike effort, providing post-launch top-off tankings to inbound attack aircraft, and then emergency refuelling cover for returning fighters. A handful of sea search missions were undertaken, especially during a period when it was thought that some al-Qaeda terrorists were escaping by boat, having crossed into Pakistan. Shown here are S-3Bs from VS-21 aboard Kitty Hawk (above and below left), and VS-32 aboard Roosevelt (below).

Carrierborne support

With the Afghan air defences nullified within hours and with no appreciable naval aspect to the operation, the Navy's Viking and Prowler fleets were assigned support roles. The Vikings, in particular, were extremely busy, flying intensely on refuelling missions, while the Prowlers engaged in their secondary roles of electronic intelligence gathering and comms jamming.

Above: A VAQ-137 'Rooks' EA-6B returns to Theodore Roosevelt (CVN 71) at the end of a mission. In the early days of the campaign EA-6Bs fired a handful of AGM-88 HARM missiles, but for most of the conflict were involved in Sigint missions, working closely with EP-3E Aries II land-based aircraft operating from Bahrain and EC-130H Compass Calls. Much of the Taliban's command and control was effected by mobile phone, and it is thought that EA-6Bs could monitor this traffic.

capture of two of Afghanistan's major cities was the first tangible success for the anti-Taliban coalition, and marked the beginning of the end.

Week 6 – 11 November/17 November

Buoyed by their successes at Mazar-e-Sharif and Herat, and with much of the northern area under their control, the Northern Alliance continued their push on Kabul across the Shomali Plain. Following relentless bombing from US aircraft, notably B-52s, Taliban forces pulled back from the front lines, and then slipped out of Kabul altogether on Tuesday 13 November. Despite pressure from the international community to remain outside the city until some semblance of an interim government had been formed, Northern Alliance forces drove victoriously into the capital, to the evident pleasure of the population.

In the wake of the fall of Kabul, Taliban forces regrouped around their strongholds of Kandahar in the south and Jalalabad in the east. Both were heavily targeted throughout the week by US bombers, attention being paid especially to the many al-Qaeda facilities in the mountains surrounding Jalalabad. During raids on 14/15 November the US claimed to have killed Osama bin Laden's right-hand man, Mohammed Atef.

On 14 November the coalition's surveillance fleet received a boost by the arrival of an RQ-4 Global Hawk UAV, which flew non-stop from RAAF Edinburgh, New South Wales, to Al Dhafra in the UAE for operations over Afghanistan. Meanwhile, on 12 November the carrier John C. Stennis left its home port of San Diego on 12 November, bound for the war zone. It would not arrive in the northern Arabian Sea until mid-December, in time to participate in the final combat missions.

On the ground, the first overt deployment of coalition forces came when 100 elite troops from the UK's SBS (Special Boat Service) were flown into Bagram airport to the north of Kabul, their mission being reconnaissance for

Above: A Marine keeps close watch as part of the defensive perimeter around early ground operations at Bagram. Marine Hercules, like this VMGR-352 KC-130R, were used for mid-air refuelling, transport and for ground refuelling of forward-deployed helicopters.

USS John C. Stennis (CVN 74) sails out of San Diego on 12 November 2001. The vessel and its air wing (CVW-9) had just completed a fleet training period, and were turned round and dispatched with great haste. Stennis arrived in the northern Arabian Sea in mid-December and flew operations in the last days of the campaign.

Harriers at sea

Boeing/BAE AV-8B Harrier IIs joined the attack force on 3 November, when four aircraft from the 15th MEU(SOC) launched from USS *Peleliu*, each armed with four 500-lb (227-kg) Mk 82 bombs. The sorties lasted for about four hours and involved two refuellings. *Peleliu*'s Harriers are Night Attack aircraft. Later, AV-8Bs from the *Bataan*'s 26th MEU(SOC) joined the attacks, including radar-equipped AV-8B Harrier II Plus aircraft. One of these aircraft is seen above being directed on deck. It carries one Mk 82 which was brought back to the ship after the mission. Marine KC-130 Hercules tanker/transports were used for refuelling purposes, and also to provide transport. They were involved in moving Marines into Dolangi airfield after the initial heliborne insertion.

An AV-8B launches from Peleliu *(LHA 5) on the type's first combat mission on 3 November (above), carrying four Mk 82s.* Peleliu *was subsequently joined by* Bataan *(left, LHD 5), carrying the 26th MEU(SOC) and its full ACE (Air Combat Element).*

ensuing peacekeeping/humanitarian operations. Their deployment was not entirely welcomed by the Northern Alliance, although their presence was tolerated due to the humanitarian nature of their mission.

The week ended with the beginning of the Moslem holy month of Ramadan, but bombing continued, with attention focusing on the town of Kunduz, where Taliban/al-Qaeda forces retained one last foothold in the north of the country.

Week 7 – 18 November/24 November

Throughout the week US aircraft bombed Taliban/al-Qaeda positions in Kunduz as the Northern Alliance closed in on the city. Resistance was fierce: many of the defenders were non-Afghan fighters driven by religious beliefs, while those attempting to desert or surrender were killed. However, at least 1,000 fighters, mostly Afghans, surrendered

Throughout the week international support for the anti-terrorism coalition continued to grow. The first French ground forces arrived in Mazar-e-Sharif to aid and support the humanitarian effort in the region. On 23 November USAF KC-135R tankers from the 100th Air Refueling Wing flew their first Enduring Freedom missions from Burgas in Bulgaria. On 24 November US bombs inadvertently fell on Pakistani territory, causing no damage.

Week 8 – 25 November/1 December

Enduring Freedom moved into its ground phase at the start of the week. In an operation entitled Swift Freedom, US Marines from the 15th MEU (*Peleliu*) and 26th MEU (*Bataan*) began to deploy to a base in Afghanistan. The Marines first went to a shore base before being airlifted forward into Afghanistan to set up an operating location at Dolangi airstrip. Dubbed Camp Rhino by Central Command, this airstrip was located in a remote region to the west of Kandahar. It had originally been established to serve a game-hunting reserve, but had latterly been used by Taliban and al-Qaeda leaders. The first Marines, from Charlie Company, Battalion Landing Team 1/1 of the 15th

Tanker support

A massive refuelling effort was required, especially for the US Navy aircraft flying from carriers and for the USAF's tactical jets. KC-10s (below right) were based at Diego Garcia and other locations, while KC-135s (below left) operated from Middle East bases and from Incirlik in Turkey and Burgas in Bulgaria. RAF VC10 and TriStar tankers operated from Seeb, and were praised for their willingness to accompany strike aircraft deep into Afghanistan.

RAF VC10 and TriStar tankers worked closely with US Navy tactical aircraft. Here a No. 101 Sqn VC10 refuels two JDAM-carrying Hornets from VFA-22.

'Viper' over Afghanistan

F-16s operated on long-range missions, believed to have originated from Doha in Qatar. The 366th Wing's 389th Fighter Squadron was involved, as were aircraft from Air Force Reserve units, including the 419th FW from Hill AFB, Utah. The F-16 was used for precision attacks using GBU-12 bombs designated by the Litening pod. F-16s also accompanied B-1s during attack missions.

Above: Carrying a baggage pod, an F-16CJ Block 50 from the 366th Wing arrives at its Middle East operating location after deploying from the US.

Right: A pilot from the Air Force Reserve Command's 419th Fighter Wing performs a walk-round check of his F-16 prior to an Enduring Freedom mission. The F-16 carried the Litening targeting pod, ALQ-131 ECM pod and GBU-12 laser-guided bombs.

MEU(SOC), were flown in by CH-53E under escort of AH-1 and UH-1 gunships, and under cover of AV-8s and carrierborne jets. In less than an hour they had swept the immediate vicinity, cleared buildings, established a security perimeter and installed runway lights along the sand strip. KC-130s arrived to bring in Alpha Company and tactical vehicles, and a constant flow of troops and equipment followed. With the airstrip prepared for heavier aircraft, C-17s were able to fly in heavy equipment.

Eventually over 1,000 Marines were deployed to Camp Rhino, which provided a secure base from which to set up roadblocks around Kandahar, and to undertake reconnaissance patrols into the surrounding countryside in search of the Taliban and al-Qaeda leadership.

On their third day ashore, the Marine force came under threat from a Taliban vehicle column which approached Dolangi. The vehicles were swiftly dealt with by a force of USMC AH-1 gunships and LGB-armed F-14s and F/A-18s from US Navy carriers.

In the north, fierce fighting continued around Kunduz, until the city finally fell to the Northern Alliance forces of General Daoud on 26 November.

After the fall of Mazar-e-Sharif, Taliban prisoners had been held in a fort by their Northern Alliance captors. On 26 November the prisoners instigated an uprising that took three days to put down. Armed with concealed and stolen weapons, the Taliban put up a fierce fight which resulted in most being killed. US and UK special forces were heavily involved in quelling the revolt, one CIA agent – Johnny Spann – becoming the first US combat death of the war. Airpower was brought to bear during the uprising, with around 30 sorties targeting the fortress with precision weapons. Several US and UK

Airfield attack

Although the Taliban's air force was tiny and ineffective, it had to be neutralised early in the war. Even abandoned wrecks were targeted.

Shindand airfield (above) was hit with great precision, bombs striking parked aircraft, the runway and taxiways (below). The latter were hit to deny them being used as alternative runways.

The results of an earlier strike against runways and taxiways at Kandahar (above) can be seen, while MiG-21s in the revetments were taken out in a subsequent raid (below).

Herat in western Afghanistan boasted a line-up of derelict fighters and an An-24 transport. Although it was inconceivable that they could play any part in the war, they were destroyed.

Kabul military airfield had already been hit (above), but US bombers returned to destroy two hangars which were being used to store Taliban military equipment (below).

Strike Eagle

F-15Es were widely used, flying from Ali Al Salem air base in Kuwait. A range of precision weapons was employed, including the TV-guided AGM-130 missile (below) and laser-guided bombs. The first unit in action was the 391st 'Bold Tigers' from the 366th Wing. F-15Es occasionally worked in co-operation with RQ-1 drones which provided detailed target information.

Although used in small numbers, and facing a long journey from their base in Kuwait, the 391st EFS F-15Es were very useful on missions requiring precision and flexible 'on the hoof' targeting. The GBU-12 500-lb (227-kg) LGB was widely used – a small weapon useful against armour and vehicle targets.

Above: An inbound F-15E refuels during the long haul to Afghanistan, clutching a GBU-10 2,000-lb (907-kg) LGB to its starboard wing pylon. Despite the lack of any air threat, F-15Es routinely carried Sidewinders and AMRAAMs.

Below: Armed with GBU-12s, a 391st EFS F-15E launches on the type's first Enduring Freedom mission on 17 October.

Two views show an RQ-1B operating on an Enduring Freedom mission. The UAV is operated by the 11th Reconnaissance Squadron of the 57th Wing, normally based at Indian Springs Air Force Auxiliary Field, near Las Vegas, Nevada.

RQ-1B Predator – the UAV goes to war

RQ-1B Predators, operated by both the CIA and USAF, were widely used for surveillance and, for the first time, in direct action against key targets. At least three were lost, two of which crashed in icing conditions, which are known to affect the pitot tube and cause the UAV to fly into the ground. In general, the Predators were used to provide detailed targeting information, often expanding on electronic intelligence from sources such as the Nimrod R.Mk 1, RC-135 Rivet Joint and EP-3E Aries II. In this role they worked alongside U-2s and, later, RQ-4 Global Hawks and E-8 J-STARS. Predators fed live imagery back to their ground stations, revealing intimate detail of targets, whose precise location was then uplinked to loitering strike aircraft. Predators worked closely with, among others, AC-130 gunships and F-15Es. In the direct attack role the Predator carried two AGM-114 Hellfire missiles (above), which could be guided by operators at ground stations in real time. In one instance they were used against a specific al-Qaeda target in a residential area, where the use of precise aiming and a small warhead were imperative.

A pair of MH-53M Pave Low IVs flies over the Afghan desert. Such aircraft are likely to have been used for special forces insertion missions, and to have provided a combat search and rescue alert capability.

servicemen were reported wounded by a JDAM explosion.

Further nations joined the coalition effort. On 25 November an Iranian AF An-74 arrived at Herat, reportedly with British special forces aboard, while the following day a large Russian contingent arrived at Kabul in a fleet of Il-76s to prepare for humanitarian work. On 30 November the USAF's C-17s dropped their two millionth HDR package.

The week ended with the noose closing around Kandahar and Jalalabad, and with US aircraft concentrating their attacks in these regions to support local opposition groups.

Week 9 – 2 December/8 December

The fight for Kandahar, the last city in Taliban hands, continued throughout the week with Pashtun anti-Taliban ground forces being supported by US aircraft. US aircraft also began

raids in the Tora Bora mountain region, where many al-Qaeda fighters were believed to be grouping in the myriad of caves. Anti-Taliban forces took control of Kandahar on 8 December, although pockets of Taliban stayed on and tension remained high for many days after. The Marines began major interdiction operations from their base at Dolangi.

Despite the success at Kandahar, the week was marred for the US by the deaths of two US special forces and the wounding of 20 more when a JDAM dropped by a B-52 exploded close to their position. A third troop died of his injuries during the medevac flight.

Week 10 – 9 December/15 December

During the week US aircraft concentrated on the last major area under Taliban/al-Qaeda control: the Tora Bora mountains south of Jalalabad. Large numbers of bombs were

dropped against forces found on the move, and against cave and tunnel entrances. AC-130 gunships were much in evidence, while on 9 December an MC-130H dropped a BLU-82 in the region. Afghan ground forces and special forces followed up after air attacks, inspecting and clearing caves. Additional Taliban resistance was also met near Shindand, around Kandahar and in Helmand province to the northwest of Kandahar.

On 12 December a B-1B was lost due to a major systems malfunction, the four crew ejecting at 15,000 ft (4572 m) into the Indian Ocean north of Diego Garcia. A KC-10 tanker orbited overhead, directing the destroyer USS *Russell* to pick up the crew.

Humanitarian air drops continued, delivering HDR ration packs, blankets, wheat and dates. They were due to end on 13 December, as road traffic from Uzbekistan (across the Friendship Bridge) and Pakistan (from Quetta and through the Khyber Pass) had been fully re-established. However, air drops continued for a few more days after to remote areas. Commando Solo and leaflet-dropping sorties also continued.

On 14 December US Marines secured Kandahar airport as a major hub for humanitarian aid and interdiction/reconnaissance operations.

Supplying the fleet

CH-46s from both HC-8 (illustrated) and HC-11 helped shuttle supplies from ship-to-shore, and around the battle groups.

C-2 Greyhounds kept up a shuttle between the carriers and the 5th Fleet base at Bahrain. Here a VRC-40 Greyhound unloads vital supplies aboard Enterprise.

MH-53Es are better known as mine-sweepers, but are also used for transport tasks. Here one lands on Roosevelt to deliver supplies.

Enduring Freedom

An MC-130H Combat Talon II refuels from a KC-135 (above), while at right one is seen operating at a base in Afghanistan. MC-130Hs paradropped Special Forces into Kandahar during the setpiece raid on 19 October and also dropped at least three 15,000-lb (6804-kg) BLU-82/B bombs.

Week 11 – 16 December/22 December

On Sunday 16 December the Tora Bora region was cleared, marking the end of the main air campaign: on the following day no ordnance was expended. Air operations continued at an intense pace, however, with 70-90 sorties being mounted daily to provide air cover for mopping up operations on the ground. By this time the US forces had been joined by Italian navy AV-8B Harriers flying from the carrier *Giuseppe Garibaldi*.

Scheduled humanitarian air drops ended during the week, with nearly 2.5 million HDRs delivered throughout this major effort. C-17s remained on call to provide further air drop capability should it have been required.

On 22 December Hamid Karzai was sworn in as prime minister of the interim government, 77 days after the launch of Enduring Freedom operations.

Aftermath

Air operations continued throughout late December and into the new year as ground forces embarked on the process of mopping up the last pockets of resistance, sifting through caves and former Taliban/al-Qaeda facilities for intelligence concerning the whereabouts of Osama bin Laden and Mullah Omar, and in repairing and re-establishing transport infrastructure to facilitate the influx of humanitarian aid and peacekeeping forces.

Taliban/al-Qaeda fighters were hit when they were found: bombs were dropped on 25 December, 26 December, 28 December and on several days in early January. The latter attacks were against al-Qaeda fighters who had regrouped at the Zawar Kili facility hit by cruise missiles during the 1998 US retaliation for the embassy bombings.

On 30 December an RQ-4 Global Hawk UAV crashed on return to its base, while on 9 January a USMC KC-130R crashed while attempting to land in Pakistan, killing all seven on board. ***David Donald***

Right: Marines from the 26th MEU watch as equipment is loaded on to a C-17. Having deployed ashore from Bataan, the Marines were airlifted forward to the base at Dolangi.

Camp Rhino – Marines ashore

Above: AH-1Ws operating from Peleliu (illustrated) and Bataan escorted assault helicopters as they landed Marines at Dolangi. They were subsequently used to protect Marine ground operations.

Left: A Navy Seabee maintains vigilance while equipment is offloaded from a C-17 at Camp Rhino (Dolangi). Seabees were instrumental in preparing the rough strip for operations by heavy transport aircraft. Initially Marine KC-130s were used, before C-17s joined the airlift effort.

On the night of 25 November CH-53Es flying from *Peleliu*, escorted by AH-1s and UH-1s, landed at the remote Dolangi airstrip. The first troops to disembark were Marines from Charlie Company, Battalion Landing Team 1/1, 15th MEU (SOC). With the few buildings cleared and the field secured, KC-130s brought in the Marines of Alpha Company. Over the next few days the Marine force built up to over 1,000, with the aim of providing a stronghold for search and destroy missions into Taliban/al-Qaeda territory. A Taliban armoured column approached the base on 27 November, but was attacked and destroyed by a combination of Navy F/A-18s working with F-14 FACs, and AH-1W attack helicopters. On 14 December some of the Marine force moved forward from Dolangi to the airport at Kandahar.

F-15A Improved Baz 678 is seen here fitted with four Python 3 missiles – a weapons load more associated with the basic Baz configuration. The patch worn by Improved Baz pilots is shown above.

Israeli F-15 upgrade

Improving the Baz

Only a decade after the McDonnell Douglas F-15 Eagle, known locally as the Baz (Buzzard), entered Israel Defence Force/Air Force (IDF/AF) service in 1976, the possibility of an upgrade programme was investigated. Two options were initially evaluated: the USAF Multi-Stage Improvement Program (MSIP) and the IAI-promoted Baz 2000. The two fundamental elements of MSIP were the introduction of the APG-70 radar with multi-target engagement capability and the AIM-120 Advanced Medium Range Air-to-air Missile (AMRAAM). Baz 2000 involved the integration of avionics from the IAI Lavi multi-role combat aircraft, development of which was cancelled by the Israeli government on 30 August 1987. After initial studies, both upgrade programmes were rejected on the grounds of budget restraints and priority of budget allocation.

The IDF/AF F-15A/B/C/D fleet was forced to soldier on with only minor improvements for another decade, before the catalyst for a major upgrade progamme finally emerged in the shape of the 1994 purchase of the F-15I Ra'am (Thunder), a derivative of the F-15E Strike Eagle. Known as Baz Meshopar (Improved Baz), the upgrade programme actually incorporated advanced 1990s technology F-15I avionics elements into the proven F-15A/B/C/D airframes that at the time were still equipped with 1970s vintage technology avionics systems.

The two major elememts of the Improved Baz programme are the modernisation of the avionics systems and the introduction of new weaponry options. The integration of the F-15I mission and weaponry computers along with Multi-Function Display (MFD) cockpit architecture, as well as the replacement of the original F-15A/B/C/D stick with a Hands On Throttle And Stick (HOTAS), similar to that fitted to the F-15I, all contribute to the improved combat efficiency, while also allowing the introduction of new air-to-air weapons such as the Rafael Python 4 short-range and AIM-120 medium-range air-to-air missiles. Air-to-ground weaponry options have also been considerably improved, while Global Positioning System (GPS) equipment is part of the upgrade package to improve navigation, as well as to offer delivery of GPS-guided weapons.

The initial value of the Improved Baz programme was US$40 million from US Foreign Military Support (FMS) funding and 50 million Israeli Shekels (around US$12.5 million) for local Israeli expenditure. The first major contract (valued at 18 million Shekels) was awarded in 1995 to IAI Lahav Division and to Elbit for the joint development and construction of an avionics integration laboratory at the IDF/AF Air Maintenance Unit (AMU) at Tel Nof air base. Elbit was also awarded the MFD contract, while the two other major Israeli contractors involved in the Improved Baz programme are Rafael (weapons integration) and Rada with its Data Transfer Equipment (DTE) and debriefing system. US contractors comprise Boeing (redesign of the existing weapons hardpoints and the addition of two weapons stations), Hughes (radar improvements), Honeywell and Loral.

F-15D 706 was selected as the upgrade programme prototype, emphasising the importance attached to the conversion of the relatively large number of IDF/AF F-15B/D two-seaters into multi-role combat aircraft with Precision Guided Munitions (PGM) capability. The rear cockpit was redesigned with two colour MFDs and HOTAS technology, allowing the back-seater (still termed navigator in IDF/AF terminology) to operate offensive and defensive systems as a complimentary element to the pilot's inputs.

Left: Each individual IDF/AF F-15A/B/C/D has its name painted on the port side of the nose. F-15A Improved Baz 678 is named Hayoreh (The Shooter) and its two kill markings are the result of a single air combat against Syrian air force MiGs on 11 June 1982.

Right and below right: F-15D 706, named Kochav Hazafon (The Northern Star), was rolled-out of the IDF/AF AMU in November 1998 as the 'prototype' Improved Baz. The IDF/AF golden anniversary badge was applied within a blue band on the outer surface of the port vertical satbiliser, with the outer surface of the starboard fin carrying the unit marking and, as with all upgraded F-15s, the legend Baz Meshopar (Improved Baz) painted in Hebrew beneath the tail number on the starboard vertical stabiliser.

Left: The carriage of Python 4 air-to-air missiles is one of several clues that give away the upgraded status of this ex-USAF F-15B, which was one of a number delivered to Israel in the wake of the 1991 Gulf War.

The first upgrade for each F-15A/B/C/D model is treated as a 'prototype', although only F-15D 706 was officially rolled out as the first Improved Baz in November 1998. Initial upgrades were to Block A standard with new computers, new displays, additional hardpoints and modified existing hardpoints. HOTAS capability was introduced on Block B standard aircraft, after exhaustive inputs from squadron pilots at the avionics integration laboratory were analysed.

Programme completion is expected within three or four years, by which time the entire fleet will have been upgraded to Improved Baz standard, placing these old warriors at the forefront of air combat technology for another 20 years.

Shlomo Aloni

The IDF/AF has Conformal Fuel Tank (CFT) sets for its entire F-15A/B/C/D fleet. However, the CFTs were sometimes not carried by Baz aircraft as Python 3 missiles could not be carried on inboard pylons due to their relatively large aerodynamic surfaces. Baz aircraft, when fitted with CFTs, would normally carry AIM-9Ls on the inner stations, while Improved Baz F-15s can carry Python 4s in this position as illustrated here.

Operation Noble Eagle

Defending continental USA

Following the 11 September 2001 terrorist attacks on Washington and New York, the USAF rapidly implemented Noble Eagle – a widespread homeland security operation involving fighter, tanker, transport and airborne early warning assets throughout continental USA. Tasked with defending major cities, nuclear power stations, bridges, military installations and other potential targets for terrorist attack, Noble Eagle has involved over 11,000 air force personnel generating over 8,000 sorties since the 11 September attack.

Under the control of Norad, the bulk of the defensive counter-air missions have been conducted by Air National Guard units flying fully-armed F-15s and F-16s from 26 locations across the USA, providing 24-hour coverage to high-risk areas and less extensive coverage of other areas. Some 120 fighter aircraft, supported by 75 KC-135 tankers and 20 airborne early warning aircraft, have been involved in the operation, with support missions conducted by some 45 C-130s.

Over 80 per cent of the aircrew flying fighter patrol and support missions have been from Guard or Reserve units and, with daily flying hours rising from an average of 15-20 to over 50, the strain on both aircrew and engineers rose to unprecedented levels in the post-Cold War era.

Within days of the attack the USAF was aware that its 32-strong fleet of E-3 AWACS would not be sufficient to provide the coverage of US airspace demanded by Norad, while committing assets to the build-up of forces for Operation Enduring Freedom and maintaining its long-term committment to enforcing the northern and southern no-fly zones over Iraq. In an unprecedented move, political leaders requested the help of NATO, and the first of five NATO E-3As from Geilenkirchen, Germany arrived at Tinker AFB, Oklahoma, on 9 October. These, along with US Customs service P-3 AEW&Cs and the remaining USAF E-3s, have maintained 24-hour coverage of New York, Washington and other potential targets, and guided fighter intercepts of many inadvertent incursions by civilian aircraft into 'areas of interest'.

Above: A NATO E-3A takes on fuel from an ANG KC-135 tanker during a Noble Eagle sortie. NATO and USAF E-3s received Norad mission taskings as if they were a homogenous fleet, with high-sortie rates requiring the crews to fly on alternate days, with each sortie typically lasting 11-15 hours.

Left: A Vermont ANG F-16C Block 25 from the 134th FS, 158th FW flies a CAP over the smouldering remains of the World Trade Center, Manhattan, in the days following the 11 September attack. Guard units maintain a continuous 24-hour coverage over the city.

Below: Among the ANG units tasked with flying defensive patrols over the western seaboard and west coast cities are the 'Red Hawks' of 134th FS, Oregan ANG. A pair of the unit's F-15As are seen here carrying live AIM-9, AMRAAM and AIM-7 missiles.

'Night Stalkers'

160th Special Operations Aviation Regiment

One of US Special Operations Command's (USSOCOM) most important aviation units is the 160th Special Operations Aviation Regiment (Airborne), nicknamed the 'Night Stalkers'. Units of the 160th are based at Fort Campbell, Kentucky, Hunter AAF, Georgia, Roosevelt Roads, Puerto Rico, and Taegu, Korea. The 'Night Stalkers' recently celebrated their 20th anniversary. This is a review of the helicopter types in service.

Ever since the 'Night Stalkers' unit was conceived in 1981, the Hughes (McDonnell Douglas) H-6 'Little Bird' has been in service in some form or other. Originally the 160th flew OH-6As with added Miniguns and rocket pods. This version received the designation AH-6C in recognition of the modifications. The MH-6E, a militarised Hughes 500 model, was introduced in 1981. Another modified Hughes 500 version, the AH-6F, followed in 1985 to replace the older AH-6C models. In 1993, the MH-6Es and AH-6Fs received mission-equipment packages that upgraded them to MH-6J and AH-6J, respectively.

There are three models of the H-6 currently operated by the 160th – the AH-6J, MH-6C and MH-6J. The AH-6Js are used for attack, the MH-6J is used for light assault, and the MH-6C is a trainer. The Vietnam-era OH-6A was officially named the Cayuse, but the helicopter series universally became the 'Loach', a product of the original LOH (Light Observation Helicopter) acronym. In the 160th community, they are referred to as 'Little Birds'. The AH-6Js are sometimes referred to as 'Gunships' and 'Killer Eggs'. All of the 'Little Birds' sport a flat-black paint scheme.

MH-6C – training mount

This version is only operated by the SOATC (Special Operations Aviation Training Company), and is employed to teach new pilots basic H-6 skills, basic NVG (Night Vision Goggles) navigation using maps, and for staff aviator support. In appearance, the MH-6C looks very similar to the original OH-6A and is called the 'V-tail' since the newer AH/MH-6Js have a 'T-tail'. While the MH-6C has the more powerful 250-C30 engine of the AH/MH-6J, rather than the OH-6A's 250-C20, it still retains the original four-bladed rotor system.

According to the SOATC commander, "We train the new personnel to the 160th for duty in the aviation regiment. The SOATC training programme is known as Green Platoon and we have all three models of 'Little Birds' in the SOATC. They are very nimble, responsive and fun to fly. Both the AH-6J and MH-6J crews are initially trained in parallel, and start out flying the 'V-tail' for three or four weeks. They learn basic navigation, how to plan and brief, and then become qualified in one of the variety of

The 160th SOAR(A) operates a mix of three basic helicopter types (H-6, H-47 and H-60) to fulfil a variety of tasks and to offer a spread of size and range capabilities. The types are represented here by the MH-6J (main picture), MH-47E (above), MH-60L (left) and AH-6J (below, firing rockets and guns). To a greater or lesser degree all are equipped with additional communications equipment, and with sophisticated sensors and navigation aids to allow them to support Special Forces in most conditions. Typically the aircraft work under the cover of darkness, using FLIR, night-vision goggles and/or terrain-following/avoidance radar to fly their missions at low level. Insertion, resupply and retrieval of Special Operations Force troops is the main role, along with armed support.

helicopters flown at the regiment. Later they go to the tactical training phase."

MH-6J – nimble transport

The 160th operates two sub-variants of the MH-6J, dubbed the 'J-light' and 'J-heavy'. The small number of MH-6J-heavy aircraft, also designated A/MH-6J, differ by being wired for the carriage of a FLIR system, and have an additional radio and extra circuit breakers. Compared to the original OH-6A, the MH-6J has a five-bladed rotor system, uprated transmission and a Rolls-Royce/Allison 250-C30 engine. Unlike most other helicopters, the 'Little Bird' controls are all mechanical linkages and there arc no hydraulics.

NVGs. Navigation aids include GPS, which is backed up by a LORAN that uses the Trimble TNL-3100. Also included is the Argus 7000 – which displays aircraft flight path and way points, airfields, special use airspace etc – and TACAN.

MH-6Js are single-pilot rated, but usually operate with two pilots on board. While there is a small space aft of the pilots, no passengers are carried there. All MH-6J passengers are carried externally on the EPS (External Plank System) – the attachments that protrude on each side. Commonly called 'pods', and some-

Empty weight is about 2,000 lb (907 kg) and the gross take-off weight 3,950 lb (1792 kg). The MH-6J can carry up to 800 lb (363 kg) of fuel, 400 lb (181 kg) in the main tank and a similar amount in an auxiliary tank mounted in the cabin behind the pilots. The aircraft typically burns about 240 lb (109 kg) of JP-8 fuel per hour, and has an endurance of 3 hours 20 minutes, with a 20-minute reserve. Range is over 300 nm (345 miles; 555 km).

While the 'Little Birds' are not IFR-rated or all-weather aircraft, the crews can use OMNI-4 NVGs that have a slightly wider field of view, higher resolution and better low-light characteristics compared with the popular ANVIS-6

One of the many skills taught by the SOATC is manoeuvring in enclosed spaces – a necessary talent for Special Operations support pilots. The MH-6C provides a useful vehicle for teaching new pilots the 'basics', including NVG flying and precision navigation, before they move on to the front-line types. It is also used for staff continuation flying.

Principal role for the MH-6J is the transport of small groups of special forces troops, who sit on benches either side of the cabin, restricting the helicopter's speed and manoeuvring when they are carried. A handful of the fleet are in MH-6J-heavy configuration, with FLIR for nocturnal missions.

times 'benches', they are detachable but are normally permanently fitted. The aircraft has provision for carrying up to three passengers on each side. Cargo hooks were removed from the aircraft when they were upgraded to J models, so they can no longer carry slung loads. Other than personal weapons such as side-arms and a carbine carried by the crew, the MH-6J is unarmed. The MH-6Js are FRIES (Fast Rope Insertion Extraction System) capable.

When carrying passengers externally on the 'pods', speed is affected by the significant drag caused by the troops. Also, flying any faster than 80 or 90 knots would present an intolerable amount of wind blast for the passengers. The MH-6J typically flies at about 300 ft (90 m) AGL and at 80 kt (92 mph; 148 km/h), although the max VNE is 152 kt (175 mph; 281 km/h). With 'pods' removed and doors installed the MH-6J has a fast cruise of about 135 kt (155 mph; 250 km/h). It has a 16,000-ft (4880-m) service ceiling, but never has a need to go that high as most operations are conducted at low level. Placarded maximum bank angle is 60° and pitch limit is 30°, but these can be exceeded if required.

MH-6J training

When a pilot transitions to the MH-6J, they spend about three months in the SOATC Green Platoon. Included in the training is ground school, academics, aircraft qualification, basic navigation and how to be on-station within plus or minus 30 seconds at over 60 nm (69 miles; 111 km) distance. Other training involves single-pilot navigation, desert and mountain training, water survival and deck landing training. The pilots receive between 100-110 hours of flight time in the MH-6J before they are assigned to a line unit.

When asked about the MH-6J, a Flight Lead and Instructor Pilot with 5,200 hours commented, "The Little Bird is very agile, handles well, and can land in a lot of places – unlike the larger helicopters. It can be airlifted quickly and is very mobile. For its size, it can carry pound-for-pound as much as some of the bigger aircraft. One of the aircraft's biggest strengths is its small size.

"You won't find better maintenance in the military anywhere compared to what we have right here in the 160th. Before joining the 160th, I was an Air Cav OH-58 pilot and also flew UH-1s. I liked the light observation role and wanted to fly the MH-6 in the 160th. It is both challenging and rewarding – and the fact that I have been here 10 years speaks for itself. In the 160th, we all work very hard and constantly train. We get all the rockets and weapons we need for training, we don't need to watch the number of our flying hours, and as a result we are very well prepared."

AH-6J – armed 'Little Bird'

Used for attack missions, the AH-6J has a weapons plank which can mount the 7.62-mm 4,000-rpm M134 and GAU-2A/B series Miniguns, and the three-barrelled 1,200-rpm GAU-19 0.50-in calibre machine-gun (usually mounted outboard). It can also carry 2.75-in (70-mm) Hydra rockets in seven-shot M260 or 19-shot M261 pods. The AH-6J can also carry the lethal AGM-114 Hellfire missile. The small space behind the pilots almost always has a 400-lb (181-kg) auxiliary fuel tank installed. The space is also where 7.62-mm ammunition is located for the Miniguns. AH-6Js are FLIR-capable, and are technically designated A/MH-6J, same as the MH-6J FLIR aircraft. The FLIR, manufactured by Raytheon, incorporates a laser designator.

Four special forces troops ride on the EPS benches during a training mission. Due to its small size the MH-6J cannot carry the same level of equipment as the larger helicopters, but NVG compatibility and GPS navigation allow it to be used in good weather at night.

Right: This view of an AH-6J illustrates the weapons plank which is mounted across the rear cabin. Beneath the nose is the pedestal for the FLIR turret. This also houses a laser for target designation.

Below right: Ammunition boxes and 70-mm rockets surround an AH-6J during a weapons training exercise. The aircraft can be rearmed with great rapidity in the field.

Rotor system, engine, transmission and navigation aids are identical to that of the MH-6J. The AH-6J is also single-pilot capable, but for most missions it will have a pilot and co-pilot on board. Empty weight ranges between 1,980 to 2,100 lb (762 to 952 kg), depending on the particular airframe. The maximum gross weight of the aircraft is 3,950 lb (1792 kg) and it normally carries between 600 and 800 lb (272 and 363 kg) of weapons payload.

With the auxiliary tank fitted, the AH-6J has a 280-nm (322-mile; 518-km) range with a 20-minute reserve. Fast cruise is typically 110 kt (127 mph; 204 km/h), but speed is dependent on aircraft configuration. Refuelling and rearming can be accomplished in about 10 minutes.

A Project Officer in SIMO (Systems Integration Management Office) is a 160th AH-6J pilot who has 4,700 hours. He is responsible for new weapon upgrades and testing within the AH-6J fleet, including a new digital cockpit upgrade currently under way. "The AH-6J is extremely manoeuvrable, agile, is a small target and hard to hit. It is easily transportable and is very effective in its role – we like to call it our 'Urban Fighter'. I think the digital cockpit will be a great enhancement to the 'Little Bird' community."

MELB – upgrading the 'Little Birds'

'Little Birds' are expected to be in service with the 160th for a long time to come. A new upgrade now under way is the MELB (Mission-Enhanced Little Bird) and the first example is slated for completion in 2002. The MELB programme includes a new and more powerful 250-C30-R3M engine with FADEC (Full Authority Digital Electronic Control), and stronger skids. It will have an enhanced digital cockpit with VOR and ILS capability, and some of the avionics will be compatible with other Special Operations assets. The new cockpit will provide much better situational awareness and simplified tasking for single-pilot operations.

MELB aircraft will have an upgraded rotor system with six main blades and a four-bladed

The crew of the MH-47D has a comprehensive
navigation and communication suite at its command,
and some aircraft have HUDs fitted. Here a crew is
using FLIR imagery to aid flight in twilight operations.

X-shaped tail rotor, similar to that of the AH-64,
replacing the current two bladed design. With
the stronger skids and new engine, the gross
weight will rise from 3,950 lb (1792 kg) up to
4,700 lb (2132 kg), allowing the 'Little Bird' to
carry additional weapons and even an external
fuel tank. The MELB internal fuel tanks will be
improved and made fully crashworthy.

All of the AH-6J and MH-6J airframes will
become common so that they can be outfitted
to perform either the AH or MH role. After the
upgrade, the end product will officially be
designated as the A/MH-6M. The 160th will
have a mixed fleet of J and M models until
about 2005. Each J model will take about three
months to be converted into an M model. The
160th SIMO has been instrumental in oversee-
ing the MELB programme.

MH-47D – modified Chinook

Chinooks form an important part of the
160th's operations. The 'Night Stalkers' first
started out with modified CH-47Cs that were
eventually replaced in 1983 by the more
powerful CH-47D. In the mid-1980s, some of

*Most 160th missions are performed at night and at low
level, conditions for which the MH-47D is ideally suited.
The Chinook is often used for 'FARPing' – the
establishment of a forward refuelling point for use by
other helicopters.*

the 160th CH-47Ds received enhanced cockpit
avionics, a fully coupled flight control system, a
weather radar and a digital intercom system.
They also received FLIRs, M134 Miniguns and
inflight refuelling probes. The end product was
designated as the MH-47D. The handful of
remaining unmodified CH-47Ds attached to the
160th became known as CH-47D SOAs (Special
Operations Aircraft) due to minor enhancements and their mission.

MH-47Ds were converted from either
CH-47As or CH-47Ds. There remains a minor
difference between the two, as the converted A
models have a slightly shorter cockpit while the
converted CH-47D models are stretched 2 in
(5 cm). Thus, some MH-47Ds have more leg
room than others.

Cockpits were upgraded in 1989 to include a
common software architecture with the
MH-60L. The MH-47D also had an integrated

*The refuelling probe is the main feature which
distinguishes the 160th's Chinooks from regular Army
machines, allowing the aircraft to refuel from HC-130,
KC-130 and MC-130 Hercules tankers.*

GPS and a trio of 800-US gal (3028-litre) internal auxiliary tanks installed. Other enhancements have included new engines, an EGI
(Embedded GPS Inertial) system (which
includes INS), newer cockpit avionics,
enhanced ASE equipment, and a new rescue
hoist. Other navigation aids includes AHARS
(Attitude and Heading Reference System),
Doppler, TACAN, VOR, and ADF with DME.
The seven radios on-board include FM, VHF,
UHF, HF, SATCOM, SINCGARS, PLS UHF
system, and Saber.

MH-47Ds are equipped with an AN/AVS-7
HUD that displays the airspeed, altitude, msl,
pitch and roll, engine torque, barometric altitude, turbine inlet temperatures, ground speed,

Among the methods used by the MH-47D for rapidly inserting or extracting troops are the fast-rope (above), FRIES (Fast Rope Insertion Extraction System, below left) and the SPIES (Special Patrol Insertion Extraction System, below right). For rapid extraction the troops clip themselves on to the FRIES or SPIES rig before being lifted away.

located at each forward side opening and manned by the crew chiefs. For each gun, a 4,000-round ammunition can is normally used. Two cans can be carried for each weapon if the mission dictates it.

Crew composition

Typically the crew consists of two pilots, a flight engineer and three crew chiefs. The minimum crew is four (two in front and two in the back). When flying at night, they use OMNI-4 NVGs. Seating is provided for 33 fully-equipped troops, but with a seat belt waiver 76 troops can be carried. Rumour has it that in Iraq, 176 prisoners of war were airlifted in one. The MH-47D has both single- and dual-point hoists, and frequently carries HMMWVs ('Hummers') externally.

Empty weight is 28,500 lb (12928 kg) and gross weight is 50,000 lb (22680 kg), although a waiver can allow the gross weight to be increased up to 54,000 lb (24494 kg) in special circumstances. Internal fuel load of JP-8 is 6,800 lb (3084 kg) or 1,000 US gal (3785 litres) without using auxiliary tanks. The three 5,400-lb (2449-kg)/800-US gal (3028-litre) tanks raise total capacity to 23,000 lb (10433 kg)/ 3,400 US gal (12870 litres). Inflight refuelling is routinely undertaken.

Maintenance access around the aircraft is good and the various components are mostly easy to get at. The standard Army Chinook phase inspection is 200 hours, while the 160th aircraft inspections are done at 300 hours. An engine can be changed out in about six hours by six people, including ground and air testing.

On cross-country transits, the MH-47D will fly at altitudes between 300 ft (91 m) and 8,000 ft (2438 m), and at about 140 kt (161 mph; 259 km/h). With three auxiliary tanks and flying at 120 kt (138 mph; 222 km/h) the maximum range is about 800 nm (920 miles; 1480 km) and endurance is around seven hours. Maximum *g* limits are between 0.5 to 2.5, and the maximum allowed angle of bank is 60°. The MH-47D is rated as an amphibious aircraft, and it can land in water and completely shut down for 30 minutes. The minimum landing zone size is 150 by 120 ft (46 by 37 m).

For new MH-47D pilots that have not previously flown Chinooks, they must first become qualified on the regular CH-47D at Fort Rucker, Alabama. The AQC (Aircrew Qualification Course) must be completed and then they will go through the Green Platoon at Fort Campbell. The 3-160 has a CMI (Civilian Mission Instructor) at Hunter AAF, where the aircraft-specific training is completed for the type. After a pilot completes three days of ground school and five hours in the MH-47D, they are qualified to fly the aircraft. Later comes an array of various mission qualifications.

An MH-47D Flight Lead and AMC (Air Mission Commander) pilot who has been in the regiment for seven years and has over 2,000 hours described the versatility of the type. "The MH-47D has an external hoist and excellent range. There are many seating arrangements including litters for medevac and various assault configurations. We can conduct fast-roping, SPIES, FRIES (Fast Rope Insertion Extraction System), FARPing (Forward Area Refueling Point), and more – which makes the MH-47D extremely versatile. Our maintainers here are very good, and in the five years I have

heading, navigational information and way points, radar altimeter, FLIR data, hovering symbology, and more. It has an HSVD (Horizontal Situational Video Display) system, which displays the HSI, navigational information, colour radar and additional information. The aircraft has an RDR-1300C weather search radar and a storm scope with a dedicated display. Both pilot and co-pilot have CCDs (Cockpit Control Devices) which are used for selecting radios and setting up the nav, and are the basic controlling input modules of the aircraft.

Hughes AN/AAQ-16 FLIR (with a dedicated display) is fitted, used for avoiding hazards en route to a target. The MH-47D also has the PLS (Personnel Locator System) which works in conjunction with the PRC-112 survival radio. The PLS homes in on the signal transmitted by

a downed aircrew member, and is an instrumental tool for CSAR. A data cartridge (called the '10-Meg') containing the navigation way points, radio frequencies, ELT and PLS codes, and other data is loaded into the aircraft before flight. The information is loaded into the cartridge during the mission planning phase.

Some MH-47Ds are powered by a pair of T55-L-712 turboshafts, although now some of the aircraft are receiving the more powerful T55-GA-714A engines, along with the com/nav3 upgrade. The com/nav3 upgrade adds a new computer, flat-plate digital screen, a new HF antenna, and adds an external rescue hoist. The new engines have FADEC, which gives them a more efficient fuel burn, reduced spool up time, and increased horsepower.

Painted in an all-black scheme, the MH-47D often carries a pair of M134 Miniguns on board,

been here I have only had to drop two missions with an aircraft that broke.

"Going from a single-pilot aircraft to the Chinook with a crew of six was eye-opening. Compared to the CH-47D, the MH-47D is very advanced and was intimidating with all of the digital systems. The MH-47D is a lot more powerful and capable than I expected. While large, the Chinook is fast, is very manoeuvrable and a very capable aircraft. For its role it is both viable and effective. In the 160th, we get to fly a lot and I average between 30-40 hours a month."

In the future, the 160th will receive the MH-47G, which will have 2,000-US gal (7571-litre) side tanks like the MH-47E and a next-generation multi-mode radar which will have a weather capability. The G models will all have the newer engines that come with the com/nav3 upgrade. The 160th is slated to receive its first MH-47G in December 2002, and the G model will be even more capable than the MH-47E.

MH-47E – Special Ops Chinook

While the MH-47E shares the same Chinook name as its CH-47 cousin, the 'Special Ops Chinook' is an altogether different and more capable beast. The 160th began flying the MH-47E in 1993 and received its final aircraft in May 1995.

The MH-47E has a fully integrated cockpit with a mission management system. It has an MDG (Map Display Generator) and, when combined with the DTM (Data Transfer Module), can display charts, maps, photos and 3D modes of operation. There are two CDU (Control Display Units) and four MFD (Multi-Function Displays) in the cockpit.

An Integrated Avionics Subsystem (IAS) permits global communication and navigation, and includes a passive AN/AAQ-16 FLIR (Forward-Looking Infrared). The IAS gives on-board redundancy, dependability and enhances survivability. The navigation suite employs an EGI system, Doppler and AHARS for pinpoint navigational accuracy. The aircraft is also equipped with the AN/APQ-174A MMR (Multi-Mode Radar), which is a forward-looking airborne radar with TF (Terrain Following) and TA (Terrain Avoidance) capabilities.

Radios and communication equipment include VHF, FM (with SINCGARS VHF-FM), UHF (with HAVE QUICK II), HF, SATCOM and

King of the 160th fleet in terms of range and load-carrying is the MH-47E. Oversized fuel tanks scabbed on to the cabin sides combine with auxiliary tanks inside to provide an impressive 1,450-mile (2333-km) range. All of the tankage is refillable inflight.

the Motorola Saber. These allow communications with a wide variety of other forces and ground agencies. Many avionics components are common to the MH-60K, enhancing the spare parts situation and maintenance. All crew

As evidenced by the forest of antennas, the MH-47E is extremely well-equipped in terms of communications. This is vital to the Special Operations role, allowing the crew to be in contact with a wide range of ground or air assets, and with commanders at several levels.

An MH-47E scoots across the Californian desert at high speed during a training exercise. The long range and high transit speed of the type makes it ideal for the long-range combat search and rescue role (CSAR), as well as deep infiltration missions.

members are trained to use OMNI-4 NVGs, and specialise in arriving at a particular target or location within plus or minus 30 seconds.

Power is provided by a pair of Lycoming T55-L-714 engines, which each produce 4,168 hp (3109 kW) sustained power. The engines are equipped with FADEC that gives the engines maximum power and efficiency. The fibreglass blades can be manually folded for shipboard operations.

Maximum gross weight is 54,000 lb (24494 kg) and the useful load of fuel and cargo is 25,000 lb (11340 kg). The MH-47E carries more internal fuel than any other species of Chinook, and has extended side fairings which house fuel. The MH-47E can carry 2,068 US gal (7828 litres)/13,442 lb (6097 kg) of JP-4. Up to three CCERFS (Cargo Compartment Expanded Range

The pod strapped to the side of the MH-47E's nose houses the APQ-174 radar, which provides terrain following/avoidance, among other functions. The front door Minigun has a large chute fitted to eject spent cases safely away from the aircraft.

Fuel System) self-sealing and ballistic-tolerant 780-US gal (2953-litre) tanks can be installed and are refillable during aerial refuelling.

There are three weapons stations in the left forward window, the cabin entrance on the right, and on the rear cargo ramp. The two forward positions mount M134 7.62-mm Miniguns, while a 7.62-mm M60D gun is used from the cargo ramp. The Minigun can be fired at 2,000 or 4,000 rpm, and initially has 8,000 rounds before having to be reloaded. The weapons are primarily used for self-defence and fire suppression.

A typical crew consists of pilot, co-pilot and flight engineer in the cockpit, and two crew chiefs in the cabin, which offers seating for 44 troops. Without seats, the Chinook can carry an impressive 65 troops, and for medevac operations it is capable of carrying 24 litters.

External loads are carried using either single- or tandem-point attachments using the External Cargo Hook System. Tandem-point loads are preferred since they allow greater weight

capacities and the loads are more stable, allowing the aircraft to fly faster and carry up to 25,000 lb (11340 kg) slung below. An external rescue hoist is fitted for SAR work, load-rated at 600 lb (272 kg) and with a 245-ft (75-m) cable.

Despite the first impression of a cumbersone machine, the Chinook is one of the fastest helicopters around (in the same class as the CH-53E) and can dash up to speeds of 170 kt (196 mph; 315 km/h), faster than the AH-64 Apache and H-60 Blackhawk. However, the aircraft typically cruises between 120 and 140 kt (138 mph; 222 km/h and 161 mph; 259 km/h). Although most missions are conducted at low/medium altitudes, and while light icing conditions are usually avoided, the MH-47E may operate in that regime if required. The Chinook can go to altitudes up to 20,000 ft (6096 m), which makes it a great asset for working in mountainous areas. The Special Operations Chinook has an impressive range of 1,260 nm (1,450 miles; 2333 km) without refuelling.

MH-47Es can work alone, or in the company of other special operations helicopters such as 'Little Birds' and MH-60s. The Special Operations Chinooks also often work with a variety of other helicopters such as AH-1s, UH-1s, CH-46s, CH-47s, CH-53s, UH-60s, AH-64s and foreign aircraft for various missions and joint operations. In addition to being used for overt missions, the MH-47E is also used for deep penetration missions. The all-weather helicopter mostly performs missions under the cover of darkness using night vision devices. It routinely carries Rangers and Seal Teams for infiltration and exfiltration, and is also instrumental in resupplying troops at forward operating locations.

Above: 160th helicopters from the constituent units routinely work together: here an MH-60L and MH-47E undertake a training mission.

Right: Razor's Edge is an MH-60L DAP, carrying a 19-round M261 70-mm rocket launcher on its stub wing, one of several weapon options that include AGM-114 Hellfire missiles and 30-mm Chain Guns.

Other operations that may be conducted include shipboard, platform, urban, water, FRIES, parachute, FARP, mass casualty and SAR operations. Using the plethora of sensors and systems, the MH-47E excels at nap-of-the-earth, low-level flight in adverse weather at night.

A pair of MH-47Es can be transported to distant locations in a C-5 Galaxy, or one in a C-17A Globemaster III. The preparation time for each MH-47E is about eighteen hours. Long-term plans for the 160th fleet will probably include the MH-47Es being upgraded to MH-47G standard.

MH-60L – multi-role hawk

The first Blackhawks used by the 'Night Stalkers' were UH-60As that came from the 101st Airborne. The UH-60As were modified with internal long-range fuel bladders, long-range navigation equipment and over-the-horizon communication equipment. FLIRs were later added, while the aircraft carried M60D machine-guns in the cabin doors, and the cockpits were eventually integrated to minimise pilot workload. After the cockpit upgrade, the aircraft were re-designated as MH-60As. The MH-60As were dubbed 'Velcro Hawks' due to the somewhat piecemeal way they had been put together.

Eventually the M60Ds were replaced by the 4,000-rpm M134 Minigun, and the internal long-range fuel system was enhanced. 1989 brought the MH-60L to the 160th, which has more powerful engines, a colour weather radar, FLIR, GPS, FRIES and an external rescue hoist. The MH-60As were upgraded to the MH-60L DAP (Defensive Armed Penetrator) configuration and received stub wings that carried additional armament. DAP birds can carry a fixed M230 30-mm Chain Gun, AGM-114 Hellfire missiles,

2.75-in (70-mm) Hydra rockets, 0.50-in machine-guns, M134 Miniguns and air-to-air Stinger missiles. Thus, the MH-60L can be used for both the assault and attack roles, giving it a dual-role capability.

Minimum crew is two, comprising a PIC (Pilot In Command) and a co-pilot. However, typically a pair of crew chief/door gunners is carried. The MH-60L, with internal fuel bladders installed, has seats for seven troops. Troops may also be placed on the floor of the cabin, allowing it to carry up to 12 troops. The all-black aircraft has a pair of T700-GE-701C engines backed with an improved durability gearbox.

Not all MH-60Ls have yet been fitted with refuelling probes, and some that have the capability do not carry the probe at all times. This is a Defensive Armed Penetrator-configured aircraft, modified from MH-60A 'Velcro Hawk' standard.

Using the 30° pitch angle limit to the maximum, an MH-60L crew brings its mount to a rapid hover after a high-speed approach to the LZ. The two crew chiefs man the Miniguns in the cabin side windows, ready to return fire if the landing is opposed.

Maintenance is good and an engine can be changed out in three to four hours. After the engine is replaced, it must be vibration-tested and go through ground tests. A check flight is also required, and the entire process could be done in half a day if treated with priority. Every 500 hours, a Blackhawk goes through a major phase inspection and is torn down and thoroughly checked.

MH-60Ls have a storm scope, and navigation aids include TACAN, DME, VOR, ADF, ILS, GPS, AHRS and an air data computer. Recent MH-60L upgrades have included a comm/nav upgrade that adds better communication equipment, new MFDs (Multi-Function Displays), and EGI(INU). The end result is an EFIS cockpit with enhanced situational awareness. Other improvements include a variety of improved ASE equipment, weapons management system, and the addition of inflight refuelling capability.

Fast-roping from a low hover is a routine method of inserting troops in quick time. MH-60s and MH-47s are well protected against missile threats with IRCM turrets, low signature exhaust shrouds and flare dispensers.

The probe upgrade is now in process, but not all MH-60Ls will have the probes installed all of the time, as that will be mission-dependent. For a time the probe was the easiest way to externally identify a MH-60K from the L model, but that is no longer true.

Empty weight of the L is about 14,500 lb (6577 kg) and gross weight is 22,000 lb (9979 kg). 2,400 lb (1089 kg) of JP-8 fuel is carried in the internal tanks, although two 185-US gal (700-litre) auxiliary tanks are often installed in the cabin. External fuel tanks can also be installed using the ERFS (Extended Range Fuel System) but this is rarely done. The MH-60L has an external cargo hook that can handle an 8,000-lb (3629-kg) load. It also has a rescue hook on the starboard side. It is day/night capable, has both de-ice and anti-ice systems, and is all-weather rated. Crews use OMNI-4 NVGs and routinely fly at night.

Refuelling technique

MH-60Ls can refuel in flight from HC-130s, KC-130s and MC-130s. At night, hook-ups are conducted using NVGs. The refuelling is performed at 110 kt (127 mph; 204 km/h) and the MH-60 usually goes to the left side of the Hercules as there is less turbulence there. The pre-contact location is five to 10 feet aft of the drogue, and the Blackhawk pilot will inspect the C-130 and drogue for any wires and unsafe items. Then the pilot performs the contact phase by plugging the drogue and moves forward and outboard into the refuel position.

For cross-country flights, the MH-60L can usually be found flying between 300 and 500 ft (91 and 152 m). With internal auxiliary tanks, the aircraft has a range of about 400 nm (460 miles; 740 km) and an endurance of about 4 hours and 20 minutes. The service ceiling is about 18,000 ft (5486 m), although for that altitude the crew would use oxygen.

A slick MH-60L model is slightly faster than the MH-60K since it weighs less and has less drag, although the DAP model is an exception due to the extra drag of its weapon pylons. A fast cruise is about 130 kt (150 mph; 241 km/h). The landing zone size should be no less than 80 sq ft (7.43 m²).

A new MH-60L pilot will spend between two to three months with the Green Platoon before becoming qualified in the aircraft. Now with EFIS cockpits in L models, it takes new pilots additional training time. After the pilot has completed the Green Platoon training by SOATC, the pilots will be assigned to a line company.

MH-60K – sophisticated Blackhawk

MH-60Ks entered into service with the 160th in 1994, possessing inflight refuelling capability from the outset. The K model is generally similar to the L model, except for having a different type of radar and only carrying the M134 Miniguns (no DAP capability). Most of the data regarding the MH-60L is pertinent to the K model as well, including the power train. Used for assault, the MH-60K has an AN/APQ-174B multi-mode terrain-following/terrain-avoidance radar in place of the weather radar. The terrain-following radar can be set for altitudes of 100, 300 or 500 ft (30, 91 or 152 m).

Marking the rise in importance of Special Operations within the US Army, the MH-60K and MH-47E were the service's first helicopters to be built from new for the role, and they share similar equipment. Below the nose radar is a FLIR turret.

Externally, the easiest way to distinguish the MH-60K from the L model is the radar nose, which is a little smaller in diameter. The MH-60K has both ASE and mission-equipment package upgrades. The K model received heavy gauge sheet metal and stringers when they went through refit, and thus has an empty weight of 15,000 lb (6804 kg). The aircraft also has a higher gross weight of 24,500 lb (11113 kg). The MH-60K can also mount external fuel tanks strapped on using the gull-wing ETS (External Tank System), with 230-US gal (871-litre) tanks. Recent MH-60K upgrades have included an enhanced SATCOM radio, various avionics, new software and a flat black paint job. In the long term, MH-60s will most probably receive some of the same upgrades that the UH-60s receive when they are upgraded into UH-60M models.

MH-60K pilots must spend about six to seven months in the SOATC Green Platoon before being qualified in the aircraft. The first few weeks include ground school and flying the full-motion simulator. Thereafter time is divided between academics and flying. Pilots accrue about 80 flight hours before they are assigned to a line company.

An MH-60K instructor pilot, with over 3,000 hours and 1,800 in the MH-60K, was asked about his job and the MH-60K. "I have been flying MH-60s since 1988, they are very safe aircraft and were built for the pilot. Originally I flew UH-60As, UH-60Ls and EH-60As before I joined the 160th. The maintenance in the 160th is superb – as a Special Operations unit, we get a higher priority for parts and everyone here is hand-selected.

"Transitioning from an aircraft that is fully analog, such as the UH-60, to a digital cockpit MH-60K seems overwhelming at first. With the MH-60K digital cockpit, there is a lot to learn. It takes a pilot one to two years to become fully mission-qualified in a Company. The MH-60K is a good handling aircraft and is very responsive.

"The 160th has been involved in Desert Storm, Panama, Somalia. Haiti, and we are often the first in. We get a lot of flight hours in our unit and we conduct realistic and safe training constantly. We immensely enjoy what we do and are a cohesive team. The 'Night Stalkers' are a great group of people and I will be here until I retire. The people are dedicated at doing their jobs and you do not have to ask them to do things. For what we use it for, the Blackhawk is the best helicopter out there. The MH-60 was built for the crew and to safely get the people in the back to their destination."

With more third-world conflicts now than ever, and with the US declaring war on international terrorism, the 160th crews and their helicopters are vital in meeting US military demands. The 'Night Stalkers' and their capable Special Operations helicopters stand ready to be wherever needed within hours, and to be on target within plus or minus 30 seconds!

Ted Carlson

7.62-mm cases fall as an MH-60K door-gunner fires his M134. The MH-60 is the most numerous of the Special Ops helicopters and has seen service in several conflicts, including Afghanistan.

93 Zrakoplovna Baza (Air Force Base 93) at Zadar-Zemunik, Croatia is home to the training arm of the *Hrvatsko Ratno Zrakoplovstvo* (HRZ, Croatian air force). The air force academy shares the base with the country's airborne fire-fighting force, controlled and operated by the HRZ.

Zadar Air Base's origins reach back over 50 years: during World War II it was used by the Reichsluftwaffe (specifically the 3rd Reich Air Force) and the Italian Regia Aeronautica. In 1950 the Yugoslav air force (JRV) academy moved to Zadar after some years on the island of Vis. In the early years academy courses lasted two years and began after the 10th school grade. The academy later accepted non-commissioned officers (NCOs) and, in 1968, the academy course was extended to three years. Today the course is four years in duration – and has been since 1975.

Although Zadar was the academy headquarters, only parts of the flying syllabus – screening and basic training, including aerobatics and navigation, for both fixed-wing aircraft and helicopters – were taught there. Advanced aspects of the fixed-wing syllabus, including IFR training, were carried out at Pula (now in Croatia), whereas all tactical training was done at Podgorica (formerly Titograd, now capital of Montenegro) and advanced helicopter training took place at Mostar (now in Bosnia-Hercegovina). Regular JRV academy operations continued at Zadar until 1991.

Zadar was also home to three squadrons of SOKO G-2 Galebs, each with 20 aircraft, and one squadron of SOKO J-1 Jastrebs, also with 20 machines. Basic training aircraft at the base included 14-16 examples of both Zlin and UTVA types. For countrywide fire-fighting duties, six Canadair CL-215s were also based at Zadar. In 2001, 93 Zrakoplovna Baza housed four squadrons, three from the Peresin academy and one fire-fighting unit.

The aircraft of Zadar-Zemunik Air Base, Croatia, photographed by Heinz Berger and Erich Strobl

93 Zrakoplovna Baza, Zadar-Zemunik
Croatian training and fire-fighting base

Rudolf Peresin Air Force Academy

After successfully passing the screening process, Croatian cadets serve a total of nine semesters, five at Zagreb's Traffic Sciences University (TSU), where theoretical instruction takes place and two summer camps provide basic military training, and four at the Rudolf Peresin Air Force Academy at Zadar Air Base.

At Zadar, trainee pilots begin with a 40-hour basic training course on the UTVA-75 (above) before proceeding to either the Pilatus PC-9M (right) or Bell 206 JetRanger (below).

Upon graduation and the presentation of his/her wings, the young pilot holds the rank of 2nd Lieutenant and is assigned to one of the HRZ's active units. There is also the possibility of continuing on to the instructor pilot school, also based at Zadar.

Between 15 and 18 cadets graduate from the Academy each year; about one-third of these are helicopter pilots. According to Croatian sources the overall drop-out rate per intake is around 30 per cent. The academy is also open to female cadets, one or two of whom join each year's intake. I Squadron (EzTO) had a female instructor pilot in 2001, another female graduate was flying with III Squadron (EzOPH), while two more were in training, on the PC-9 course.

UTVA-75 *(above and right)*

The nine remaining UTVA-75s at Zadar are the oldest and only non-Western aircraft in the academy's fleet. Drawn from Croatian aero clubs at the beginning of the 1991 Homeland War, they served as trainers and liaison aircraft; some are known to have been camouflaged at this time. In 2001 all are in the same white/red training livery and are civil-registered. The UTVA-75s serve with the *Eskadrila za Temeljnu Obuku* (EzTO, Basic Training Squadron), alongside Pilatus PC-9Ms, and are employed for all screening and basic flight training. Cadets generally go solo after about 10 hours with an instructor and progress to the PC-9M after 40 hours.

Now approaching the end of their operational life, the UTVA-75s (on average, all about 20 years old) have only one or two years remaining in service. At the time of writing a decision on a new basic trainer type was due; among the contenders are the Aermacchi SF.260, FFA Bravo and Zlin 243.

Pilatus PC-9M *(left, below and opposite page)*

The search for an advanced trainer began at a time when arms sales to the region were subject to embargo. Croatia decided that a European type should be selected and, despite the embargo, began negotiations with Pilatus. Twenty PC-9Ms were ordered in 1995/96, though approval was only given after the Croatian government undertook not to arm the aircraft.

The PC-9Ms are used by the EzTO (Basic Training Squadron) as well as the *Eskadrila za Visu i Namjensku Obuku* (EzViNO, Advanced Training Squadron), though tactical training is not possible given the enforced limitations on the type's configuration. The HRZ is considering fitting a number of its PC-9Ms with a HUD to aid training, though they would remain unarmed. PC-9M training is supported by 50 hours in a simulator, one of which is operated by the air force academy, mainly for IFR and night flying instruction.

Bell 206B-3 JetRanger *(below)*

In late 1996 10 Bell 206B-3s were ordered from Bell Helicopter Textron at a reported total cost of US$15 million. These helicopters, manufactured in early 1997, equip the third squadron in the academy – *Eskadrila za Obuku Pilota Holikoptera* (EzOPH).

Helicopter trainees generally go solo after 16 hours of instruction. After two years at the academy cadets have usually logged some 100 hours on the JetRanger and go on to accumulate a further 50 hours converting to the specific helicopter type that they will fly when posted to on operational squadron. In 2001, the Croatian air force operated Mil Mi-8s, Mi-17s (Mi-8MIV-1s) and Mi-24s. Two JetRangers have been lost in accidents on training sorties, though thankfully without loss of life.

In the near future, if funding permits, a separate helicopter-training compound will be built on the opposite side of the airfield to the current facility in order to reduce interference with fixed-wing traffic, especially during auto-rotation exercises.

Airborne Fire-fighting Squadron

Fire-fighting from the air was previously undertaken in the former Yugoslavia using a number of different aircraft types, including four Canadair CL-215s obtained in 1981. It was not until 1994 that the Croatian *Protupozarna Zrakoplovna Eskadrila* (PZE, Airborne Fire Fighting Squadron) was established, using a pair of CL-215s, three CL-415s (acquired in 1997) and two AT-802Fs (ordered in 2000). The Zadar-based squadron is responsible for fighting fires along the Dalmatian coast and nearby islands; these break out on a virtually annual basis.

Fire-fighter pilots must log a minimum of 500 hours on an air force type before becoming eligible to fly the Air Tractor; after two years and a minimum of 1,000 hours he/she may convert to the CL-415. In order to place all fire-fighting operations under one umbrella, all

assets (both ground and air) are controlled by the HRZ, instead of a number of government ministries. In 2001, plans existed to purchase further aircraft as funds permit; an optimum force would comprise 10 AT-802Fs and between six and eight CL-415s.

Air Tractor AT-802F *(above)*
In 2000 Croatia selected the AT-802F as a scouting aircraft for its new fire-fighting squadron. After logging some 500 hours on the PC-9, interested pilots convert to the AT-802F. Two years' experience is required before an Air Tractor pilot becomes a good fire scout.

AT-802F crews aim to be at the fire scene as soon as possible so that an assessment can be made and decisions taken regarding the co-ordination of ground and air fire-fighters. The aircraft carry 3000 litres (792 US gal) of water, though larger fires are the domain of the Canadairs. With their considerable endurance, the AT-802Fs can fly two non-stop fire-spotting missions if required (between 08:00 and 13:00 hours and from 14:00 hours until dusk) daily during the fire season.

Canadair CL-215/CL-415

In 2001 the HRZ's Canadair amphibian fleet comprised two CL-215s (serialled 822 and 833) and three CL-415s (844, 855 and 866). Since 1 January 2001 all the fire-fighting aircraft have been operated by the HRZ and, accordingly, during the summer Croatian roundels and serials were applied in place of civil registrations.

CL-215 822 is currently the oldest Canadair still in use anywhere; it is planned to trade-in both remaining CL-215s and purchase a new CL-415 in the near future, and take options on two further examples.

Zadar's 'museum'

During Operation Oluja (Tempest) in August 1995, when Croatian forces re-captured Slavonia in the the eastern part of the country, a number of aircraft were captured at Udbina airfield. Examples of five types are displayed at Zadar.

Among these are SOKO G-2A Galeb (Seagull) '661' and SOKO J-1 Jastreb (Hawk) '601' – both types that have served at Zadar in the past – and SOKO J-20 Kraguj '701', an example of an aircraft that saw action during the Homeland War (as the 1991 conflict is known in Croatia).

The **SOKO G-2A Galeb** (right) was heavily used by the former Yugoslav air force academy at Zadar, as a basic jet trainer. '661' was the only example to see service in the newly-formed HRZ, which employed UTVA-75s as its only other trainers. It was flown only briefly due to a lack of parachutes for its ejection seats

Equipping five wings in the JRV (in which it was known as the J-21), the **SOKO J-1 Jastreb** (far right) also served in the former JRV academy.

When hostilities broke out in Yugoslavia in 1991 the **SOKO J-20 Kraguj** (top) was obsolete but, nevertheless, saw service with both Slovenian and Croatian forces during the two wars of independence. '701' is the believed to be the only example in existence in Croatia.

All three aircraft have recently been painted in

Croatian colours, though none is airworthy. All are regularly towed from their hangar at Zadar for display at open days and air shows.

Perhaps the rarest aircraft in Zadar's collection in an **UTVA-60** (bottom left), formerly YU-DBD. This aircraft is displayed near to the air base's headquarters

building and is the last remaining example of the type. Only four UTVA-60s were built before the design was superseded by the UTVA-66.

Also displayed as a gate guard is '195', a former JRV **MiG-21UM 'Mongol'** (bottom right). The HRZ has a number of active MiG-21UMs in its inventory.

Dassault Rafale

Omnirole fighter

Along with Sweden's Gripen, the Rafale is almost certainly to be the last major combat aircraft to be developed in Europe by a single nation. Arising from Dassault's ACX programme revealed in 1982, what matured as the Rafale was briefly one of three contenders to form the basis of the Eurofighter, until France left the discussion table in 1985, at which point the two programmes diverged. By going it alone, the Rafale has suffered from few of the political problems and attendant delays which have dogged its rival, and also allowed the French aviation industry to flourish in areas of cutting-edge technology. The result is a highly advanced aircraft which is already in service and assured of a large home market. To highlight its capabilities and export potential, its makers have dubbed it the 'omnirole' fighter – able to perform any task required of it in the current and future battlespace.

In the 1980s, when Rafale was under development, it was considered that current and projected avionics capabilities would alleviate much of the pilot's traditional workload, allowing him/her to concentrate on the important aspects of conducting the mission at hand. In the light of Gulf War experience, France became one of the first to admit that the workload on a single pilot would be too much in a sophisticated strike mission scenario in a heavily-defended environment, with the result that the order was revised to include a majority of two-seaters. At the time, this decision was derided by many, but other programmes have subsequently arrived at a similar conclusion: Saab (Gripen), Boeing (Super Hornet), Mikoyan (MiG-29M2), Sukhoi (Su-30) and even Eurofighter are either pursuing or studying fully 'missionised' two-seaters, able to handle the most complex of mission taskings, which are expected to include command and control of other combat aircraft and of UCAVs.

The imposing form of a Rafale M rises above the steam from the catapult of Charles de Gaulle. When it came to Rafale procurement, the needs of the Aéronavale were considered more pressing than those of the Armée de l'Air, with the result that the first operational aircraft were delivered to Landivisiau to form Flottille 12F. In so doing, a true air defence capability was restored to the carrier air group.

Having dropped the 'Super' from its appellation, the Mirage 4000 company-funded prototype for an F-15-class 'superfighter' re-emerged in 1986 as a trials aircraft for the Rafale programme. Among the areas it was used to research was the performance in turbulent air of an aircraft equipped with close-coupled canard foreplanes.

In May 2001 the creation of Flottille 12F at Landivisiau Naval Air Station marked the entry into service of the Rafale, the latest combat aircraft designed and produced by Dassault Aviation. With the Rafale, the French Armed Forces have entered the 21st Century with one of the most modern fighters in use anywhere.

The history of the Rafale programme can be traced back to the mid-1970s, when the Aéronavale and the Armée de l'Air (AdA) started to look at various options to replace aircraft then in service or about to enter service. In an effort to reduce costs, the Armée de l'Air and the Aéronavale agreed on a common requirement and issued a common request for proposals.

"The stringent French Ministry of Defence requirements called for a true swing-role fighter which could operate by day or night in all weather conditions, covering the whole spectrum of air-to-air and air-to-ground operations which, until now, required a fleet of different types of aircraft: Jaguar, Mirage F1CT and Super Etendard attack fighters, Mirage F1CR and Etendard IVPM reconnaissance aircraft, F-8P Crusader, Mirage F1C and Mirage 2000 interceptors, plus Mirage 2000N nuclear bombers," explains Jean Camus, a Dassault Operational Expert and former Test Pilot in charge of the Rafale programme when serving with the French Air Force. "Moreover, the design had to offer affordable life-cycle costs through low fuel consumption, low maintenance requirements, and long airframe and engine lives."

European co-operation was seriously considered as it was felt that it would help reduce costs, especially those associated with the development of the fighter, but the agreement France had entered into with Britain, Germany and Italy stalled over the weight issue. French armed forces wanted a fully swing-role aircraft, in the 9-tonne class for aircraft-carrier compatibility, whereas the other European nations were committed to buying a heavier air defence fighter, in the 10-tonne class, which was to emerge as the Eurofighter Typhoon.

Programme development

Development of the new French aircraft was remarkably swift, and the first Rafale – the white-painted demonstrator Rafale A – was unveiled at Saint-Cloud in a ceremony headed by the late Marcel Dassault in December 1985. From then on, things proceeded quickly: the aircraft first took to the air from the Dassault flight test centre in Istres, in the south-east of France, on 4 July 1986. To limit the risks traditionally associated with a first sortie, the aircraft had been fitted with two proven General Electric F404-GE-400 turbofans, the same engine which powers the US F/A-18A/B Hornet. The Rafale A used all-moving canard foreplanes coupled with an advanced fly-by-wire (FBW) flight control system, with the large delta wing, canards and FBW system all ensuring maximum manoeuvrability.

Rafale A was first displayed at Farnborough in September 1986, where it demonstrated its agility to good effect. A Snecma M88 turbofan replaced the left F404 from May 1990 onwards. Supercruise in dry thrust was achieved with this powerplant when the aircraft reached Mach 1.4. Rafale A was retired in January 1994, after 865 sorties.

A severe blow to the programme came with the fall of the Berlin Wall and the disappearance of the old Soviet threat. This event, in turn, led to a considerable drop in French defence budgets, with a corresponding substantial modification to the French Air Force order of battle, and a reassessment of future needs. The first consequence was a drastic reorganisation of the Armée de l'Air, resulting in the nearly immediate withdrawal of the Mirage 5F fleet, and the decision to modernise 55 Mirage F1C fighters into Mirage F1CT tactical fighters. The Mirage 5Fs had initially been built for Israel, but the aircraft had been brought into

French service when an armament embargo was put in place. Those two events had a considerable impact on Rafale, absorbing budgets which could have been used to accelerate the new fighter programme. Instead, it was then decided to slow down research and development. One of the consequences was the need to enhance Armée de l'Air air defence capabilities with other advanced aircraft, and the French Air Force decided to update some of its Mirage 2000Cs to Mirage 2000-5F standard. This allowed the Rafale to be further delayed, and gave time to the competitors to partially catch up with Dassault.

Air Force versions

To carry out the different missions envisioned (air defence/superiority, precision attack, nuclear strike and reconnaissance), the Armée de l'Air required two Rafale variants: the single-seat Rafale C (C for Chasseur, fighter) and the two-seat Rafale B (B for Biplace, two-seater). The black-painted Rafale prototype C01 first flew in May 1991 at Istres but, to reduce costs, the planned second single-seat prototype was in the event not built.

Rafale C01 differed significantly from the earlier Rafale A demonstrator: the overall configuration had been retained but the C prototype was fractionally smaller and lighter. Furthermore, considerable changes had been implemented to reduce radar cross-section (RCS). They included a gold-coated canopy, reprofiled fuselage/fin junction, radar-absorbing materials and more rounded wing-root fairings. Composite material content had been considerably increased, contributing to a simultaneous reduction in weight and RCS. "In the prototypes and production aircraft, engineers have used high-tech materials to drive down weight, increase fatigue life and reduce production costs," explains Jean Camus. "For instance, the Rafale's canards are made of superplastic forming diffusion-bonded titanium, and the wings are constructed of carbon fibre."

Dassault engineers have developed a very simple fighter with fixed air intakes and no dedicated airbrake, helping to reduce maintenance requirements. With the Rafale, Dassault pioneered the use of revolutionary advances such as the fully redundant, very high-pressure (350-bar/5,000-lb/sq in) hydraulic system and the variable-frequency alternators which all boost reliability and safety. These advances have proven so successful that they have been adopted as standard by other aircraft manufacturers and, for instance, they have been specified for the Airbus A380.

Yves Kerhervé climbs the Rafale A away after an approach to the deck of either Clémenceau or Foch. Just visible is the tail of an F-8 Crusader, the type which Rafale M was initially procured to replace. As well as approaches to the two carriers, Rafale A was also tested on dummy decks at Nîmes and Istres.

Rafale M's undercarriage is considerably strengthened, while the nosewheel (left) is extendible to increase incidence at launch. The nosewheel mounts a launch strut and hold-back system, compatible with US carriers.

Like the Mirage 2000, the Rafale was conceived from the outset for NATO interoperability, and Dassault engineers strictly adhered to the various NATO Stanags (Standard Agreements). All the Rafale's major systems are fully compatible with NATO systems: its radios have Have Quick secure capabilities, its inflight-refuelling equipment is fully compatible with NATO probe and drogue systems, its IFF system is completely interoperable, and the MIDS-LVT (Multifunction Information Distribution System – Low Volume Terminal) system has been designed hand-in-hand with other NATO countries. Interoperability extends to the armament: the GBU-12, one of the most widely available air-to-ground stores, has already been qualified on the Rafale, and the fighter's 14 fully NATO-compatible hardpoints offer excellent flexibility, allowing a wide variety of stores to be carried. It is anticipated that other widely available weapons such as the Mk 82/83/84 'dumb' bombs and the GBU-22/24 Paveway III laser-guided bombs will be qualified on the Rafale in the not too distant future.

Rafale C01 was mainly used to open the flight envelope, and to test the M88-2 engines. Later, it took part in weapon firing/separation testing – with Magic II air-to-air missile and gun – and participated in the man-machine interface validation. At some stage it was felt that C01 was getting too old and that it should be retired, but it has been decided instead to retain the aircraft for further engine testing, and it will participate in the M88-3 development programme.

Tactical two-seater

Since the successful introduction of the two-seat Mirage 2000N in the mid-1980s, the French Air Force has always been keen to operate two-seat tactical fighters. Initially, it was intended that only 25 two-seat Rafales were to be procured for conversion training but, after the Gulf War, studies showed that two-seaters were far better adapted to complex attack missions, as the workload in single-seat Jaguars or Mirage F1s was deemed too high in poor weather or in demanding combat environments. The experience with two-seat Mirage 2000Ns and 2000Ds had proved that the concept of 'burden-sharing' between a pilot and a weapon system operator (WSO) was well suited to the air-to-ground role in high-threat scenarios. Another key driver was the need for increased operator functionality. As a result of these ideas, the AdA procurement plan was amended, and now includes 95 single-seat and 139

two-seat Rafales, although 16 aircraft were deleted from the overall Air Force requirement at this time (from 250 to 234 fighters). It is worth noting that the US Navy has followed a similar path with the F/A-18E/F Super Hornet, increasing the percentage of two-seaters on order.

Rafale B01, the only two-seat prototype, first flew in April 1993, and was in charge of the fire-control/weapon system testing, including the RBE2 radar and the SPECTRA electronic warfare suite. Later, B01 took part in weapon separation and heavy load trials, and the fighter regularly flew with three 2000-litre (528-US gal) drop tanks, two Apache/Scalp cruise missiles and four air-to-air missiles. It even took part in numerous flying displays in that configuration, demonstrating the agility of the aircraft even in the most extreme load conditions. Compared with the single-seat variant, the two-seater is 350 kg (771 lb) heavier, while fuel capacity is reduced by 400 litres (106 US gal).

Nevertheless, it is fully combat-capable. "The front and rear cockpits of the Rafale B are similar, so that either the pilot or the WSO can perform any of the tasks associated with the different types of missions, thus enhancing flexibility," explains Pierre Delestrade, a Dassault Flight Test Engineer. "However, it is generally agreed that the pilot in the front cockpit would deal with the air-to-air modes whereas the WSO in the back seat would take care of air-to-ground functions."

Naval variants

The Aéronavale has had a long-standing requirement for a new carrierborne fighter to supplant ageing Dassault Etendard IVPMs, Vought F-8P Crusaders and, ultimately, Dassault Super Etendard Modernisés. At some stage, the then-McDonnell Douglas (now Boeing) F/A-18 Hornet was seriously considered, but it was felt that a naval variant of the new fighter envisioned for the French Armed Forces was deemed the most likely candidate, and a carrier version was developed. However, the aircraft was so much delayed because of financial constraints that the service had to modernise its ancient Crusaders in an attempt to

M02 launches from Charles de Gaulle during deck trials. Along with B01, M02 was extensively used as an avionics testbed, being fitted with RBE2 radar and FSO. It was also the first Rafale to fly with a working SPECTRA suite.

Opposite page: Although closely matched in overall capabilities, Eurofighter's Typhoon and the Rafale offer individual advantages in certain areas – Rafale is perceived as offering more air-to-surface options than its rival. Both the Aéronavale and Armée de l'Air have opted for more two-seat versions than single-seaters, a weapons system officer being seen as vital for complex attack and reconnaissance missions.

M01 was essentially an aircraft systems and aerodynamic trials platform for the 'Sea Rafale', undertaking several trials campaigns on the dummy deck at NAS Lakehurst, New Jersey, and aboard Foch. This view highlights the clean lines of the Rafale when not encumbered with pylons and the 'lumps and bumps' associated with SPECTRA and FSO.

Navalisation changes introduced on the Rafale M (M for Marine) include a strengthened airframe, a massive tail hook to bring the aircraft to a halt even during power-on, no-flare, carrier landings, a reinforced main undercarriage to absorb much higher vertical velocities, a power-operated built-in ladder to improve cockpit accessibility and to reduce the amount of ground support equipment needed, a new fin-tip Telemir system which allows the aircraft's inertial navigation system to exchange navigation data with the carrier's navigation suite, and a carrier-based microwave landing system. The most obvious modification introduced on the Rafale M is the imposing longer and stronger nosewheel gear, which gives the aircraft its noticeable nose-up attitude.

Because of the oversized nose gear, the front centreline weapon pylon had to be eliminated, but this does not have any significant impact on the Rafale's load-carrying capability as this pylon cannot be used when a centreline fuel tank or a Scalp stand-off missile is carried. To increase interoperability with US Navy aircraft-carriers, the nosewheel leg incorporates a launch-bar coupled with a 'hold-back' system, allowing Aéronavale Rafales to operate from American 'flat-tops'. The consequence of all these modifications is a Rafale Marine heavier than its Air Force counterpart, about 500 kg (1,102 lb) more, but this figure is slightly less than anticipated when the fighter was designed. The new combat aircraft represents a considerable improvement over the old F-8P Crusader, and will allow French Navy pilots to deal with any threat likely to be encountered for a considerable amount of time as the design has warfighting relevance beyond 2030.

Rafale N

Initially, it was planned that the Navy would receive 86 single-seat Rafale Ms, but budget cuts soon limited the overall number of fighters to be procured to 60. After an in-depth study, the Aéronavale procurement plan was recently amended to include a two-seat variant, the Rafale N (N for Naval), which will have a very high commonality with the single-seat Rafale M version, but will carry slightly less fuel: 4485 kg (9,888 lb) instead of 4700 kg (10,362 lb).

Dassault engineers have begun development of the new version, and a few modifications will have to be introduced: the adoption of the beefed-up airframe and nosewheel has already imposed the withdrawal of the 30 M 791 cannon to make room for relocated equipment, and reinforced canopy hinges will be required to handle stronger winds encountered on carrier decks. The final split between single-seat and two-seat aircraft is still to be announced but is likely to be either 20 Ms and 40 Ns, or 25 Ms and 35 Ns. It was initially anticipated that the first two-seat Rafale N for the Aviation Navale would make its first flight in 2005, and that the first aircraft would be delivered to an operational unit in 2007, but the recently announced budget cuts could have an impact on the planning.

Carrier-suitability trials were carried out quite early in the development programme when the Rafale A demonstrator flew approaches to the now-retired aircraft-carrier *Clémenceau* to ensure that the fighter behaved as it should in the carrier circuit. Although not carrier-capable, the

The Aéronavale received its first four Rafale Ms (M2 to M5) in Standard LF1 – without guns and only able to fire Magic 2 missiles, and with limited RBE2 and SPECTRA modes. From M6 onwards, deliveries have been to the full Standard F1. The initial aircraft were subsequently raised to this standard, and Flottille 12F should have all 10 of its allocated F1 aircraft by mid-2002. Subsequent deliveries will be to Standard F2 to begin the replacement of the Super Etendard.

increase their efficiency and maintain a modicum of air defence capability.

A number of changes had to be introduced for Rafale carrier operations, but the single-seat naval variant retains a high degree of commonality with its Air Force counterpart. As a result, the multi-spar wing cannot be folded, reducing complexity and weight but limiting the number of aircraft which can be stored on the deck or in the hangar. "This compromise is not really a problem because of the significantly larger size of the new *Charles de Gaulle* nuclear-powered carrier compared with the older *Clémenceau* and *Foch*, and the *Charles de Gaulle*'s air group will be significantly more powerful than that of its forebears," explains Capitaine de Frégate Planchon, Commanding Officer of Flottille 12F, the first Rafale squadron.

Above: Flottille 12F Rafale Ms participated in their first major exercise (Trident d'Or) in late May 2001, when four joined Charles de Gaulle's air group. The squadron officially reformed on 18 May.

Right: M1 was the second production Rafale to fly, handed over for testing in July 1999. As well as operational use, it flies from Istres on Standard F2 development duties.

Rafale A conducted further testing with *Foch* to confirm that, although heavier than the Super Etendard, the aircraft could be flown more slowly than its Super Etendard and Crusader predecessors.

Two dedicated navy prototypes were built, Rafale M01 and M02, and the delivery of M01, in December 1991, signalled the start of the naval development programme. This aircraft first underwent catapult testing at NAS Lakehurst, New Jersey, USA, in the summer of 1992. This trip to America was imposed by the lack of special European facilities since the closure of the catapult test bench at RAE Bedford, UK, following the withdrawal of HMS *Ark Royal*. A second campaign was carried out at Lakehurst in January-February 1993 by M01 before the first actual landings on *Foch* by Yves Kerhervé, Dassault Chief Test Pilot, in April 1993.

Rafale M02 first flew at Istres in November 1993, while M01 travelled to Lakehurst for a third series of tests in November-December 1993. The subsequent campaigns, both at Lakehurst and on *Foch*, allowed the aircraft to be tested under extreme conditions, including take-offs and landings with heavy loads such as 1,250-litre (330-US gal) and 2000-litre (528-US gal) external fuel tanks. The Rafale M has excellent bring-back capabilities, and can 'trap' with heavy unexpended ordnance. The first 'traps' and launches from the new aircraft-carrier *Charles de Gaulle* were performed in July 1999 by Yves Kerhervé, flying Rafale prototype M02. This paved the way for further successful trials campaigns on board the new French flagship. Today, M02 is still used for Standard F2 development programme, but M01 has now been withdrawn from service. In June 2001, for the week of the le Bourget air show, it was on

static display in Place de la Concorde in central Paris.

During the test programme, all Rafale prototypes have been fitted with a fixed refuelling probe to the right of the nose, ahead of the windscreen, and inflight refuelling from co-located C-135FRs was often used to top up fuel during long-duration sorties. The development programme of such an ambitious fighter by a single country was certainly a challenge but was ultimately successful, with not a single major problem involving the traditionally difficult-to-develop fly-by-wire flight controls.

Differing French standards

To reduce research and procurement costs, and to limit development risks, a stepped approach was adopted by Dassault and the French MoD. The first three Air Force

Top: M01 overflies the 'porte-avion' Foch during carrier tests. Foch was partially modified to handle the Rafale, including changes to the catapult shuttle, but it was sold to Brazil before the Rafale became operational. The first carrier landing was undertaken by M01 in April 1993.

Above: M02 lands on 'CdG' (deck-letter 'G'). Rafale M's first landings on the new carrier were carried out by Yves Kerhervé in July 1999.

aircraft (two-seaters B301 and B302, and single-seater C101) and the first 10 naval fighters (M1 to M10) are being delivered in the so-called Standard F1, dedicated to air-to-air combat and air defence missions. The F1 variant is armed with MICA EM radar-guided missiles and the long-serving Magic II short-range AAM, and its RBE2 (Radar à Balayage Electronique 2 plans) electronically-scanned radar has no air-to-ground modes. Standard F1 Rafale Ms entered service in 2000, and the first unit to be equipped is Flottille 12F, at Landivisiau, in Brittany. Rafales B301, B302 and M1 are used for the Standard F2 development programme, and mainly operate from Istres. However, from 2002, it is anticipated that M1 will spend about half of each year at Landivisiau with Flottille 12F.

The improved Standard F2 will allow air-to-ground attacks to be performed with advanced weapons such as the powerful Scalp cruise missile and the low-cost AASM (Armement Air-Sol Modulaire, modular air-to-surface armament). The Front Sector Optronics (FSO), a Link 16 MIDS-LVT (Multifunction Information Distribution System – Low Volume Terminal), and air-to-ground modes for the RBE2 radar will also be introduced in Standard F2 Rafales. Moreover, a high-resolution three-dimensional digital database will permit automatic terrain-following at low level.

Finally, the MICA IR will replace the Magic II, and an inflight-refuelling pod will be adopted for naval aircraft to allow them to adopt the 'buddy' tanker mission. Full authorisation for the Standard F2 development programme was signed on 26 January 2001, and a total of 33 Rafales 'Air' for

the Armée de l'Air and 15 Rafale M/Ns will be delivered to this standard. The first Standard F2 aircraft was due to enter service in 2004 and the first squadron was planned to be fully operational in 2005. However, this programme could slip by a year.

Swing-role fighter

Delivered from 2008 onwards, the final 198 Rafale B/Cs and 35 Rafale M/Ns will be in the swing-role Standard F3. These aircraft will be able to carry out other specialised tasks: nuclear strike with ASMP-A missiles, anti-ship attacks with Exocets or ANFs, reconnaissance missions with the Pod de Reconnaissance Nouvelle Génération (Pod Reco NG, or New Generation Reconnaissance Pod), and inflight-refuelling sorties with a new 'buddy' tanking pod which, unlike the earlier design to be utilised by Standard F2 Rafales, will contain fuel as well as the hose/drogue.

French Navy Rafale F3s will replace the Super Etendard, which will be used in these three crucial roles until 2009, at least. Today, all Super Etendards can fire Exocets, and some Standard 4 Super Etendard Modernisés are fitted with a belly-mounted reconnaissance system, but the Standard F3 Rafale will offer considerably improved capabilities. Similarly, French Air Force Mirage F1CR reconnaissance aircraft will progressively retire from 2005, and will be supplanted by the new podded system. Progress in the field of reconnaissance has been rapid in recent years, and the Pod Reco NG will significantly boost capabilities. The pod will be fielded by Mirage 2000Ns from 2006, followed by Rafales from 2008.

In an effort to standardise the fighter fleet, all Standard F1 and F2 aircraft will be updated to the latest Standard F3 while undergoing inspection maintenance. A subsequent Standard F4 is already envisioned, and could feature yet to be announced improvements. Apart from the already budgeted MBDA Meteor long-range air-to-air missile, it is highly likely that the conformal fuel tanks, active-array electronic scanning RBE2 radar, and the uprated M88-3 engines being developed by Dassault, Thales (formerly Thomson-CSF) and Snecma for the export market will be adopted for this Standard.

Engine reliability

In order to reduce costs of ownership and to improve performance, the M88 has been designed to achieve the optimum combination of operational readiness and reliability, and to facilitate maintenance in harsh conditions. The engine comprises 21 modules, interchangeable without the need for balancing and re-calibration. Module exchange allows rapid engine repair and minimises spares holdings. Some of these modules can even be changed without removing the engine from the Rafale airframe and, in any case, an M88 can be replaced in under an hour. After maintenance, there is no need to check the turbofan on a bench before it is installed back in the aircraft.

As a consequence, the Armée de l'Air has decided not to procure any test bench for the M88, but the Aéronavale has chosen instead a more traditional approach and, after maintenance or repair, its Rafale engines will be checked in new facilities at

Landivisiau Naval Air Station and/or on board the carrier *Charles de Gaulle*. M88 reliability is very good and, even for sustained combat operations, it is anticipated that only limited quantities of spare parts and spare engines will be required. The first deployments on board *Charles de Gaulle* have confirmed the maintenance requirements.

"When introducing into service such an advanced engine, you have to be very cautious at first," explains Jacques Desclaux. "For the M88, we have selected new technologies such as powder

metallurgy, and we want to be certain that problems do not appear. This is why the M88-2 Stage 1 engine initially had to be inspected every 150 hours, but in January 2001 this interval was raised to 500 hours, corresponding to roughly two years of operational use. As experience builds up with the M88-2 Stage 4, this interval will be progressively extended to 800 hours or 1,000 hours, depending on the components. By comparison with the Rafale, when the Mirage 2000 entered service, the M53 had to be checked every 75 hours."

Right: An M88-2 is mounted in a test rig. The engine is very compact and its accessories are easily accessed for maintenance.

Below: An M88-2 is removed from a storage canister.

Smart sensor fusion

Like every modern combat aircraft, the Rafale is dependent on technology to achieve air dominance. The advanced avionics suite of the new Dassault fighter is divided into different types of systems, all closely integrated to improve the pilot's situational awareness: sensors, electronic warfare suite, navigation and identification equipment, displays etc.

"On the Rafale, there is no primary sensor: the radar, the FSO and the SPECTRA electronic warfare suite all contribute to situational awareness, and the data obtained by the different means is fused into a single tactical picture displayed on a central eye-level screen collimated to infinity," says Jean Camus. "All the sensors have inherent advantages and drawbacks: the passive FSO has excellent countermeasures resistance, and its angular resolution is better than that of the radar. On the other hand, the radar is very accurate in range at long distances, and can track more targets than the FSO. The SPECTRA suite can analyse enemy radar emissions to precisely identify an emitter. The data fusion system combines and compares the data gathered by those sensors, and can accurately position and positively identify targets. It's much more than simple correlation as it allows the pilot to build up an accurate and unambiguous tactical picture."

Test pilots and front-line pilots all praise the new concept, which will give them the upper hand. "Multi-channel target acquisition/tracking, associated with smart sensor fusion, is a key-enabler which will radically change the face of air warfare," says Yves Kerhervé, Dassault Chief Test Pilot. "Until now, pilots only had their brains to process the information obtained by their radars/eyes and to build a mental image of the evolving situation. With the Rafale, the system has taken over the processing role, considerably reducing pilot workload, and allowing aircrew to devote more time to tactics management. The pilot now concentrates on fight, not on flight. Additionally, the Rafale's multi-channel weapon system can simultane-

Above: Rafale excels at low-level flight with heavy loads. Automatic terrain-following is possible thanks to a digital terrain database, removing the need for an emitting TF radar which may give away the aircraft's position.

Right: This test rig at Istres integrates all of the cockpit systems, including HUD, MFDs and VTAS.

Below: Another cockpit test rig at Istres allows test pilots and engineers to simulate highly complex mission profiles and scenarios.

Right: Rafale production
will be undertaken
alongside that of the Mirage
2000 for some time. Indeed,
the earlier generation of
Dassault fighter continues
to secure important fighter
contracts, aimed by
Dassault at nations who
cannot afford the Rafale.
Key to this success is the
development of the Dash 5/9
family, with RDY radar,
state-of-the-art cockpit,
MICA armament and new
EW systems. Here Rafale
B01 flies with the first
Mirage 2000-5.

Opposite page, top: In some
respects the C01 prototype
has been overtaken by
events, notably the desire to
equip the Aéronavale's
fighter squadron as a matter
of considerable urgency,
and the decision to
restructure the overall
Rafale buy greatly in favour
of two-seat aircraft. Never
fitted with representative
avionics systems, C01 was
initially worked extremely
hard on envelope expansion
duties. Now it is the main
engine testbed, involved in
M88-2 Step 4 work and,
soon, trials of the all-
important M88-3.

ously deal with airborne and ground threats. This is a crucial advantage over the nearest competitors because, from now on, pilots will be able to attack targets on the ground while engaging the enemy fighters presenting the greatest threat. For example, even with the radar in an air-to-surface mode, the FSO will be fully capable of detecting and tracking hostile interceptors, and the pilot can instantly engage an emerging threat."

At the core of the Rafale's capabilities is the Modular Data Processing Unit (MDPU) composed of line-replaceable modules. Built from commercially available off-the-shelf elements, the MDPU enhances avionics/armament integrations – thanks to its redundant, open and modular architecture, the system is highly adaptable, and new avionics and new ordnance now under development can be easily adopted. The system has been conceived with growth in mind, so that modifying the aircraft from one standard to another will not be a problem. Noteworthy is the fact that the MDPU is not fitted to Standard F1 Rafale Ms, which are equipped with older technology systems. However, the MDPU will be fitted to the first production Air Force Standard F1 Rafales – B301, B302 and C101.

RBE2 radar

The adoption of specialised phased array radars for the B-1B Lancer and for the MiG-31 Foxhound had not gone unnoticed, and the French authorities quickly recognised that this revolutionary technology was the way forward. This trend was confirmed by the announcement by the US military that the F-22 Raptor and other future fighters (which gave way to the F/A-18E/F and JSF programmes) were also to be equipped with phased array radars. French radar specialists soon embarked on an ambitious research and development programme to develop indigenous electronically scanned (e-scan) radars to equip a wide variety

of systems, including warships and new combat aircraft. Obviously, the Rafale was the first aircraft to benefit from this massive research effort.

"Compared with outdated mechanical planar antenna radars, electronic scanning radars offer a quantum leap in efficiency: as they do not need complex actuators to point the antenna, they are inherently more reliable and more stealthy," points out Philippe Ramstein, Thales Director of the Rafale programme. "The beam-shifting of an electronic scanning radar is extremely precise and nearly instantaneous in both vertical and horizontal planes, ensuring a very high revisiting rate on detected targets in the search-while-track mode." Modern air combat tactics have been devised to counter mechanical scanning radars operating in the track-while-scan mode, with a pair of fighters usually splitting to confuse the interceptor. In this scenario, some of the dispersed enemy aircraft will inevitably fall off the screen. However, against a fighter equipped with an e-scan radar operating in search-while-track, these tactics would prove totally ineffective.

"Even more important is the capability to share time between modes, thus carrying out different tasks simultaneously," stresses Philippe Ramstein. "Powerful data processors and unmatched beam agility allow the Rafale to fully interleave functions within a given mode: the radar combines search, track and missile-guidance functions, processing them simultaneously to assist the crew in achieving air dominance. It also features a superior fighter/missile datalink which gives better fire-control capabilities in adverse environments, thus increasing the overall lethality of the Rafale's weapon system. Finally, fixed arrays considerably reduce radar returns towards enemy aircraft. All these factors contribute to the enhancement of the Rafale's combat efficiency and electromagnetic stealthiness compared with fighters fitted with classic radar sets."

Developed by what is now Thales Airborne Systems, the RBE2 (Radar à Balayage Electronique 2 plans, two-axis electronic scanning radar) is the first airborne look-down/shoot-down multi-mode electronic scanning radar designed and produced in Europe. Developing such a compact, high-performance radar system was a real techni-

Left: In the early 1990s
Dassault was describing the
AdA versions of the aircraft
as the Rafale D (D = Discret,
discreet) to highlight the
low radar cross-section and
IR signature of the type. This
view shows the carefully
blended contours of the
forward fuselage and
intakes, and also displays
the gold-tinted canopy
applied to the C01 prototype.
Clearly visible is the wide-
angle HUD screen, which
dominates the pilot's
forward view. As well as
providing short-term flight
information and targeting
data, the head-up display
can also project FLIR or TV
imagery to aid low-level
flying at night or in bad
weather.

Right: Ninety-five of the
AdA's 234-aircraft order are
being procured as single-
seat Rafale Cs. The pilot-
only aircraft will be geared
towards air defence duties,
but retain full air-to-surface
capability, perhaps under
the direction of two-seat
aircraft. The first front-line
unit to form will be EC 1/7 at
Saint-Dizier, with whom the
Rafale will replace the
Jaguar.

cal challenge for the French industry: the RBE2 had to have a long range while fitting in the relatively small nose of the Rafale. Besides, the radar and all of the related electronics had to be able to withstand the shock of a carrier landing.

Flight testing started in Mystère 20 s/n 104 in July 1992 and, at one point, no fewer than five Centre d'Essais en Vol aircraft – three Mystère 20s and two Mirage 2000s (s/n 501 and 504) – were involved in the RBE2 development programme before further tests were conducted aboard Rafales B01 and M02. Today, Rafales M1, B301 and B302 also participate in radar development. The first production RBE2 was delivered in October 1997, and the radar is already in operational service with the French Naval Aviation. The first RBE2s for Standard F1 Rafales are only capable of air-to-air modes, but subsequent radar sets will have improved capabilities, culminating in the Standard F3 radar which will introduce comprehensive air-to-surface modes, including automatic terrain-following. The RBE2 is now totally qualified for air-to-air combat, and air-to-ground functions are actively being developed in preparation of the Standard F2.

Radar modes

"By embracing open architecture and commercially off the shelf (COTS) technology, Thales has designed a highly evolutive multi-mode radar which will satisfy even the most stringent requirements," states Philippe Ramstein. The RBE2 multi-mode, electronic scanning array radar enables Rafale pilots to accurately detect, track and engage airborne and ground threats from very long range.

Thanks to its unique waveform design and electronic

scanning management, the RBE2 radar performs long-range detection and tracking of up to 40 air targets in look-down or look-up aspects, in all weathers, and in severe jamming environments. Interception and firing data are calculated for eight priority targets which can be engaged with MICA BVR/air combat active radar seeker missiles fired in quick succession at the rate of one every two seconds. With its electronic scanning antenna, the radar is fully capable of tracking the other 32 targets while updating the MICAs with the dedicated, mid-course, secure, radar-to-missile link

Below: An impressive SAR capability is being developed for the RBE2, as demonstrated by this 'patch map' obtained by a development radar. Using such images, the WSO can designate targets into the system, which slews other sensors to peer at the designated location.

which enables very long-range multiple firings with an exceptionally high probability kill rate, even against manoeuvring enemy fighters. This gives the Rafale a unique combined situational awareness and combat capability/efficiency, while considerably reducing aircrew workload, especially in complex tactical situations.

For air-to-ground strikes, the radar has dedicated functions for low- and high-level navigation, target-aiming, searching and tracking of moving and fixed targets, ranging and terrain-avoidance/following. In the terrain-avoidance/following mode the RBE2 looks ahead to build a constantly changing, wide-angle, three-dimensional profile of the terrain to be overflown. With the electronic scanning technology, terrain-avoidance is optimised to improve survivability while flying at extremely low altitude and very high speed.

With its open architecture, the RBE2 has been designed for growth. For instance, a Synthetic Aperture Radar (SAR) mapping mode is actively being developed for Standard F3 Rafales. It will allow Rafale aircrew to 'paint' from stand-off distances, providing high-resolution maps of surface targets in any weather, day and night, and to designate a precision aiming point to the fighter's weapon system. Airborne trials have already started on board a flying test bed, and the SAR mode will be fully qualified on the Rafale in 2006.

Anti-ship attacks require specific modes, and the RBE2 will be able to detect, track and engage ships, even in high sea states. An air-to-sea radar surveillance mode will be introduced on Standard F2 Rafales, and the fighter's weapon system will be capable of firing anti-ship missiles with the advent of the Standard F3.

Optronics suite

To complement the radar, the Rafale is fitted with a comprehensive optronics suite divided into three systems: the Front Sector Optronics (FSO, or OSF in France), the Damoclès laser designation pod, and the aforementioned Pod de Reconnaissance Nouvelle Génération (Pod Reco NG, or New Generation Reconnaissance Pod).

Mounted on top of the nose, ahead of the windshield, the FSO allows an uninterrupted view of the forward sector. Operating in different infra-red wavelengths, the FSO provides discreet long-range detection, multi-target angular tracking and range-finding for air and surface targets, considerably enhancing the Rafale's stealthiness as the fighter can covertly detect and identify enemy aircraft without using its own radar which would betray its presence.

The FSO comprises two modules: the infra-red sensor (Infra-Red Search Track) and the TV system, coupled with an eye-safe laser rangefinder. The functions of the two systems are clearly complementary: surveillance and multi-target tracking by the starboard IR surveillance module; target tracking, identification and ranging by the port TV/laser module. Whatever the rules of engagement restrictions, the FSO minimises the risks of fratricides ('blue on blue'), and it allows instantaneous battle damage assessment to be performed. Although this unique surveillance and identification system has been thoroughly tested onboard a Dassault Falcon 20 and Rafale prototypes M02 and B01, plus series aircraft B301 and B302, it will not be introduced into production aircraft until the first Standard F2 Rafale is delivered.

Produced by Thales, the new-generation Damoclès laser designation/targeting pod has been designed to be used in conjunction with existing and future laser-guided ammunitions such as Paveway LGBs and AASM precision weapons. The 250-kg (551-lb) Damoclès represents the third generation pod after the ATLIS which equips Armée de l'Air Jaguars and Aéronavale Super Etendards, and the PDL-CT/PDL-CTS of the Mirage 2000D. The new infra-red (third-generation staring array detector) and laser technologies chosen for the Damoclès ensure extended detection and recognition ranges, permitting laser-guided armament to be delivered at substantially greater ranges and from higher altitudes, considerably reducing the aircraft's vulnerability to short/medium-range air defence systems.

Two fields of view are available to aircrew: wide

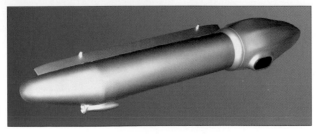

Left: This image shows a Rafale and Mirage 2000 at extreme range, as captured by the Front Sector Optronics (FSO) system. The FSO suite is fully integrated with the radar and other aircraft systems, allowing it to be cued by the radar at long range for identification or raid assessment, and for continued passive tracking of targets. It also incorporates a laser rangefinder, highly useful for feeding slant range and offset point information to the weapons computer for bomb release.

Right: This is a model of the Pod Reco NG which will provide Rafale with a state-of-the-art digital reconnaissance capability. The nose section rotates to allow horizon-to-horizon operation, and the pod also has a function which allows it to automatically track linear features such as roads.

Deployability – autonomy and rapidity

Over the past three decades, the French Air Force has participated in countless deployments to numerous trouble spots in Africa (Chad, Djibouti, Central African Republic, Gabon etc.), the Middle East (Gulf War) and Europe (Bosnia and Kosovo). Accordingly, the Rafale has been designed to impose a minimum logistic footprint, facilitating deployments far from traditional support infrastructures or from dispersed locations. The number of Rafale external ground support equipment is extremely low, and will prove less constraining for strategic airlift assets during long-distance deployments.

"To ease up operations from friendly bases, the Rafale's cross-servicing procedures have been simplified, and the aircraft is fully compatible with standard NATO tools and support equipment," explains Carl Chevillon, Rafale International Support Manager. "A pressure refuelling system is fitted as standard for both internal and external fuel tanks, and internal tanks can be refilled in four minutes only. The French Navy has already validated hot refuelling procedures. For full self-supportability, the fighter is equipped with an OBOGS (On-Board Oxygen Generation System), giving infinite oxygen endurance, and helping reduce costs through a diminution of necessary ground equipment and maintenance. It will also contribute to the reduction of the already very short turnaround time between flights, and will ease up deployments by eliminating the need to transport liquid oxygen, a highly volatile and dangerous product. Similarly, the Rafale is equipped with a closed-loop cooling fluid system for onboard coolanol and nitrogen circuits. Finally, it is fitted with a Microturbo Auxiliary Power Unit for autonomous operations and engine start."

"The Rafale is very powerful and can take off in a very short distance," says Jean Camus. "It has been designed for carrier operations, and its approach speed is remarkably low. Although the type is not fitted with a brake chute, its landing roll is incredibly short. It has been regularly demonstrated at events such as Farnborough or the Paris Air Show that its landing distance is shorter than those of all its competitors. These characteristics mean that the fighter can routinely operate from short runways, and it has been determined that pilots can avoid using afterburner for take-off from a NATO standard runway in all configurations, even with a heavy load. Alternatively, the fighter can operate from short civilian runways or from damaged military airfields. The Rafale's navigation precision is astounding, and autonomous initial and final approaches can be performed, even in extremely bad weather."

Inflight refuelling is a prerequisite for a modern fighter intended to undertake rapid-response deployments to anywhere in the world. The AdA's primary tanker is the C-135FR, which serves with ERV 93 alongside three ex-USAF KC-135Rs. The tanker unit is conveniently based at Istres, alongside the combined Rafale test fleet. Pilot opinion differs as to which is more difficult: tanking from the very short hose attached to the boom, or refuelling from the wingtip Mk 32 HDUs and their attendant vortices

(4° x 3°) and narrow (1° x 0.5°), and the pod is fitted with an eye-safe (wavelength 1.5 µm) laser rangefinder, with a laser designator fully compatible with NATO's Stanag 3733 (wavelength 1.06 µm), and with a laser spot tracker (wavelength 1.06 µm). Additionally, its greatly improved resolution means that it can be used for stand-off reconnaissance and battle damage assessment. The Damoclès pod has been conceived for considerably lower maintenance requirements and costs than earlier designs, and it can withstand the shocks associated with carrier landings. It is currently being qualified on Mirage 2000-9 and Super Etendards, and is to be qualified on the Rafale in 2003.

Damoclès has already been ordered by the Aéronavale for its Standard 5 Super Etendard Modernisés, and is highly likely to be adopted by the French Armed Forces for their Rafales. Additionally, it has been selected by the United Arab Emirates for recently ordered Dassault Mirage 2000-9s (30 of which have been bought, plus 33 earlier Mirage 2000s brought to the Dash 9 standard). From 2010, a joint British-French electro-optic targeting pod, called JOANNA (for JOint Airborne NavigatioN and Attack), will be introduced on Rafales to supplant Damoclès. Flight-testing of JOANNA is due to start in 2005.

For reconnaissance missions, electro-optics is the way of the future, and the Pod Reco NG will make good use of this technology. Designed and produced by Thales, this system will be fielded by suitably modified Mirage 2000Ns and Rafale tactical fighters. The performance of this system is still classified, but it is understood that the imagery provided will be exceptionally good, and will allow high-resolution images to be acquired from stand-off distances. For maximum efficiency, the sensors use different wavelengths, and the pod features state-of-the-art digital recorders. Additionally, it will be equipped with a datalink to relay data in a timely and accessible way for real-time interpretation. For targets of opportunity, it is envisioned that the pilot will be able to slew the pod's sensors using his helmet-mounted display. A total of 23 Pod Reco NGs will be acquired, including eight for the Aéronavale.

Voice control and helmet-mounted sight

Rafale's man-machine interface has been tuned to reduce aircrew workload: the aforementioned sensor fusion system is not the only novelty introduced in the fighter, as the helmet-mounted display (HMD) and voice control system will also considerably enhance aircrew situational awareness. The direct voice input control system and the HMD will be adopted for Standard F3 Rafales.

Rafale weapons

Rafale is equipped with 14 hardpoints: two wingtip rails and two outer wing pylons for small AAM carriage (MICA, AIM-9, Magic etc), four intermediate and inner underwing pylons for heavy stores, two rear fuselage-side pylons (for MICA or small bombs), two forward fuselage-side points (primarily for podded sensors) and two centreline hardpoints (only one on the Rafale M because of the oversized nose gear). The rear centreline hardpoint and four heavy-store underwing pylons are 'wet' for drop tanks. Maximum external load is quoted at 9500 kg (20,925 lb).

GIAT 30 M 791 cannon

Recent combat experience has shown that cannon are still invaluable in a number of scenarios when missiles are either too expensive or unable to hit targets at very close range. As a direct consequence, guns are still considered ideal for the interception of low- and slow-flying targets, such as helicopters or transport aircraft. Additionally, guns are increasingly regarded as being useful and highly-effective weapons for use against ground and naval targets.

For the Rafale, GIAT Industries of France has developed the new 30 M 791 seven-chamber revolver cannon, the world's only single-barrelled 30-mm weapon capable of firing at a rate of 2,500 rounds/minute. This cannon is designed to offer maximum efficiency in air-to-air combat, and its very high firing-rate and high initial velocity (1025 m/3,362 ft per second) optimise hit probabilities. The 120-kg (265-lb) gas-powered gun is autonomous, and its effective air-to-air range is 2500 m (8,200 ft). The firing-rate is reached instantaneously, and 21 rounds are fired in 0.5 second (the normal burst duration). The powerful 30 M 791 is mounted on the side of the starboard engine duct in all Rafale versions, apart from the two-seat naval Rafale N. It is worth noting that, to minimise maintenance requirements, the internal cannon will only be carried for training sorties on firing ranges, or for actual combat missions.

The technically advanced 30 M 791 gun fires the powerful 30 x 150 range of ammunition designed specifically for the Rafale. These new rounds have high penetration and incendiary effects, and provide an excellent compromise between their splinter effects and detonation power. A total of 125 rounds is carried, and ammunition ignition is electrical. The gun is equipped with a pyrotechnical rearming device which ejects a faulty round after a short safety time period.

The first prototypes were completed in 1991, and the first trials were conducted with CEV Mirage III s/n 605 carrying a suitably modified pod containing a 30 M 791. Single-seat Rafale prototype C01 undertook the first aircraft/gun compatibility firing trials in 1993 and, since then, the 30 M 791 has been extensively tested in extreme conditions – including 9-*g* turns – to check that it can resist corrosion, shocks and quick temperature changes. The trials also proved that the Rafale's airframe and electronic systems could withstand the extreme loads and vibrations associated with a 2,500-round/minute firing-rate. Numerous firing campaigns have been conducted at the Cazaux Flight Test Centre, in the southwest of France, with the last campaign carried out by Rafale B301 between September 2000 and February 2001. Final approval for operational use has now been granted, and the 30 M 791 is in full-scale production. It will enter operational service with the French Naval Aviation in 2002.

MATRA Magic 2

Although the gun remains a useful last-ditch weapon, the armament of choice for modern fighters are missiles, especially in the air-to-air arena. The Rafale will initially be armed with two types of air-to-air missiles: the MBDA Magic 2 and the MBDA MICA (Missile d'Interception, de Combat et d'Autodéfense, Interception, Combat and Self-Defence missile).

The short-range IR-guided Magic 2 entered operational service in 1985 on Armée de l'Air Mirage 2000B/Cs. Compared with its Magic 1 predecessor, the new variant could carry out all-aspect engagements, including head-on firings. So far, 11,270

Magics have been purchased by 19 nations, and the type has been qualified on 18 different aircraft types, including the F-16. The Magic 2 is extremely agile, and can be fired at up to 8 *g*. The Magic will arm only French Standard F1 Rafales, and it is not anticipated that any export customer would purchase the type as the MICA IR offers much improved overall performance.

MATRA MICA

To replace both the ageing
Magic II and the Super 530D
arming the Mirage 2000B/C/E,
it was decided at the start of
the Rafale programme to
design a single multi-target multi-mission missile. First studies began in 1978, and
full-scale development was launched in 1982. The advanced MICA is capable of
both interception missions beyond visual range and close-up dogfight combats.
This lightweight (112-kg/247-lb) weapon has excellent manoeuvrability, confirmed
by approximately 100 test firings carried out so far. Two variants of the MICA will
be used by the French Air Force and Navy for the Rafale – the radar-guided MICA
EM (Electromagnétique) and the infra-red guided MICA IR – ensuring a reduction in
direct costs as the airframes, warheads and motors are the same for both variants,
the only difference being the seeker. This concept of interchangeable seekers was
first pioneered in France with the Matra 511 and the Matra 530 missiles, the latter
arming Mirage IIIs, Mirage F1Cs and Crusaders. The availability of two guidance
systems will offer enhanced tactical flexibility, and will hamper enemy
countermeasures selection. As expected, both MICA seekers have excellent
counter-countermeasures capabilities.

   "Thanks to its active radar homing head, the MICA EM is fully autonomous after
launch so that a pilot can either engage several targets simultaneously or turn
away after a shot, reducing the time spent in a potentially dangerous area or
denying the enemy aircraft any firing possibility", explains Jean Camus. "Its very
high impulse motor ensures very long range, and the fighter/missile datalink
permits beyond-visual range interceptions with a remarkable kill probability
percentage." The MICA EM has been tested in very demanding environments, and
the trial programme culminated in the engagement by two MICAs fired from one
fighter of two widely separated targets using countermeasures. First firings of the
MICA EM were carried out in 1992, and the missile was qualified on Standard F1
Rafales in July 2000, following 27 successful test firings conducted at the Centre
d'Essais en Vol/Centre d'Essais des Landes at Cazaux/Biscarosse. A first batch of
225 MICA EMs was ordered by France for its Mirage 2000-5s and its Rafales in
December 1997. The MICA EM will be fully operational on Aéronavale Rafales in
early 2002, and is already in service with the Armée de l'Air, Republic of China Air
Force (Taiwan) and Qatar Emiri Air Force Mirage 2000-5s. It is on order for the
United Arab Emirates Air Force for its Mirage 2000-9s, and for Greece for its
Mirage 2000-5 Mk IIs.

   Successive development of various generations of infra-red sensors for the

*MICAs are shown mounted on underwing (above left) and fuselage (above) pylons.
The EM version has a pointed nose housing an active radar seeker, while the IR
version has a rounded nose with a window for the dual-band seeker.*

Matra 530, Magic 1 and Magic 2 missiles has allowed French specialists to design
a highly-effective dual-band imagery IR seeker for the MICA IR. The extremely
manoeuvrable MICA IR will supersede the long-serving MBDA Magic 2 short-range
IR-guided air-to-air missile in Standard F2 Rafales. In fact, the MICA IR will initially
be restricted to short-range combat modes and, when Standard F3 Rafales are
delivered, a long-range interception function will be introduced. When used in
conjunction with the helmet-mounted sight, the MICA IR will allow lock after
launch, off-axis shots to be performed. The IR seeker has many advantages for
such a long-range missile. It has excellent angular resolution and countermeasures
resistance – thanks to dual-band imagery – and is totally stealthy: when used in
conjunction with the FSO, the passive homing head enables absolute 'silent'
interceptions without tell-tale radar emissions. First MICA IR firings were recorded
in 1995, and the missile is now in pre-production. A first batch of 70 MICA IRs has
been ordered by the French Armed Forces in 2000, and the type might eventually
be selected for the Armée de l'Air Mirage 2000-5F and Mirage 2000D to fulfil the
MACK (Missile d'Autoprotection et de Combat Rapproché, or Self-defence and
Close-Combat Missile) programme aimed at finding a successor for Magic 2.

   On the Rafale, the MICAs are fitted to two hardpoints under the sides of the
rear fuselage, and to wing/wing-tip points. With the planned dual pylons mounted
under the wings, the capacity of the fighter will grow to 10 missiles while still
retaining three 2000-litre (528-US gal) or 1250-litre (330-US gal) drop tanks, giving
an outstanding operational range.

   A vertical-launched naval/land variant of the MICA, the VL MICA, is on offer. The
basic naval configuration comprises eight missiles housed in off-the-shelf vertically-
mounted containers for ship defence against saturation attacks. The land version is
adapted to a truck, and could prove highly efficient in defending high-value static
targets such as air bases or refineries. Using either IR or active EM seeker, the VL
MICA system is claimed to be capable of engaging up to eight different targets
spread over a 360° arc in less than 12 seconds.

Meteor/MIDE

The appearance of other long-range active missiles on the market (Raytheon
AIM-120 AMRAAM and Vympel R-77 (AA-12 'Adder') forced the French Ministry of
Defence to rethink its strategy, and a requirement for an even longer-ranged
beyond visual-range air-to-air missile has recently appeared. The new missile would
need a no-escape zone several times greater than that of today's missiles.
Accordingly, the MIDE (Missile d'Interception à Domaine Elargi, Extended Domain
Interception Missile) programme was launched, with a desired in-service date said
to be 2010-12. In June 2001, at the Paris Air Show, the French Ministry of Defence
officially announced that it was joining the Meteor missile programme, and an
agreement was signed with the UK and Sweden. Other European partners have
elected to participate in the project, and Italy confirmed its commitment in
September 2001. At the time of writing, only Spain and Germany were still to
commit to the programme. Eventually, three aircraft types – the Eurofighter

Typhoon, the Saab Gripen, and the Dassault Rafale – will be armed with the
Meteor.

   The 185-kg (408-lb) ramjet-propelled Meteor was initially selected for the Royal
Air Force EF2000 Eurofighters as part of a programme dubbed Beyond Visual-range
Air-to-Air Missile (BVRAAM), but it was soon recognised that it would be ideal for
the Rafale. This extremely fast missile – Mach 4+ – is designed to retain sufficient
energy at end-game to defeat hard manoeuvring targets. When fitted to the
Standard 4 Rafales, the new Meteor/MIDE missile will create a formidable
combination of weapon/sensor/airframe. According to the latest information
published, a first batch of 256 Meteors will be ordered in 2005 with first deliveries
planned for 2012. Looking even further ahead, the Meteor could be equipped with
an anti-radiation seeker to participate in DEAD (Destruction of Enemy Air Defences)
missions. With its very high speed and very long range, a dedicated variant of the
Meteor would prove ideal for the role, allowing a Rafale to attack known enemy
emitters or to react against a pop-up threat.

Free-fall weapons

With 12 of the 14 hardpoints available for bomb carriage,
the Rafale can carry a heavy load of free-fall bombs – up
to 22 bombs in the 250-kg or 500-lb class. It is also
capable of carrying a wide range of other unguided
weaponry, such as rockets, Durandal and Belouga.
However, the Rafale is a sophisticated aircraft with the
emphasis on precision attacks. Therefore, the first free-
fall weapon to be qualified is the GBU-12 Paveway II
laser-guided bomb (illustrated), which can be carried on
twin or triple launchers. Further varieties of laser-guided
weapons are scheduled to be cleared to increase the
aircraft's air-to-surface repertoire and its exportability.

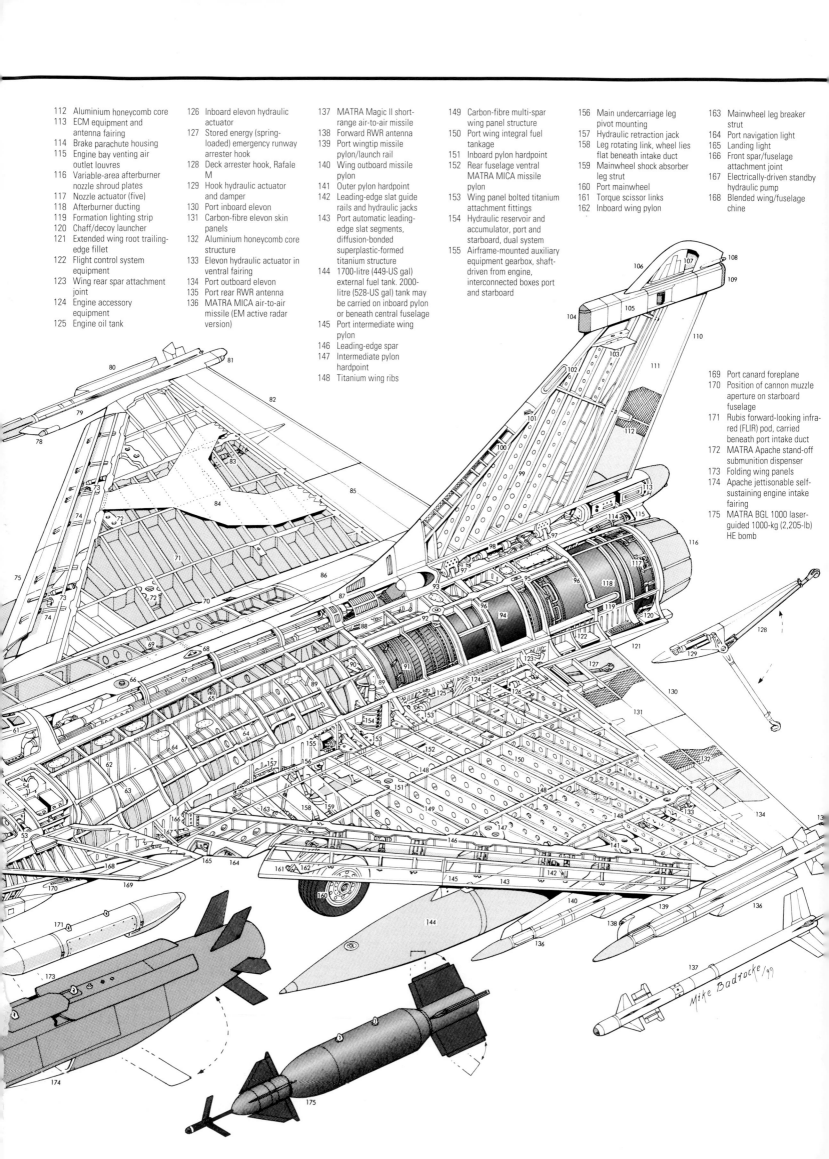

112 Aluminium honeycomb core
113 ECM equipment and antenna fairing
114 Brake parachute housing
115 Engine bay venting air outlet louvres
116 Variable-area afterburner nozzle shroud plates
117 Nozzle actuator (five)
118 Afterburner ducting
119 Formation lighting strip
120 Chaff/decoy launcher
121 Extended wing root trailing-edge fillet
122 Flight control system equipment
123 Wing rear spar attachment joint
124 Engine accessory equipment
125 Engine oil tank

126 Inboard elevon hydraulic actuator
127 Stored energy (spring-loaded) emergency runway arrester hook
128 Deck arrester hook, Rafale M
129 Hook hydraulic actuator and damper
130 Port inboard elevon
131 Carbon-fibre elevon skin panels
132 Aluminium honeycomb core structure
133 Elevon hydraulic actuator in ventral fairing
134 Port outboard elevon
135 Port rear RWR antenna
136 MATRA MICA air-to-air missile (EM active radar version)

137 MATRA Magic II short-range air-to-air missile
138 Forward RWR antenna
139 Port wingtip missile pylon/launch rail
140 Wing outboard missile pylon
141 Outer pylon hardpoint
142 Leading-edge slat guide rails and hydraulic jacks
143 Port automatic leading-edge slat segments, diffusion-bonded superplastic-formed titanium structure
144 1700-litre (449-US gal) external fuel tank. 2000-litre (528-US gal) tank may be carried on inboard pylon or beneath central fuselage
145 Port intermediate wing pylon
146 Leading-edge spar
147 Intermediate pylon hardpoint
148 Titanium wing ribs

149 Carbon-fibre multi-spar wing panel structure
150 Port wing integral fuel tankage
151 Inboard pylon hardpoint
152 Rear fuselage ventral MATRA MICA missile pylon
153 Wing panel bolted titanium attachment fittings
154 Hydraulic reservoir and accumulator, port and starboard, dual system
155 Airframe-mounted auxiliary equipment gearbox, shaft-driven from engine, interconnected boxes port and starboard

156 Main undercarriage leg pivot mounting
157 Hydraulic retraction jack
158 Leg rotating link, wheel lies flat beneath intake duct
159 Mainwheel shock absorber leg strut
160 Port mainwheel
161 Torque scissor links
162 Inboard wing pylon

163 Mainwheel leg breaker strut
164 Port navigation light
165 Landing light
166 Front spar/fuselage attachment joint
167 Electrically-driven standby hydraulic pump
168 Blended wing/fuselage chine

169 Port canard foreplane
170 Position of cannon muzzle aperture on starboard fuselage
171 Rubis forward-looking infra-red (FLIR) pod, carried beneath port intake duct
172 MATRA Apache stand-off submunition dispenser
173 Folding wing panels
174 Apache jettisonable self-sustaining engine intake fairing
175 MATRA BGL 1000 laser-guided 1000-kg (2,205-lb) HE bomb

Mike Badrocke/99

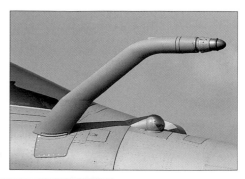

Three styles of external tanks have been developed: 1150-litre (304-US gal) conformal fuel tanks (left), 1250-litre (330-US gal) supersonic tanks and 2000-litre (528-US gal) subsonic tanks (below). The latter are similar to those originally developed for the Mirage 2000N, albeit with a revised tail section lacking auxiliary fins.

Above: The refuelling probe can be removed easily, but is not retractable for space and complexity reasons. The domed FSO turret to starboard houses the infra-red search/track module. TV and laser are in the port unit.

Rafale C cutaway

1 Kevlar composite radome
2 Thales RBE2 electronically-scanned look-down/shoot-down multi-mode radar scanner
3 Fixed (detachable) inflight refuelling probe
4 Front sector optronics (FSO) – Infra-red scanner/tracker (IRST)
5 FSO – Passive visual sight, low-light television (LLTV)
6 Forward-looking optronic system module
7 Airflow sensors, pitch and yaw
8 Total temperature probe
9 Radar equipment module
10 Dynamic pressure probe
11 Cockpit front pressure bulkhead
12 Instrument panel shroud
13 Rudder pedals
14 Canopy emergency release
15 Electro-luminescent formation lighting strip
16 Alternative nose undercarriage assembly, Rafale M
17 Catapult strop link
18 Deck approach and identification lights
19 Drag strut
20 Hydraulic retraction jack
21 Nosewheel bay
22 Port side console panel
23 Engine throttle lever with display imaging controls, and hands-on throttle and stick (HOTAS) control system. Sidestick controller for digital fly-by-wire control system on starboard side
24 Elbow rest
25 Pilot's wide-angle holographic head-up display (HUD)
26 Frameless windscreen panel
27 Canopy, open position
28 Thomson-CSF ATLIS II laser designator pod, carried on starboard intake pylon
29 ATLIS II mounting pylon adaptor
30 Rear-view mirrors (three)
31 Pilot's helmet with integrated sight display
32 Cockpit canopy, hinged to starboard
33 Pilot's SEMMB (licence-built Martin-Baker) Mk 16F 'zero-zero' ejection seat
34 Forward fuselage/cockpit section all-composite carbon-fibre structure
35 Lateral equipment bays, port and starboard
36 Nose undercarriage pivot mounting
37 Nosewheel door-mounted lower UHF antenna
38 Taxiing lights
39 Hydraulic steering jacks
40 Twin nosewheels, forward-retracting
41 Hydraulic retraction and lock strut
42 Port engine air intake
43 Boundary layer splitter plate
44 Ventral intake suction relief door
45 Port forward oblique SPECTRA ECM antenna
46 SPECTRA RWR antenna
47 Onboard oxygen generation system (OBOGS)
48 Canopy centre arch and support frame
49 Embedded electrically charged canopy emergency breaker
50 Circuit breaker and diagnostic panel
51 Avionics equipment bays
52 Canard foreplane hydraulic actuator
53 Foreplane hinge mounting
54 Environmental Control System (ECS) equipment bay
55 Canopy emergency release
56 Cockpit pressurisation outflow valves
57 Canard foreplane hinge fitting
58 Starboard canard foreplane
59 Carbon-fibre foreplane structure with honeycomb core
60 Starboard navigation light
61 Air system heat exchanger exhaust
62 Centre-fuselage aluminium-lithium primary structure
63 Intake ducting
64 Fuselage integral fuel tankage, total internal capacity 5325 litres (1,407 US gal)
65 Port main longeron
66 SATCOM antenna
67 Dorsal spine fairing housing systems ducting
68 Anti-collision beacon
69 Starboard fuselage integral fuel tank
70 Kevlar composite wing/fuselage fairing panels
71 Starboard wing integral fuel tank
72 Wing pylon hardpoints
73 Leading-edge slat hydraulic jacks and position transmitters
74 Slat guide rails
75 Starboard two-segment automatic leading-edge slats
76 Starboatd external fuel tank
77 GIAT 30 M 791 30-mm cannon located beneath starboard wing root
78 Forward RWR antenna
79 Wingtip fixed missile pylon/launch rail
80 MATRA MICA air-to-air missile (IR version)
81 Rear RWR antenna
82 Starboard outboard elevon
83 Elevon hydraulic actuator
84 Wing carbon-fibre skin panelling
85 Inboard elevon
86 Fuselage aluminium-lithium skin panelling, carbon-fibre ventral engine bay access panels
87 Auxiliary Power Unit (APU) intake grilles
88 Microturbo APU
89 Wing panel attachment forged and machined fuselage main frames
90 Engine compressor intake with variable guide vanes
91 Snecma M88-2 afterburning turbofan engine
92 Forward engine mounting
93 APU exhaust
94 Carbon-fibre engine bypass duct
95 Rear engine mounting
96 Fin attachment main frames
97 Fin root bolted attachment fittings
98 Rudder hydraulic actuator
99 Carbon-fibre multi-spar fin structure
100 Carbon-fibre leading edge
101 Flight control system airflow sensor
102 Formation lighting strip
103 VOR localiser antenna
104 Forward ECM transmitting antenna
105 SPECTRA integrated ECM system equipment housing
106 Fin-tip antenna fairing
107 VHF/UHF communications antenna
108 Rear position light
109 Aft ECM transmitting antenna
110 Rudder
111 Carbon-fibre rudder skin panelling

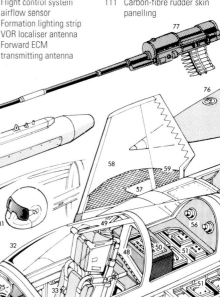

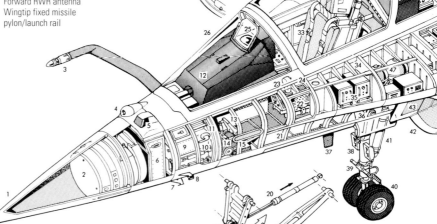

MBDA Scalp EG/Apache

"The recent participation of the French Armed Forces in the operations over Kosovo has confirmed that the use of 'dumb' bombs will be more and more limited in future conflicts," explains Yves Robins. "As a consequence, smart weapons will be increasingly needed to neutralise high-value targets while avoiding collateral damage. This is the reason why the Rafale will be primarily equipped with guided weapons."

For long-range strike attacks, the main air-to-ground weapon will be the MBDA Scalp EG (Emploi Général, General Purpose) stealth cruise missile of the Apache/Scalp EG/Storm Shadow/Black Shaheen family. It has now been decided that the Apache (Arme Propulsée Antipiste à Charges Ejectables, Propelled Anti-Runway Weapon with Ejectable Charges) designed to attack enemy air bases will only be carried by Mirage 2000Ds, but this weapon is available for Rafale export contracts. Powered by a lightweight (60-kg/ 132-lb) Microturbo TRI 60-30 engine rated at 5.32 kN (1,195 lb), the Scalp EG is a long-range stealthy cruise missile fitted with a powerful 400-kg (882-lb) unitary Broach conventional penetration warhead. It is intended to be used in pre-planned attacks against strongly defended, hardened high-value static targets. "After launch, the Scalp is fully autonomous thanks to its global positioning/INS/terrain reference navigation system," explains

Jean Camus. "Its passive IR imagery homing head is activated during the final target approach. Automatic target recognition algorithms compare the actual scene with memorised scene, identify the designated target, and select the impact point in order to hit with very high precision." Although Dassault or MBDA would not confirm its maximum range, a French Parliamentary Study reveals that it can strike targets after a 400-km (249-mile) flight.

More than 100 captive flight tests with the Scalp seeker mounted on a Puma have been carried out to check the closed-loop performance of the image processing, and the first full-scale trial of a Scalp EG was conducted in poor weather conditions in December 2000 at the Centre d'Essais des Landes (Landes Test Centre/firing range): the missile, fired by a Mirage 2000N flying at Mach 0.9 at 609 m (20,000 ft), dropped down to low level, and autonomously navigated over a 250-km (155-mile) trajectory at Mach 0.8 before striking a purpose-built target.

Scalp/Storm Shadow has been ordered by France, the UK, Italy and Greece, and a specific variant, the Black Shaheen, has been ordered by the United Arab Emirates Air Force for its Mirage 2000-9s. Up to 500 Scalps are planned to be purchased by France, including 50 for the *Aéronavale*, and the first production Scalp will be delivered to the *Armée de l'Air* in 2003, initially for Standard R2 Mirage 2000Ds. French Air Force Rafale B/Cs will carry two Scalp EG missiles under wing pylons, whereas Rafale Ms will accommodate a single missile on the centreline station. It is worth mentioning that the Rafale M will be able to land back on the *Charles de Gaulle* carrier with an unexpended Scalp EG.

A land-attack derivative of the Scalp is being developed to satisfy French Navy needs, and will enter service in 2011. The 500-km (311-mile) range variant will equip the six new 'Barracuda'-class submarines and the 17 future multi-purpose frigates. The Naval Scalp will be fired vertically from the Sylver launching system or from torpedo tubes, and will share about 80 percent platform commonality with the Scalp EG.

Sagem AASM

Scalp EG is a very expensive weapon which will be acquired in relatively small quantities. As a result, the French Ministry of Defence's specialists have decided that an affordable armament must be procured, and that requirement led to the AASM (Armement Air-Sol Modulaire, Modular air-to-surface armament) concept of modular all-weather attack weapons. Three contenders were shortlisted by the French Ministry of Defence – Aérospatiale/Matra Missiles, Sagem and Matra/BAe Dynamics (now MBDA). After a comprehensive study, Sagem was declared the winner in September 2000. In all, 3,000 AASMs will be ordered, including 750 for the Aéronavale, and the first weapons will enter service in early 2005. The AASM kits ordered by the French Armed Forces will initially be adapted to 227-kg (500-lb) general-purpose bombs (typically American Mk 82s), but several variants of various weights and powers will eventually join the French inventory, culminating in a 907-kg (2,000-lb) version for use against hardened or buried targets. The first weapons to enter service will be fitted with a global positioning/inertial navigation guidance system which uses known co-ordinates to hit the target. This variant has the same capabilities as the American JDAM (Joint Direct Attack Munition) series of precision weapons.

From 2006, an infra-red imagery terminal-phase seeker will become available to enhance precision in some variants. As such, the low-cost AASM will only be able to attack stationary targets. The AASM will have a

The AASM family is initially planned as a French equivalent of the GPS-guided JDAM family, but will expand to add EO terminal guidance. Pop-out fins give extra glide distance.

very long range – up to 60 km (37 miles) when launched at 13716 m (45,000 ft) – and will be able to hit targets with pinpoint accuracy (precision of about 9-14 m/30-45 ft for the INS/GPS version, and 1-3 m/3-10 ft for the IR-guided variant). Six 227-kg (500-lb) AASMs on two triple ejector racks will normally be carried by a Rafale for long-range strikes, allowing six widely separated targets to be attacked in one pass.

Future armament

For the Standard F3 Rafales, a full range of next-generation air-to-surface ordnance will become available. The ANF missile (Anti-Navire Futur, future anti-ship), which was due to supplant the proven sea-skimming Aérospatiale AM39 Exocet, was to be adopted. However, the ANF was put on hold in 2000, but it is understood that the programme will be revived at a latter stage, and it is widely anticipated that the missile will find its way into the Rafale inventory. The long-range (150- to 200-km/93- to 124-mile), supersonic (Mach 2.5), sea-skimming, fire and forget ANF is propelled by a ramjet engine, and will have a high operational efficiency in terms of range and penetration, in all weathers. According to specialists, its high terminal agility coupled with a very high speed will be sufficient to escape all the enemy anti-missile weapons currently in service or under development. The ANF is the first member of a new multi-mission supersonic missile family derived from the Vesta (Vecteur à Statoréacteur, ramjet vector) aerodynamics and propulsion programme. Vesta has been under development since 1996, and bench tests have now been completed. Three test flights are due to be performed in 2002. The Vesta programme will help

reduce the technological risks associated with the ANF, and will help develop solutions which will limit the installation and ownership costs.

The recently funded ASMP-A (Air-Sol Moyenne Portée-Amélioré, Improved Medium Range Air-to-Ground) stand-off nuclear weapon will also benefit from the technological spin-off from the Vesta programme. This new pre-strategic nuclear missile will supersede the current Aérospatiale ASMP which is in service on the French Air Force Mirage 2000Ns and French Navy Super Etendards. The ASMP-A is based on the current ASMP's general architecture, but is propelled by a new-generation liquid-propellant ramjet with a longer burning time which will allow a considerable range extension and enable new, more aggressive trajectories. It will also make use of a new generation of components and advanced, more precise navigation/ control/guidance equipment. According to figures recently published by the French National Assembly Defence Committee, its maximum range is about 100 km (62 miles) at low level and 500 km (311 miles) at high level. The pre-feasibility phase was completed in 1996, and the feasibility/definition phase in mid-1999. Full-scale development started in 2000, and initial operational capability with Mirage 2000Ns and Rafales is expected in 2007 and 2008, respectively.

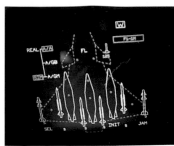

The left and right touch-sensitive MFDs are typically used for aircraft status displays. This display format shows a simulated combat load of two Magics (in white, at the wingtips), four MICA EMs (in white, under the wings and fuselage), three 1250-litre supersonic drop tanks (yellow), plus 125 30-mm rounds.

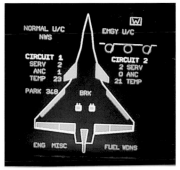

The pilot can call this display format to check the hydraulic systems. For the Rafale, Dassault has chosen an innovative approach with the choice of fully redundant, very high-pressure (350-bar/5,000-psi) hydraulic systems which boost reliability and safety. This choice has proved so successful that it has also been made by other aircraft manufactures. For instance, these systems have been adopted for the Airbus A380.

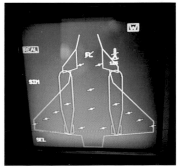

The Rafale is fitted with a centralised weaponry safety system which totally replaces safety pins usually inserted into the weapon pylons. The pilot is now the only person in charge of weapon safety, and the engineers do not have to remove the safety pins associated with older fighters. The system has been tested in stringent conditions, and has been cleared for aircraft-carrier flight deck compatibility. The two 2000-litre (528-US gal) drop tanks appear in red because, at the time the picture was taken, they were being fitted by engineers. When they become operational, their colour switches to yellow.

Shown above is the cockpit of Rafale M01 which, although showing the main features and layout, differs in detail from production aircraft. Rafale's cockpit is a highly advanced work-station, incorporating cutting-edge VTAS (voice, throttle and stick) technology, and touch screens. Tactical and flight information is mainly displayed on the wide-angle HUD and the central head-level display. Symbology is collimated to infinity, so that the pilot does not have to refocus as he/she switches between the two displays and the outside world. Non-critical control of menus is by touch, for which special seamless gloves are worn, but critical control, for instance during combat, is by DVI (direct voice input) or by switches on the single throttle lever and side-stick (left). Arm rests are provided on both sides of the cockpit for pilot comfort.

Dassault Rafale B (302)
Combined Rafale Test Team
Istres Flight Test Centre

**This aircraft is one of two production-standard Rafale
Bs delivered to the Armée de l'Air. Although initially
configured as basic Standard F1 aircraft, the pair is
used for Standard F2 development work.
Consequently, several F2 features are incorporated,
such as Front Sector Optronics, RBE2 air-to-ground
modes and MIDS-LVT terminal capability. To
maximise efficiency, the aircraft are operated by a
joint test team which includes Dassault, Snecma,
Thales and various CEV and AdA agencies.**

Construction
In the interests of weight, cost and
reliability, Rafale employs a high
degree of advanced materials in its
construction. Around 24 percent of
the aircraft's weight is of
composite materials, which
account for about 70 percent of
the 'wetted' surface area. SPF-DB
(superplastic forming-diffusion
bonding) titanium is used for
extreme strength in components
such as the canards, and the
airframe includes SPF aluminium
structures.

Loadout
Rafale's 14 hardpoints enable a huge variety of stores
combinations to be carried, which can be further
extended by the use of twin rail launchers for MICAs
or dual/triple bomb carriers on the inboard and
intermediate underwing pylons. This aircraft is
depicted in a typical long-range attack loadout, with
Scalp EG missiles on the intermediate wing pylons,
two 2000-litre fuel tanks inboard and MICAs on the
wingtips (infra-red) and under the rear fuselage pylons
(active radar). Scalp EG has a unitary penetration
warhead, whereas the similar Apache dispenses sub-
munitions for anti-runway attack. Apache is not slated
for carriage by AdA Rafales. Air force two-seaters also
retain the 30 M 791 cannon (deleted from the Rafale
N naval two-seater).

M88 engine
The M88 which powers Rafale was the result of development of high-performance fighter
engines begun in the late 1970s as a follow-on to the M53 of the late 1960s. Studies
concentrated on new alloys and metallurgical processes to allow highly loaded blades, and on
air-cooled turbine blades. Following the decision to proceed with Rafale in 1983, development
took on a new urgency, and full funding for the M88 was finally granted in 1987. Snecma
developed the engine according to schedule, and under budget. The Rafale's M88-2 is a two-
spool turbofan with a bypass ratio of 0.3:1 and a mass flow of 65 kg (143 lb) per second. The
engine inlet is fitted with 15 variable guide vanes, downstream of which is the three-stage low-
pressure compressor (fan). Some of the low-pressure air bypasses the high-pressure
compressor, which has six fan stages and three variable stators. Overall compression ratio is
24:1. Both compressors are driven by single-stage turbines with air-cooled blades. The high-
pressure turbine features single-crystal blades capable of operating at a TGT (Turbine gas
temperature) of 1577°C (2,870°F).

ing missiles

943 lb)
version): 26000 kg

lb)
6,534 lb)
e: Mach 1.8
kt (1389 km/h; 863 mph)
213 km/h; 132 mph) at

defence mission
tack mission

iles) for a strike mission
ombs, four MICAs, one
d two 1250-litre (330-US

1 miles) for an air
ht MICAs, two 2000-litre
0-litre (330-US gal) drop

ofans rated at 50 kN
75 kN (16,854 lb) with

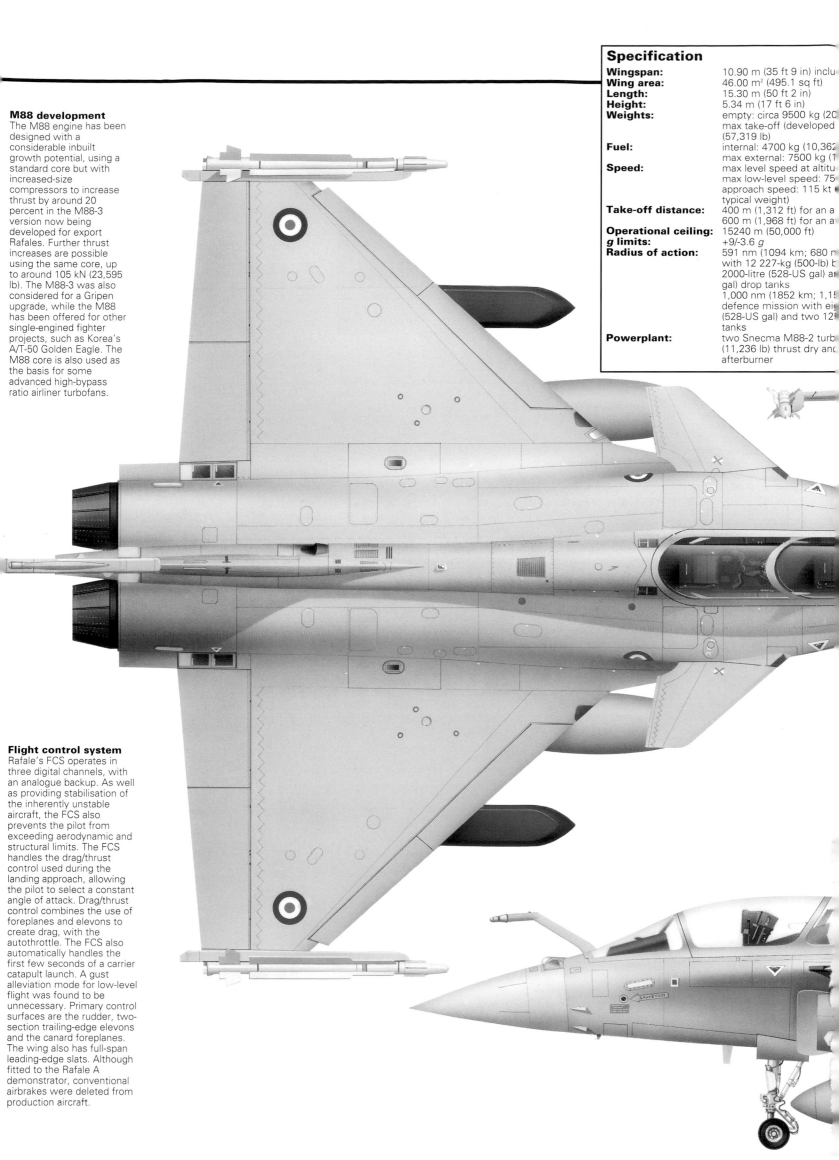

M88 development
The M88 engine has been designed with a considerable inbuilt growth potential, using a standard core but with increased-size compressors to increase thrust by around 20 percent in the M88-3 version now being developed for export Rafales. Further thrust increases are possible using the same core, up to around 105 kN (23,595 lb). The M88-3 was also considered for a Gripen upgrade, while the M88 has been offered for other single-engined fighter projects, such as Korea's A/T-50 Golden Eagle. The M88 core is also used as the basis for some advanced high-bypass ratio airliner turbofans.

Flight control system
Rafale's FCS operates in three digital channels, with an analogue backup. As well as providing stabilisation of the inherently unstable aircraft, the FCS also prevents the pilot from exceeding aerodynamic and structural limits. The FCS handles the drag/thrust control used during the landing approach, allowing the pilot to select a constant angle of attack. Drag/thrust control combines the use of foreplanes and elevons to create drag, with the autothrottle. The FCS also automatically handles the first few seconds of a carrier catapult launch. A gust alleviation mode for low-level flight was found to be unnecessary. Primary control surfaces are the rudder, two-section trailing-edge elevons and the canard foreplanes. The wing also has full-span leading-edge slats. Although fitted to the Rafale A demonstrator, conventional airbrakes were deleted from production aircraft.

Specification

Wingspan:	10.90 m (35 ft 9 in) inclu
Wing area:	46.00 m² (495.1 sq ft)
Length:	15.30 m (50 ft 2 in)
Height:	5.34 m (17 ft 6 in)
Weights:	empty: circa 9500 kg (20
	max take-off (developed
	(57,319 lb)
Fuel:	internal: 4700 kg (10,362
	max external: 7500 kg (1
Speed:	max level speed at altitu
	max low-level speed: 75
	approach speed: 115 kt
	typical weight)
Take-off distance:	400 m (1,312 ft) for an a
	600 m (1,968 ft) for an a
Operational ceiling:	15240 m (50,000 ft)
g limits:	+9/-3.6 g
Radius of action:	591 nm (1094 km; 680 n
	with 12 227-kg (500-lb) b
	2000-litre (528-US gal) a
	gal) drop tanks
	1,000 nm (1852 km; 1,15
	defence mission with ei
	(528-US gal) and two 12
	tanks
Powerplant:	two Snecma M88-2 turb
	(11,236 lb) thrust dry and
	afterburner

Taxiing his Rafale on the deck of 'CdG', a Flottille 12F pilot models the new lightweight CGF-Gallet helmet developed for the French air arms. This headgear is fully compatible with the two Helmet-Mounted Display (HMD) systems currently competing for the Rafale contract (JHMCS and Topsight E). Also evident is the Mk 16F ejection seat, which is reclined at 29° to improve high-g tolerance.

under 200 ms." As an added bonus, the system is also essential for flight safety, helping reduce pilot workload in emergency situations, and allowing aircrew to remain hands-on most of the time.

Initially, the Sextant Topsight full-face HMD with an integrated oxygen mask was designed for the Rafale, but the programme has been plagued with both development problems and budgetary constraints. As a result, the French Armed Forces are now seriously considering buying alternative HMDs. Two competitors are thought to be under consideration: Elbit, with a variant of the JHMCS (Joint Helmet-Mounted Cueing System), and Thales Avionics (formerly Sextant), with the Topsight E. The winning HMD is required to display flight reference data and to permit weapon-aiming and engagement at large off-boresight angles. This will enhance situational awareness and will give unprecedented capability to French fighter pilots – shots 'over the shoulder' will become reality, greatly increasing combat efficiency. The Topsight E system appears to be the front-runner, and could first enter service on Mirage 2000-5Fs before appearing on Standard F3 Rafales in 2008. It can be fitted to a wide variety of helmet designs, including the lightweight CGF-Gallet type recently ordered by the French Air Force and already used by Dassault aircrew and Flottille 12F Rafale pilots.

Advanced man-machine interface

"Detractors have questioned the choice made by the French Navy and Air Force to purchase a large number of two-seat aircraft, criticising the man-machine interface of the French fighter," explains Philippe Rebourg, Dassault Deputy Chief Test Pilot. "But, with its wide-angle HUD, its touch screens, and its innovative central display collimated to infinity, the Rafale possesses one of the most modern

Implementation of the Voice, Throttle And Stick (VTAS) concept will revolutionise air combat. "Development of the voice control system started in the early 1990s, and the system has been extensively tested, initially on Alpha Jets and Mirage IIIs, before full-scale testing on Rafales," says Gérard Dailloux, Dassault Flight Safety Vice-President. "Initially, word recognition was a major hurdle, especially as the cockpit noise environment considerably changes according to the aircraft's speed, altitude, and high g-load. Pilot's voice is also affected by stress or extreme g-load, and the direct voice input system has to cope with the environment. Thales and Dassault engineers progressively overcame all the hurdles, and the voice command system now works perfectly. The customer can choose between various vocabulary options with 90 to 300 words, and word recognition rates are better than 95 percent. Response time is

Ease of maintenance

In accordance with stringent French requirements, the Rafale has been conceived with ease of maintenance in mind. Moreover, the new fighter was required to operate from crowded carrier decks where maintenance is less easy than on large air bases. "Even the Air Force variants benefit from the rugged airframe, as there is a high degree of commonality between the Rafale C/B and the Rafale M/N," explains Xavier Labourdette, Dassault International Support Manager. "Corrosion protection and resistance to shocks are excellent, and electromagnetic compatibility has been extensively tested for demanding carrier operations."

The entire aircraft is monitored in real-time by a Health and Usage Monitoring System integrated into the mission computer. "This user-friendly integrated

testing system is fully embedded within the Mission Computer," says Xavier Labourdette. "It enables far more accurate diagnostics of any potential problem which might arise, and considerably shortens troubleshooting and repair times, while also reducing the amount of ground facilities needed. For instance, no external tester is required at operational level. This translates into very high mission-readiness levels and minimal operating costs."

In a modern fighter, avionics and electronics are essential for mission success and have to be easily maintainable. Rafale's on-board integrated fault-detection system is sophisticated enough to tell the engineers which line-replaceable unit (LRU) or system is unserviceable, but the Mermoz test bench goes one step further, allowing specialists to determine which component in a LRU is defective. Mermoz handles all electronic equipment, including radios,

radar and missile launchers. The first Mermoz system is already fully operational with the Aéronavale at Landivisiau Naval Air Station, and allows a considerable reduction in anticipated repair costs. The modular avionics concept means that printed circuit boards can then easily be changed instead of replacing complex and expensive LRUs. This modularity allows a considerable reduction in the spare parts inventory, and the concept extends to the engines, the mission computer and the radar processor.

With the Rafale, planned intervention rates are so low that the French armed forces are currently thinking in terms of only 60 technicians per squadron of 20 aircraft while retaining mission-readiness rates in excess of 90 percent. In the air-to-air configuration, a typical turnaround can be performed in 20 minutes with six personnel, including the fitment of six MICA missiles and refilling the tanks. This figure is based on proven achievements, and benefits from the innovative centralised weaponry system which has many advantages over the traditional safety pins found on the pylons and launchers of other aircraft types: increased personnel safety through clear and unambiguous weapon status, reduced manpower and turnaround times, independent management for each store station, and no more need for end-of-runway (last chance) checks.

Every effort has been made to facilitate maintenance. For instance, a side-opening canopy has been chosen to ease ejection seat removal. "We noticed that, on other aircraft types, engineers first had to remove the canopy before taking away the ejection seat, a labour-intensive operation," explains Xavier Labourdette. "Each year, a number of canopies were accidentally dropped to the ground – not very cost-effective! Boresighting is also a time-consuming operation which requires considerable tooling. Thanks to the unrivalled manufacturing and assembly techniques afforded by the Catia computer-aided design tool, the Mirage 2000 and the Rafale are the only fighters in the world to have eliminated physical boresighting, requiring only software boresighting."

B01 is seen during hardened aircraft shelter compatibility trials. The AdA is budgeting to operate a squadron of Rafales with 20 percent fewer personnel than is the case with Mirage 2000 units.

Rafale defences

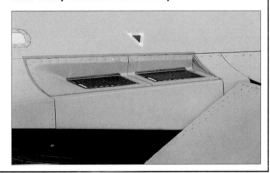

SPECTRA has a comprehensive warning capability, including laser warning receivers (above, small window on right aft of formation light strip), infra-red detectors and interferometers (RWRs). The most prominent of the latter are the forward-facing units mounted on the intake sides (below). SPECTRA gives Rafale a significant ELINT capability.

The Thales/MBDA team is justifiably proud of SPECTRA, particularly its directional ECM capabilities. This marking (above) appeared on Rafale M02 during the NATO MACE X exercise in August 2000, in which SPECTRA proved itself. Below is the fin tip, location for several SPECTRA elements, including rear-facing interferometer and laser warner on the side.

SPECTRA countermeasures include jammers mounted in the tail and just forward of the canards (above), internal chaff dispensers and two upward-firing IR flare/EO decoy dispenser boxes on each side of the rear fuselage, situated in the blended wingroot structure (below). In line with current fashion, towed radar decoys are under active study.

cockpits in service or under development anywhere in the world."

The two-seat configuration also allows new missions to be undertaken, and it has already been announced that Rafale B/Ns will be utilised as high-speed command aircraft during complex attack sorties, and as control posts for Unmanned Combat Aerial Vehicles (UCAV). The back-seater will be able to assess time-critical data and intelligence obtained through datalink, and make key decisions while the pilot handles the aircraft. "This mixed-fleet concept of fighters and UCAVs is required in a dense electromagnetic environment, when modern air defence systems have to be destroyed to achieve air dominance without exposing pilots," says Yves Robins, Dassault Vice-President for International Relations.

Dassault Aviation has started developing a range of stealthy UCAVs, and the first demonstrator, the low-cost Petit Duc (Small Duke, Duc standing for Démonstrateur d'UCav), first flew in July 2000. Powered by two AMT jet engines, the 60-kg (132-lb) Petit Duc can fly at Mach 0.5,

and boasts a 150-km (81-nm) radius of action, and the stealthy design is said to have the radar signature of a sparrow. The Dassault research programme will lead to two bigger demonstrators, called Moyen Duc (Medium Duke) and Grand Duc (Large Duke), which could give way to fully operational UCAVs. According to informed sources, a two-seat Rafale could simultaneously control up to four UCAVs for reconnaissance, real-time communication relaying, electronic warfare and even attack missions. "An innovative 3D sound system could be introduced on the Rafale to help aircrew localise their UCAVs during the whole mission," says Gérard Dailloux. "Additionally, this system would prove useful to accurately locate enemy air defence systems and fighters."

Inertial navigation

Two Sagem Spark ring laser-gyro Inertial Navigation Systems with hybridised GPS ensure very accurate autonomous navigation without the need to rely on external and vulnerable navigation aids. Their powerful and

Rafale M02 is prepared for electromagnetic testing in the anechoic chamber at Istres. The treatment on the floor, walls and ceiling of the chamber reduces electromagnetic reflections to a minimum, allowing engineers to test and measure the electronic signatures of the various sensors as if the aircraft were flying in clear air. Of utmost importance is the compatibility between the aircraft's sensors/comms and jammers, so that there is no interference between them. Furthermore, Rafale M has to be fully compatible with the naval systems to be found aboard the carrier.

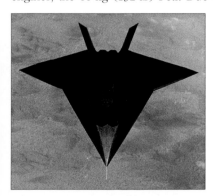

Above: Known as AVE (Aéronef de Validation Expérimentale), or Petit Duc, this is the small-scale (2.4-m/7-ft 10-in wingspan) demonstrator for a family of stealthy UCAVs being developed by Dassault. Current thinking envisages formations of full-scale vehicles being controlled from the back seats of Rafale Bs.

Above and above right: 12F Rafale Ms are seen aboard Charles de Gaulle in May 2001. The carrier has two 75-m (246-ft) stroke catapults, which typically launch the Rafale at speeds between 130 kt/241 km/h (clean) and 148 kt/274 km/h (loaded). The launch and climb-out is controlled automatically for up to 15 seconds after the 'cat' fires.

open architecture is fully capable of blending all data from various sensors (GPS, air data system, radar altimeter for terrain-matching) while performing integrity monitoring.

SPECTRA

In the last decade, the proliferation of air defence systems has put considerable pressure on airborne electronic warfare specialists. Moreover, potentially hostile fighters are equipped with ever more efficient fire-control systems, again imposing the adoption of sophisticated self-defence suites. "Modern air warfare places a severe requirement on aircraft self-defence capabilities, and only

the most advanced fighters will survive any major conflict," suggests Pierre Delestrade, a Flight Test Engineer. "Thankfully, the Rafale is equipped with highly-automated and affordable systems which provide an unprecedented level of protection against threats likely to appear in the future."

Designed and produced by Thales Airborne Systems in co-operation with MBDA, SPECTRA (Self-Protection Equipment Countering Threats of Rafale Aircraft) is a state-of-the-art self-defence system mounted on the Rafale as a complete and totally integrated electronic warfare suite. "The system, which offers a dramatic increase in survivability against modern and emerging threats, is entirely mounted internally in an effort to keep weapon stations free," says Pierre Delestrade. "It ensures efficient electromagnetic detection, laser warning and missile approach warning using passive IR detection technology, jamming and chaff/flare dispensing, even in the most demanding multi-threat environment. SPECTRA is divided into different modules and sensors strategically positioned throughout the airframe to provide all-round coverage."

The latest advances in micro-electronic technology have led to a new system which is much lighter, more compact and less demanding than its ancestors in terms of electrical and cooling powers. Thanks to its advanced digital technology, SPECTRA provides passive long-range detection, identification and localisation of threats, and allows the pilot or system to react immediately with the best defensive measures: jamming, decoy-dispensing, evasive manoeuvres

Rafale training sortie

In January 2001, the author was invited by Dassault to experience first-hand the impressive performance and capabilities of a production Rafale during a combat training mission conducted from the Dassault Flight Test Centre at Istres, in the south of France. The sortie provided a unique insight into the operational procedures which can be expected to be undertaken in the swing role for which the Rafale is optimised.

The tactical demonstration began in the OASIS (Outil d'Aide à la Spécification d'Interface Système, or System Interface Specification Tool) development simulator, where both air-to-air and air-to-ground engagements are shown and practised, helping me become familiar with the cockpit environment, the display formats and the HOTAS switchology. "This simulator is normally utilised for Rafale man-machine interface development, but it is also very useful for customer demonstrations and press briefings," explained Jean Camus, former Director of the French Test Pilot School.

Next came a visit to the equipment store for an ejection seat briefing and for flying clothing fitting. The Martin-Baker Mk 16F lightweight 'zero-zero' ejection seat will certainly set new standards in terms of simplicity, ease of use and safety, and it took only a few minutes to fully understand the normal and reversionary functioning modes. To overcome rigours associated with fast-jet flying, Rafale pilots are required to wear special clothing, including anti-*g* trousers, a lightweight CGF-Gallet helmet and an upper body jacket containing a survival kit. Special gloves with no seams at the fingertips for compatibility with the touch-sensitive displays were also provided.

A positive pressure breathing system has been extensively tested for the Rafale, and the aircraft is cleared to accommodate it, but French Flight Test Centre aviation medicine specialists have asked for a time delay to complete studies to make sure that there are no long-term medical effects on pilot health. The revolutionary Libelle hydrostatic integral *g*-suit

was tested by Dassault test pilots in the summer of 2001. The liquid-filled, self-regulating suit proved highly satisfactory and could be offered at some stage on the Rafale.

Today's sortie is in Rafale B302. This aircraft, the third production Rafale, first flew in December 1999, and is currently used for the French multi-role Standard F2 development programme. As such, it is fully instrumented, and is equipped with an RBE2 radar capable of air-to-air and air-to-ground modes, with a Front Sector Optronics (FSO) suite and, when needed, with the Spectra self-defence suite, the direct voice input system and the MIDS-LVT datalink system. However, neither Spectra nor the voice control system or the MIDS-LVT were actually fitted for this sortie. The infra-red sensor of the FSO was not installed either, but the TV sensor was fully operational. For the flight, B302 was equipped with a supersonic drop tank on the centreline pylon, two MICA EM training rounds on the aft fuselage stations, and two wingtip-mounted Magic II training missiles, giving a take-off weight of 16400 kg (36,155 lb). The 1250-litre (330-US gal) external tank, which brings the two-seater's fuel capacity to 6550 litres (1,730 US gal), is cleared to fly at speeds up to Mach 1.6.

For this mission, my pilot and mentor was Philippe Rebourg, Dassault Deputy Chief Test Pilot, who flew Mirage IIIEs when serving with the Armée de l'Air. In 1989, he was selected for test pilot training, and graduated from the US Air Force Test Pilot School at Edwards in 1990, before joining the Centre d'Essais

en Vol – the French Flight Test Centre – where he worked on the Mirage 2000D, Mirage 2000C and Rafale programmes. He was recruited by Dassault in 1995, and has now logged about 5,000 flying hours, including over 400 in Rafales.

Into the cockpit

It was soon time to walk to the flight line. When approaching the Rafale, one's immediate impression is that it is considerably larger than previous Dassault fighters such as the Mirage F1 or the Mirage 2000. Although the naval variant has a built-in ladder, access to the cockpit in the two-seat Air Force version is via an external ladder. The rear cockpit is very roomy and extremely well laid out, with every control within easy reach. Reclined at 29° to boost *g* tolerance, the Mk 16F ejection seat proved very comfortable, and strapping in was a much easier process than with earlier Martin-Baker designs such as the Mk 10, which equips numerous aircraft types. For improved safety in the event of a high-speed ejection, the Mk 16F is equipped with arm restraints to minimise the risk of injury. The traditional leg restraints found on older fighters are built into the cockpit, around the foot tunnel, allowing the pilot to dispense with wearing garters. As a result, the anti-*g* trousers do not

The undercarriage cleans up smartly, all units retracting forwards. Lift-off is typically around 130 kt for a lightly loaded aircraft. Note the deployed leading-edge flaps and deflection of the canard foreplanes.

and/or any combination of these actions to evade or defeat a diversity of airborne and ground-based threats. Even in a very dense signal environment, direction-finding accuracy is excellent, and the time taken for signal identification is extremely short (all data is classified).

Additionally, very high processing power gives excellent detection and jamming performance, optimising the response to match the threat: incoming electromagnetic signals are analysed, and the bearing and location of the emitters are determined with great precision. Exact location and types of systems detected by SPECTRA can be recorded for later analysis, giving Rafale operators a substantial built-in SIGINT/ELINT capability while minimising the need for specialised and costly dedicated intelligence platforms. In further developments, the adoption of high-debit datalinks will allow two Rafales to carry out instantaneous triangulations of threats, giving positional accuracy within a few metres. SPECTRA is also fully flight-line software reprogrammable.

Missile warning

The proliferation of new-generation weapons, such as man-portable surface-to-air missiles, has raised concern among key decision-makers, and a laser warning system has been mounted on the sides of the nose and on the tail of the fighter, providing 360° coverage, and ensuring detection and warning of incoming shoulder-launched laser beam-riding missiles. The discreet IR missile approach warner ensures high probability of detection and low false alarm rates, even against totally passive IR-guided weapons. The exhaust plume of an incoming missile can be detected at very long-range without any emission that would betray the presence of the Rafale. Four upward-firing launcher modules for various types of cartridges – flares or electro-optic decoys – are built into the airframe,

and the Rafale is equipped with internal chaff dispensers.

SPECTRA is much more than a traditional self-defence system as it is closely integrated with the primary sensors also supplied by Thales – the RBE2 multi-mode electronic scanning radar and the FSO passive front sector optronics system. As such, it considerably improves pilot situational awareness: all data obtained thanks to the various means is fused into a single tactical picture, offering the pilot a clear image of the evolving tactical situation. Lethality zones, determined by SPECTRA according to the air defence weapon types detected and the local terrain, can then be displayed on the colour tactical screen, enabling the aircrew to avoid dangerous areas. This smart data fusion

M02 was the principal SPECTRA testbed, and is seen here in August 2000 while involved in the MACE X electronic warfare trials. Evident here are the vanes fitted to the nose aft of the radome which feed accurate pitch and yaw data to the flight control system. The intakes have simple splitter plates, and exhibit excellent high-Alpha properties.

have pockets in the sides of the lower leg in order not to interfere with this restraint system.

The engineers who designed the cockpit of the Rafale concentrated on reducing the workload for the aircrew, and introduced 21st Century technology to improve situational awareness. For instance, Dassault specialists have adopted colour displays to accelerate and facilitate information assimilation. The instrument panels of front and rear cockpits are almost identical and are both NVG-compatible. They are divided into one large HUD and three multifunction screens: two touch-sensitive lateral displays (resolution 500 x 500), and a large (20° x 20°), high-resolution (1000 x 1000) head-level display. The added benefit of the new head-level display is that it is collimated to infinity, enabling the pilot to shift instantly from head-up flying to head-down mission-monitoring without a need to refocus and, furthermore, its field of view is considerably larger than that of a traditional screen. The system works nicely, and should set new standards. The front and rear cockpit displays can be operated in a tandem mode, which presents the pilot and the back-seater with the same information, or de-coupled so that crew members can carry out different tasks concurrently.

The wide-angle (30° x 22°) HUD provides short-term information, whereas the head-level display conveys medium-term information. For the benefit of the back-seater, the image of the HUD is projected on a repeater mounted on top of the instrument panel, providing the rear-seat occupant with a good view over the nose. This is a crucial advantage for flight safety during conversion training, as the instructor can now see directly ahead during low-level flying and touch-and-goes.

The Rafale has been designed with ease of operation in mind, and engine starting is a model of elegant simplicity: you just have to push two small auxiliary throttles from cut-off to idle, and then turn a rotary control left and right, or right and left, according to which engine you want to start first (there is no preferential order). The auxiliary power unit immediately came on line, and the two Snecma M88-2 turbofans were both turning and burning in under two minutes. Although the day was really sunny, the air-conditioning unit cooled the cockpit very rapidly and very effectively. Compared with the

prototypes, a new, more powerful air-conditioning system was adopted on production Rafales when the French Air Force realised that its fighters were increasingly likely to operate overseas. The new unit is 40 percent more powerful for the same volume.

This sortie was conducted as a real flight test, and two Dassault flight test engineers, Pierre-Louis Aumont and Pierre Delestrade, monitored the mission from a dedicated facility. A special radio system allowed them to listen to everything that was said in the cockpit, and to easily communicate with us. The two engineers and the test pilot exchanged data to make sure that the test instrumentation and all the systems were in working order. Philippe Rebourg then selected the Inertial Navigation System format on the left-hand touch-screen, and asked me to start the alignment, which took only four minutes (it can be done in 1 minute 30 seconds, but with reduced accuracy). With all onboard flight-test and telemetry equipment checks completed and clearance granted

Martin-Baker's Mk 16 is the state-of-the-art in lightweight ejection seat technology, used in both Eurofighter and Rafale and a strong candidate for the F-35. The Mk 16F for the French aircraft is built in France under licence. Below is a 450-kt sled test from a Rafale forward fuselage, undertaken at CEL Biscarosse.

For both home use and export potential, Rafale is being cleared for a wide variety of stores. An important addition to the aircraft's quiver is the 500-lb (227-kg) GBU-12 Paveway II laser-guided bomb, used by many nations, including France itself. Here Rafale B01 carries four of the weapons on twin racks (up to six can be carried) on a separation clearance sortie at Cazaux, while still easily hauling three of the enormous 2000-litre (528-US gal) tanks. Further clearance work will expand the LGB repertoire with Paveway III weapons, and add a variety of 'dumb' bombs. In the future, additional US weapons such as the AIM-9X next-generation Sidewinder may be cleared for Rafale use by export customers.

significantly increases mission success rates through enhanced crew awareness and improved aircraft survivability.

SPECTRA's first flight onboard a Rafale took place in September 1996, after M02 prototype had been retrofitted. Since then, the system has been thoroughly tested in very complex electronic warfare scenarios. For instance, Rafale M02 was pitted against a wide variety of the latest air defence systems during the comprehensive NATO Mace X trial organised in August 2000 in southwest France, and its self-defence suite performed flawlessly in an extremely demanding environment which comprised a very wide variety of netted systems: Crotale NGs, Aspics, DE-Hawks (Danish Enhanced-Hawks), DALLADSs (Danish Army Low-Level Air-Defence Systems), Norwegian Advanced Surface-to-Air Missile Systems (NASAMSs), and even an American SA-15 simulator and a German SA-8.

SPECTRA is now in full production, and is already operational onboard Aéronavale Rafales. The suite has been designed with growth in mind to keep the Rafale abreast of emerging threats, and further developments are envisioned, including a towed radar decoy and laser-based IR counter-measure directional turrets to defeat incoming IR-guided missiles. However, Dassault and Thales engineers are pretty confident that SPECTRA is already fully capable of dealing with current and future threats, and these systems will only be needed in the long-term, if ever.

Communication systems

To rise above the fog of war, the Rafale is fitted with an extensive communications suite which comprises four radio sets: one V/UHF and one encrypted UHF, plus two MIDS-LVTs (Multifunction Information Distribution System – Low Volume Terminals). In modern combat, information and situational awareness are essential for immediate and total success, and the futuristic network-centric warfare concept is a key enabler. One of the most significant advances in technology, the advent of this global military info-sphere will shape the future of combat operations, allowing assets to exchange and share tactical data at very high rates, and bringing together all forces in the 'battle-space' in a very efficient way.

Rafale was designed from the outset with a datalink capability and, for France and other potential NATO-approved customers, it will be equipped with the secure, interoperable MIDS-LVT Link 16 system. Jointly developed by France, Germany, Italy, Spain and the USA, the light-weight (29-kg/64-lb) LVT can transmit and receive data at a rate of 200 Kb/s. In France, the MIDS-LVT will eventually equip numerous platforms: Mirage 2000-5Fs, Mirage 2000Ds, AWACSs, A400Ms, tankers, combat SAR helicopters and various warships. With MIDS-LVT, each Rafale in a formation will have access to the sensor data of other aircraft, ground stations and AWACSs. The introduction of datalinks is recognised as a fundamental change in air warfare tactics, and Rafale pilots will have access to far more tactical information than their predecessors. Perhaps, the most obvious advantage is the capability to receive targeting information for a silent interception/attack.

Mastering digital technologies has proved essential when

---

by Istres Ground, the throttle was advanced for taxi, and we slowly left the Dassault flight line towards holding point 33.

Take-off

When ordered to do so, I armed my ejection seat using the main seat-safety handle by my right thigh. The take-off from Istres was made in excellent visual meteorological conditions: the temperature was 12°C and airfield elevation is 23 m (75 ft) above sea-level. The Rafale is so powerful that it is impossible to apply full dry power against the brakes as this would send the aircraft forward anyway, causing considerable damages to the tyres in the process. Completely trusting the automatic fault detection system, Philippe Rebourg simultaneously released the brakes and slammed the throttle forward without checking the engine parameters. The fuel-efficient M88-2 turbofan is equipped with a Snecma FADEC (Full Authority Digital Engine Control), which allows it to accelerate from idle to full reheat power in three seconds.

Acceleration was very brisk: a numerical readout of the Jx, the aircraft's longitudinal acceleration, appeared in a small box in the middle of the HUD. The obvious benefit of this feature is that it eliminates the need to carry out the traditional time/distance check to confirm engine performance and acceleration. Rotation speed was 130 kt (241 km/h), and lift-off came some 700 m (2,300 ft) down the runway. Philippe Rebourg raised the gear while we were accelerating, heading north. As soon as the wheels left the ground, the RBE2 radar automatically switched from standby to emission. In an effort to protect flight deck crew from radiation hazard, this safety feature has been slightly modified in the Rafale Marine, with the radar becoming active only when the undercarriage is retracted. When coming back from a sortie, the emissions stop when the gear is lowered. However, this can be overridden by the pilot if he/she wants or needs to maintain situational awareness, for instance during a carrier-controlled approach, when the Rafale is following another fighter in bad weather conditions using his/her radar to maintain a safety separation.

Philippe Rebourg handed me over the jet to get a 'feel' for the aircraft and become comfortable with the controls and displays. The side-stick and the

Rafale's cockpit is mounted high on the fuselage, giving an excellent view of the entire upper hemisphere. The bulged canopy aids downward vision.

massive throttle proved ideally positioned, and the visibility afforded by the large canopy is outstanding. As anticipated, the HUD repeater was invaluable, allowing me to see straight ahead of the aircraft. My only concern was that when pointing directly at the sun, the repeater whited out, and the symbology briefly disappeared. B302 is equipped with a monochrome repeater but a new colour LCD one with improved image resolution has been adopted for Armée de l'Air follow-on two-seaters, and is on offer for potential export customers.

Scalp attack

Flying at 1,000 ft AGL and 400 kt IAS, Philippe Rebourg demonstrated a cruise missile attack profile. First, he selected the air-to-surface format on the left-hand touch-screen, and created a simulated fit with two Scalp missiles, the weapon of choice for long-range attacks of defended targets. Thanks to an advanced, user-friendly mission preparation system, the missiles' firing envelope had been determined before the sortie, and the parameters clearly appeared on the head-level display. The test pilot switched to attack mode and, as explained in the simulator, asked me to initiate the alignment of the missiles' inertial navigation systems using the left-hand touch-screen. The Mil Std 1760 databus immediately transferred

Mounting the canards high on the fuselage sides impairs some downward vision in the rear seat of the Rafale B, but is less of a problem in the front.

accurate positioning data to the Scalps, and a 'Jink' order flashed in the HUD.

Responding at once, Philippe Rebourg threw the Rafale into aggressive manoeuvres so that the gyros of the INSs were fed with enough motion and g information. He stabilised again while I 'switched on' the missiles' Microturbo turbofans. You have to be careful not to start those engines too early as this reduces missile range. The Scalp firing sequence has been conceived to facilitate weapon release. All the pilot had to do was to follow the predetermined route into the crescent-shaped firing envelope, and then pull up with a positive g-load to simulate a perfect airframe/missile separation before breaking away to avoid 'hitting' the Scalps. The two simulated stand-off missiles were fictitiously fired in quick succession. The Scalp's very low IR and radar signatures help it disappear in the background clutter, preventing its destruction before the attack. For improved efficiency, the trajectory of the weapon can be adapted to the nature of the target, enabling the missile, for instance, to fly round to attack a building from a more vulnerable or less defended side.

While performing the attack, both the radar and the FSO maintained a constant surveillance of the airspace, enabling the pilot to detect and instantly engage any fighter attempting to intercept us. During the whole run, the autopilot was engaged, and Philippe Rebourg accurately flew the Rafale with the 'coolie-hat' on top of the side-stick. The advanced autopilot is fully integrated into the fly-by-wire system, designing the MIDS, and EuroMIDS and its American partners came out with the very light LVT, which also includes a TACAN. The LVT and its two associated antennas offer a 360° coverage. "Flight testing of the MIDS has already started with systems mounted on Falcon 20 and Mirage 2000 test beds," says Philippe Rebourg. "Additionally, an airborne Rafale has successfully exchanged data with a C³ simulator and an integration rig. Finally, in the summer 2001, two MIDS-LVT-equipped Rafales co-operated with an E-2C Hawkeye fitted with a JTIDS (Joint Tactical Information Distribution System) terminal, demonstrating the efficiency and the flexibility of the system." The first production MIDS-LVT for the Rafale will be delivered in 2003, and the system will be fully operational on Standard F2 aircraft. In future improvements, the Rafale's information-sharing capabilities are likely to be developed even further thanks to the adoption of an advanced satellite communication system.

For countries with no NATO approval, Thales and Dassault have designed the LX-UHF tactical datalink, which has already been selected by two customers for their Mirage 2000s. This high-tech, jam-resistant, line-of-sight system is comparable to the Link 16.

Powerplant

Developing a new engine for the Rafale was a daunting task, but the performance levels reached by the Snecma M88 are fully compliant with French needs. According to Jacques Desclaux, director of the M88 programme, "The Rafale's stringent air combat and low-altitude penetration performance requirements have imposed the adoption of an innovative powerplant with a very high thrust-to weight ratio, an extremely low fuel consumption in all flight regimes, and a very long engine life. Snecma responded with the state-of-the-art M88 twin-spool turbofan. This engine, which powers every Rafale variant, represents the third generation of French fighter engines, after the Atar family of the Mirage III/IV/V/F1, and the M53 of the Mirage 2000."

The M88 development programme was officially launched in 1986, and the first bench trials were performed in February 1989. The first flight, in the Rafale A demonstrator, was recorded in February 1990, and qualification was obtained in early 1996. The first production engine was delivered at the end of the same year and, as of September 2001, 56 turbofans had been built, including 29 series engines. "The test programme has proved highly successful, and development and production engines have accumulated nearly 23,000 functioning hours so far, including 8,000 hours of bench running, 11,000 flying hours logged by prototype engines, and nearly 4,000 flying hours by series M88-2s," continues Jacques Desclaux. "So far, Snecma has secured orders for 160 M88s, and the French MoD will eventually acquire about 700 engines for its 294 Rafales. Production at Melun-Villaroche currently runs at four a month, but will later stabilise at six a month to satisfy

Dassault's two-seat demonstrator/prototype turns for finals at a desert air base during a sales tour. The company fought hard for a large United Arab Emirates fighter contract, which was eventually won by the Lockheed Martin F-16 Block 60. Rafale's export chances have been considerably boosted by the development of advanced technologies aimed principally at export customers under the Operation Mk 2 programme, including conformal fuel tanks, active array radar and uprated engines. The latter are seen as important in silencing some of the aircraft's critics.

B01 displays a typical long-range attack loadout, with three large tanks, two Scalp EG/Apache stand-off missiles and MICA IR missiles for self-defence on the wingtips.

and can be easily engaged by a paddle switch on the stick, and deselected by a larger paddle. Similarly, the autothrottle can be quickly engaged/disengaged.

Automatic terrain-following

When clearing the firing area, Philippe Rebourg engaged the terrain-following mode, which uses a 300,000-km² (115,840-square mile) high-resolution 3D digital database to elaborate a safe trajectory. The system uses stored terrain elevation data, together with inputs from the fighter's navigation suite, to predict aircraft height above ground. This prediction, measured against the true height from the radar altimeter, enables the Kalman filters to build an error-corrected model of the aircraft's navigation system, achieving drift-free navigation, and independent and totally passive terrain-following. The pilot can choose between three ride modes – soft, medium or hard – according to the threat level.

At the moment, the Rafale is cleared to fly down to 300 ft (91 m) above ground level (AGL) with the terrain-following engaged, but it is planned that the fighter will fly as low as 100 ft (30 m) when the development programme has been completed. Compared with the Mirage 2000D, peak *g* loads during terrain-following flight are more than doubled, and the terrain-following speed envelope has been considerably expanded, meaning that the new design follows ground contours more closely, thus further reducing exposure to enemy detection and attack. Thanks to an innovative symbology displayed on the HUD, the pilot always knows what to expect when deviating left or right from the planned route when, for example, manoeuvring hard to avoid a sudden threat: the height of the surrounding terrain appears on the sight in a detailed and explicit way.

Dassault is currently developing a new operating mode which will generate predictive ground proximity warnings, preventing controlled flight into terrain. If a pilot loses attitude reference and dives towards the ground, for example when entering low clouds during an air combat training engagement, the terrain-following system will detect the danger and will warn him, giving him instructions to safely recover. When Standard F3 Rafales enter service, the aircraft will be capable of automatically switching to terrain-following mode without any pilot action. "This will be a very effective life-saving safety feature," said Philippe Rebourg.

The hilly terrain provided an excellent environment to test this totally passive terrain-following system. Before the flight, Philippe Rebourg had selected the desired flight height, 500 ft (152 m) AGL in this area of France, plus a 100-ft (30-m) safety obstruction margin, and the aircraft automatically and strictly adhered to the chosen height. The ride quality was excellent, and Philippe Rebourg explained that the gust-alleviation mode originally envisioned was eventually not needed. There was light turbulence in the area, but the aircraft remained rock-steady, even when overflying ridges.

Air-to-air role

After a few minutes at very low-level, we turned west for a demonstration of the air-to-air role. During air-to-air combat the radar searches above and below the horizon, and automatically selects high-, medium- or low-pulse repetition frequencies to optimise detection range. Targets can be IFF-interrogated automatically, and a dogfight mode is available for close combat. For air-to-air interceptions, the RBE2's long detection ranges and multi-target capabilities enable the pilot to track up to 40 targets in the track-while-scan mode, irrespective of their aspects and flying altitudes. Interception data are calculated for the eight priority targets, allowing firing of MICA missiles in quick succession in their full range envelopes. The weapon system automatically selects the nearest or most threatening target, and the pilot only has to accept the proposal and shoot (or instantly switch to another target if the tactical situation or orders received dictate another choice).

Electronic scanning radars such as the RBE2 are much more efficient than conventional radars against fighters using advanced air combat tactics. When the enemy pilots realise they are going to be intercepted, they usually split – for instance, the leader opens to the right and climbs while the wingman opens to the left and dives – to try to break radar lock or force the

French requirements. This output can be rapidly increased to respond to any export contract. Additionally, the versatile M88 could be offered for other projects such as the German Mako trainer."

M88-2

"It was clear from the outset that, compared with the M53, the M88 would have to run at much higher temperatures," says Jacques Desclaux. "This represented a challenge, but Snecma came out with innovative solutions to improve performance and durability: the engine incorporates advanced technologies such as integrally-bladed compressor disks, called 'blisks', low-pollution combustor, single-crystal high-pressure turbine blades, ceramic coatings, revolutionary powder metallurgy disks, and composite materials. Additionally, the M88 has been optimised so that its small infra-red signature does not compromise the

Rafale's overall IR signature, and its smoke-free emissions make the aircraft more difficult to detect visually."

Light, compact and fuel-efficient, the M88-2 is rated at 50 kN (11,236 lb) thrust dry and 75 kN (16,854 lb) with afterburner. It is equipped with a fully-redundant Snecma FADEC (Full Authority Digital Engine Control) which allows it to accelerate from idle to full afterburner in less than three seconds. Thanks to the FADEC, the M88-2 engines give the Rafale stunning performance: carefree engine handling allows the throttle to be slammed from combat power to idle and back to combat power again anywhere in the flight envelope. The FADEC has also proved essential for mission effectiveness, safety and reliability. For instance, it is fully capable of handling minor engine faults without a need to warn the pilot. The compressor utilises a three-stage low-pressure fan, and a six-stage high-pressure compressor. Peak engine temperature is 1850 K (1577 °C; 2,870°F) with a pressure ratio of 24.5:1, and, at maximum dry power, specific fuel consumption is in the order of 0.8 kg/daN.h, increasing to 1.8 kg/daN.h with afterburner.

For the M88-2, a staged approach was chosen from the start: the first 29 production engines were of the M88-2 Stage 1 standard, but all subsequent turbofans ordered by the French Ministry of Defence will be built to the improved M88-2 Stage 4 standard, which will feature extended time between overhauls (TBO), thanks to its redesigned high-pressure compressor and turbine. Tailored to simultaneously excel in low-altitude and air combat flight regimes, the M88 turbofan offers a combination of extremely high thrust and very low cost of ownership.

Assembly

In a ceremony held at Bordeaux-Mérignac, in the southwest of France, the first production aircraft, two-seater B301, was delivered to the French Ministry of Defence in

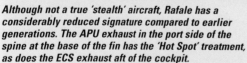

Although not a true 'stealth' aircraft, Rafale has a considerably reduced signature compared to earlier generations. The APU exhaust in the port side of the spine at the base of the fin has the 'Hot Spot' treatment, as does the ECS exhaust aft of the cockpit.

Rafale has a narrow nose which, in turn, limits the antenna area available. This was one of the many challenges encountered by the RBE2 team in devising a radar that met an ambitious specification.

interceptor to track only one target. With the vertical and horizontal separation between the enemy fighters getting wider and wider, the mechanical scanning radar will progressively have more and more difficulties to track both, and the pilot will eventually have to choose which aircraft to engage when one of the contacts is lost. Against an interceptor equipped with an e-scan radar, these aggressive engagement profiles are not considered efficient any more as the radar is fully capable of tracking all the aircraft, even those outside the chosen search volume.

On an operational Rafale equipped with a fully functioning datalink, radar, FSO and self-defence suite, all tactical data is fused in the head-level display, considerably improving situational awareness. For interception, the head-level display is divided into vertical and horizontal tactical images, offering a perfectly clear view of the evolving combat situation to the crew. The FSO image can be inserted into the main image to further improve situational awareness. Alternatively, for raid assessment, a close-up display can be selected along with the tactical images on the head-level display. As B302 was not fitted with the

Spectra self-defence suite, Philippe Rebourg chose to use the right-hand lateral screen for the FSO image.

Interception

Without specific threats pitted against us, interceptions of targets of opportunity had to be simulated. The weather had quickly deteriorated over the mountains of the central part of France, and the higher ground was in clouds. As we were coming out of a turn with the RBE2 electronic scanning radar in the search-while-track mode, the weapon system automatically acquired a lonely contact tracking south at medium level. The TV sensor of the FSO is slaved to the radar, and the system immediately produced an image of the aircraft, which turned out to be a French Air Force Transall.

Although we could not yet see our prey, the TV sensor easily tracked the contact: the optronics system seemed not to be affected by the thin cloud layer and, even at long range, the FSO clearly showed the turning propellers. As expected with such a new fighter type, no radar or FSO range will be published in this report as all data is classified. However, it should be noted that this writer was really impressed by the capabilities offered by the weapon system: the radar and FSO coverage in range, azimuth and elevation ensured excellent situational awareness. Mounted on top of the nose, the FSO has a nearly perfect field of view.

Philippe Rebourg offered to 'attack' the Transall with our MICA missiles. Using a controller on the throttle, he just had to accept the weapon system's proposal, and was ready to fire: nice, quick and simple. On the head-level screen, two lines representing the MICA maximum range and its no-escape zone clearly showed when to shoot. Thanks to this display, aircrews can choose the most adequate tactic to achieve immediate superiority: engage the target at long-range or wait to maximise hit probability. During a within-visual-range dogfight, the minimum firing range would be displayed instead.

To demonstrate the multi-target capability, the test pilot engaged another target – in this case an airliner cruising at very high level – in less than two seconds. In the HUD, the remaining fighter/missile link times were shown graphically, whereas the remaining missile flight-times were displayed numerically. Up to four missiles can be guided simultaneously by the fighter/missile datalink. The MICA is very agile, and will prove exceptionally difficult to dodge: a jet deviation system combined with efficient aerodynamic control surfaces and long-chord wings give the missile exceptional agility (load factors of

December 1998. It was followed by the first production Rafale M in July 1999. Rafale's production is split between four Dassault factories: Argenteuil (where the fuselage is built), Martignas (wings) and Biarritz (fin), with final assembly at Bordeaux-Mérignac. The extensive use of Dassault's Catia computer-aided design tool to generate three-dimensional electronic models of the Rafale has considerably eased production and will facilitate maintenance.

Production rates are currently very low, with only six aircraft to be delivered in 2001, but Dassault is still extremely busy with the Mirage 2000-5/2000-9 programme. The recent successes in the United Arab Emirates (30 Mirage 2000-9s plus 33 Mirage 2000EAD/DAD/RADs to be brought up to Dash 9 standard), Greece (15 Mirage 2000-5 Mk IIs plus 10 Mirage 2000EGs to be upgraded) and India (10 Mirage 2000Hs) mean that co-production of both fighter types is likely to go on for a considerable period of time. The first 13 Standard F1 Rafales (two Rafale Bs and one Rafale C for the Armée de l'Air, plus 10 Rafale Ms for the Aéronavale) will all have been delivered by October 2002, and 48 Standard F2 Rafales are on order. The French Armed Forces will eventually receive 294 Rafales, including 60 Rafale Ms, with final deliveries around 2020. At the time of writing in late 2001, nine production Rafales had been delivered: two Rafale Bs (B301 and B302) and seven Rafale Ms (M1 to M7).

Rafale prototypes and series aircraft are very similar, but there are nevertheless a few differences. For instance, production Rafales have a beefed-up undercarriage for operations at a higher maximum weight, they benefit from a more powerful conditioning unit, and they boast a 'Hot Spot' treatment for IR signature reduction. Rafale prototypes were restricted to a maximum take-off weight of 19500 kg (42,989 lb), but production aircraft are already cleared to 24500 kg (54,012 lb), and will eventually be able

to operate at 26000 kg (57,319 lb), quite an achievement for such a compact fighter. It is worth noting that the first two-seat Rafales (B301 and B302), plus the first naval variant fighter (M1) are still all fully instrumented for trial programmes.

Operation Mk 2

Although Rafales built for the Armée de l'Air and the Aéronavale fully respond to the French requirements, it was felt that the design could be further improved to strengthen its appeal on the export market. Accordingly, Dassault, Thales and Snecma have launched the curiously dubbed 'Operation Mk 2' for the Rafale, a phased improvement programme with systems tailored to meet potential customers' requirements. The development of new versions has been fully funded through an agreement between Dassault, Thales, Snecma and the French government, and full-scale work was officially launched in 2001.

Dassault builds the Rafale sub-assemblies at four factories, before assembling the final product at Bordeaux-Mérignac. Here M2, M3 and M4 are seen, essentially complete, before emerging for their manufacturer's test flights. Dassault production is currently running at a low rate, but will ramp up considerably. Even when operating at full rate for French deliveries, the production system has been devised to rapidly increase its rate to meet export demands should the need arise.

more than 50 *g*) and, according to Rebourg, its end-game performance seems to exceed that of the AIM-120 AMRAAM. The wing-mounted MICAs can be rail-launched at up to 9 *g* whereas the ejected fuselage missiles can be fired at up to 4 *g*.

For a long-range interception, the missile climbs to very high altitude to maximise endurance before diving at very high speed onto its target. As was the case with the radar and the FSO, the firing range of these missiles cannot be disclosed. At one stage in our flight, the Thales RBE2 radar was tracking no fewer than 21 aircraft, most of them airliners. With two simulated 'kills' credited to us, we turned south and headed towards the Mediterranean Sea.

Accurate attacks

To carry out the planned precision attacks, Philippe Rebourg again modified the simulated weapon load, replacing the two Scalp missiles with four GBU-12 Paveway laser-guided bombs (LGBs), a typical weapon fit for a strike mission. Rafale B302 was not equipped with a laser designator so an external designation/ illumination by another aircraft or a Forward Air Controller had to be simulated for the purpose of the demonstration.

Our selected target for the LGB attack was the Faraman lighthouse, in the Rhône river estuary. A suitably positioned jetty at Port Saint Louis harbour provided an ideal offset aim point (OAP), and the FSO quickly locked onto it. The eye-safe laser rangefinder was briefly switched on to update the navigation data with higher precision than with the radar. The distance and bearing from the OAP to the lighthouse had been measured during the planning session and, with these data, the mission computer was able to determine the weapon release point through simple calculation, giving excellent bombing accuracy. The attack was conducted in level-flight at 20,000 ft (6096 m), and three simulated GBU-12s were 'released' in quick succession.

The wingtip launch rails and rear fuselage pylons give the MICA missiles a 'free' ride in any configuration. The fuselage stations are limited to a 4-g launch envelope, meaning that they would normally be used to mount the active radar EM version of the missile, which is less likely to be used in hard-manoeuvring combat.

Rebourg then suggested to simulate a dive attack on the lighthouse. Rolling inverted at 18,200 ft (5550 m) and diving at 23° for the attack, he placed the designation diamond on the building, and used the laser rangefinder to improve the firing solution. With the FSO cued on the target, he zoomed in on the lighthouse to refine the designation. Then, he pressed the trigger to authorise the attack, and the remaining

GBU-12 was fictitiously dropped. Even without a laser designation, the bomb would have hit the target with near perfect precision as the Rafale's Continuous Computation of Release Point (CCRP) mode is exceptionally accurate. During the whole attack, the FSO remained perfectly locked on the target, except during the pull-up when the lighthouse disappeared under the nose of the aircraft. When the Damoclès

In contrast to the rival Eurofighter, Rafale prototypes have travelled far and wide in the hunt for export sales. Most sales jaunts have involved B01, as it can be used to give demonstration rides to interested and influential air force commanders. Here the aircraft is seen in 1998 at the Seoul international airshow. South Korea proved to be a bright prospect for Dassault as it searched for a heavy fighter to fulfil its F-X requirement. Rafale was reported to be the favoured option from a technical standpoint, but political considerations placed the Boeing F-15K as front-runner. At the time of writing the decision had been delayed due to budgetary considerations.

As is the case for the French Rafales, a stepped approach has also been adopted for export Rafales, and several variants with different levels of equipment will be available:

Block 05:	FSO
	conformal fuel tanks (optional)
	voice command
	3D digital database terrain-following system
	GBU-12 laser-guided bomb
	Mk 82 general-purpose bomb
	Scalp stand-off cruise missile
	MICA IR missile with combat modes only
	datalink
Block 10:	M88-3 engines
	active array for RBE2
	radar terrain-following capability
	synthetic aperture radar mode
	helmet-mounted display
	MICA IR with BVR interception modes
	AASM precision weapon
	Exocet anti-ship missile
	Pod Reco NG reconnaissance pod
	Inflight-refuelling pod
Block 15:	Meteor air-to-air missile
	follow-on avionics improvements

Integration of some of these items is already going ahead, and a series of flight tests was conducted by Dassault in 2000 to validate GBU-12 laser-guided bomb deliveries from the Rafale. Separation tests were carried out at Cazaux using two-seat prototype B01. The well-known GBU-12 is widely used by air forces around the world, and the type is already in service in France on Mirage F1CTs, Jaguars, Mirage 2000Ds and Super Etendard Modernisés. The introduction of the Thales Damoclès laser designation pod on French Standard F2 and export Rafales will give a true self-designation capability to the new swing-role fighter.

A typical configuration cleared during the trial programme featured four wing-mounted GBU-12s together with four MICA and two Magic 2 air-to-air missiles, plus three 2000-litre (528-US gal) drop tanks, giving enormous firing power and very long range. In this configuration, combat radius is said to be 1480 km (800 nm), and the Rafale boasts significant self-escort capability. It is anticipated that other laser-guided weapons such as the 500-lb (227-kg) GBU-22 and 2,000-lb (907-kg) GBU-24 Paveway III bombs will also be adopted for the Rafale. According to unconfirmed rumours, Raytheon is pushing Dassault to consider clearing the latest generation AIM-9X Sidewinder air-to-air missile for service on board the Rafale to increase the missile's export prospects.

Active array radar

In 1999, Thales announced that an active array for the RBE2 radar would be offered to boost export prospects. Although the innovative RBE2 already represents a giant leap forward compared with older mechanical scanning radars, the adoption of an active array will ensure that the design remains fully effective in the long term. Thales began studies on active array technology in 1990, and has

Both pilot and WSO have access to the same information from the aircraft's system, allowing them to tailor workshare to particular requirements. In general, the pilot would concentrate on the air threat while the WSO works against surface targets.

8,600 ft (2620 m), losing only 2,600 ft (795 m) in the process.

Even from the rear seat, the Rafale was a joy to fly: it was very responsive to both roll and pitch inputs. Although I did not hesitate to slam the throttle from idle to maximum afterburner and back to idle again, the M88-2 turbofans performed flawlessly, and engine response was instantaneous. "Numerous foreign test pilots have already evaluated the Rafale, and they all praise the M88's extremely short response time and pilotability," stated Philippe Rebourg. "Whatever the speed or altitude, there is no throttle movement restriction. Additionally, the M88-2 is extremely powerful. In the air defence role, you need to be able to go supersonic at will, and the two M88s allow us to accelerate very quickly."

Close encounter

On the way to our final target, Philippe Rebourg offered to demonstrate the Doppler beam sharpening (DBS) mode of the RBE2 radar. By processing the Doppler shift in the returned echo, the DBS mode created a high-definition view of a small part of the area, in this case Istres Air Base. Using the DBS mode, we obtained a high-resolution picture of our target area, and the image's quality was so good that we could easily make out the different buildings and hangars. The introduction of a Synthetic Aperture Radar mode on Standard F3 Rafales will further enhance resolution, allowing aircrews to designate targets even more accurately. The test pilot designated the control tower, and the FSO was automatically pointed at it, allowing him to refine the designation thanks to the magnification mode of the TV sensor of the FSO.

But, while preparing the attack run on the tower, Philippe Rebourg spotted a Mirage 2000N about to cross our path, slightly higher and heading towards the airfield where it is stationed. We immediately swung left to engage the 'threat'. By a single action, pressing a button on the throttle, Philippe Rebourg

pod enters service, the Rafale will be able to self-designate, track and illuminate a target with metric accuracy.

Manoeuvres

As agreed before departure, Philippe Rebourg handed over the jet to me for a few aerobatics. The Rafale's advanced digital fly-by-wire controls allow carefree handling, and there are only two types of limits according to the configuration of the fighter: heavy loaded (+5.5 g, 150°/sec roll-rate) and air combat (+9 g and 270°/sec roll-rate). However, these are 'soft limits' which can be overridden in some conditions. If need be, to avoid hitting the ground for example, the pilot can apply more pressure on the side-stick to pull up to 11 g in the air combat configuration.

Rafale's flight envelope has been explored at angles of attack (AOA) of more than 100°, and the fighter has already flown at negative speeds without departing from controlled flight. "However, we consider that a clean weapon separation at very low speed and very high angle of attack will always be problematic, whatever the fighter design or missile type fired," explained Jean Camus before the flight. "As such, we think that helmet-mounted cueing

systems and agile missiles such as the MICA IR are much better suited for within-visual-range combat."

I first tried a loop in military power, pulling 4 g with the blue Mediterranean sea providing a superb background. I started the manoeuvre at 330 kt (611 km/h) and 11,200 ft (3415 m), with the speed decaying to 94 kt (174 km/h) at the top (14,700 ft/ 4480 m). Except for a very slight buffeting, the aircraft remained very stable throughout the loop. Although not used to side-sticks and fly-by-wire technology, I was rapidly at ease with the aircraft, and found that the controls were extremely precise. I had no tendency to over-control. Aileron rolls performed at 272 kt (503 km/h) IAS at 8,000 ft (2440 m) resulted in a roll-rate of about 200°/sec.

Next, at 11,000 ft (3350 m), I retarded the throttle, and the speed rapidly decayed to 130 kt (240 km/h) as the nose progressively came up (26° AOA). Even at such a low speed, the fighter remained fully controllable and, at Philippe Rebourg's suggestion, I performed an aileron roll, a manoeuvre which would not have been feasible at such a high angle of attack with a Mirage III or a Mirage F1. At 190 kt (352 km/h) and 11,200 ft (3415 m), I rolled the aircraft inverted and pulled the stick to execute a split-S manoeuvre. The angle of attack peaked at 29°, and I stabilised at

been constantly progressing in this field, with several ongoing operational programmes for ground/naval and airborne applications. This development effort is conducted in parallel with the European collaborative AMSAR (Airborne Multi-mode Solid-state Active-array Radar) programme which will eventually lead to a production radar for Rafale and Typhoon mid-life updates.

"The new active array to be integrated into the RBE2 has many significant advantages over the current passive antenna," says Philippe Ramstein. "It is composed of about 1,000 GaAs (gallium arsenide) solid-state transmit/receive modules – embedded in the antenna assembly – which offer considerably increased power and detection range.

Above: B301 was the first production aircraft for the AdA, first flying in June 1998. Service entry is currently slated for 2005, but is likely to slip by a year due to budgetary constraints.

Right: This impression shows the RBE2 radar fitted with a Thales-developed active array, intended for export. France is also a participant in the AMSAR active array programme, which may well produce an active array antenna for French Rafales, as well as for the Eurofighter's Captor radar.

switched from air-to-ground to dogfight mode, as if we were under threat. In less than a second, the RBE2 radar acquired the target, and the FSO and Magic were instantly cued towards the unsuspecting 'enemy' fighter. The test pilot plugged in the afterburner, and threw the Rafale into a tight turn, there again demonstrating the agility of the fighter. The missiles locked onto the Mirage 2000's jet pipe, and with the Magic acquisition growl in our headsets, Philippe Rebourg 'shot' a simulated missile at short range, leaving not a single chance to the Mirage 2000N. When the helmet-mounted sight is fully operational, such a self-defence engagement will be carried out even quicker for targets at high off-boresight angle.

Circuit training

At the pre-determined fuel value – 1500 kg (3,307 lb) – the word 'Bingo' flashed on all screens, including the HUD, while a female voice whispered a warning. All too soon, it was time to return to Istres for 'circuit-bashing'. During the short transit back to Istres, Philippe Rebourg called the approach page of the left lateral display, and offered to select 15° AOA to facilitate landing from the back seat. Once downwind, the pilot selected the airbrake, which did not produce any significant pitch changes. Noteworthy is the fact that the Rafale is not equipped with a conventional airbrake. Instead, the canard leading edges come up, and the elevons on each wing deploy in opposite directions to create drag and slow down the aircraft.

Istres, which is an active fighter and tanker base, was quite busy, and we had to wait for an EC 3/4 'Limousin' Mirage 2000N and a pair of Alpha Jets to take off before turning into finals. The autothrottle was engaged to maintain 124 kt (230 km/h) with 15° Alpha and, thanks to the HUD repeater, I was able to keep the runway in sight during the whole first approach. The Rafale's autothrottle system is much more sophisticated than a traditional autothrottle: it is coupled with both the engine and the controls, which

can be automatically deflected for maximum braking effect. This so-called 'thrust/drag control mode' allows the aircraft to automatically keep a chosen Mach number (above 20,000 ft/9072 m), indicated air speed (below 20,000 ft) or angle of attack number (when the undercarriage is down) in an extremely wide flying envelope.

On short finals, Philippe Rebourg took over the controls, flared, and the Rafale gently touched down. He immediately selected maximum reheat, and we rocketed skywards. For the final landing, the test pilot changed the AOA back to 16°, a normal figure giving an approach speed of 120 kt (222 km/h). In the unlikely case of an engine failure, it is advised to choose a 14° AOA for a safe single-engine approach. In fact, this would allow a safe go-around to be performed if the landing had to be aborted for any reason.

The 16° AOA, associated with very high-performance Messier-Dowty carbon brakes, ensured a remarkably short landing run, and we came to a full stop in about 600 m (1,970 ft). This short runway capability is a crucial advantage when flying out of a

damaged air base, or when the infrastructure of the country where you operate from has not been designed to accommodate fighters on a permanent basis. Moreover, for an even shorter ground run, the pilot can select an 18° AOA giving a slower approach speed, and maximum braking effect can be obtained as soon as the wheels are on the ground (it is advised to lower the nosewheel as soon as possible). If need be, the pilot can even fully apply wheel-brakes before landing, but this is not a standard procedure. While taxiing the Rafale back to the Dassault ramp, I found that the nosewheel steering was pleasantly precise, and that the wheel brakes were equally effective.

Flying such a modern aircraft is always an honour, and this 1-hour 30-minute flight was an excellent opportunity to discover all the impressive capabilities of this outstanding French fighter. Its RBE2 electronic scanning radar, its Front Sector Optronics associated with an innovative and user-friendly man-machine interface place the Rafale in a class of its own among combat aircraft, making it a leading contender on the fighter market.

Henri-Pierre Grolleau

Casting its shadow alongside that of the Mirage 2000 chase-plane, B01 holds a steady angle of attack on the approach. A low approach speed and powerful brakes give the Rafale a credible short-field performance.

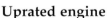

Reliability will also benefit from the introduction of the new antenna which boasts inherent redundancy: whereas a failure of the receiver or of the transceiver makes most radars useless, a percentage of an active array radar's transmit/receive modules may fail with minimal impact on radar performance. Moreover, the direction of the radiation beam from each module is very accurately controlled by computer, making it possible to scan an extremely wide area at very high speed." Thanks to the new antenna, the azimuth coverage of the RBE2 will increase from +/- 60° with the passive array to +/- 70°, bringing it on par with mechanical scanning radars and improving situational awareness. Additionally, the detection range will be considerably improved, and it is at least more than 50 percent better than with the passive array.

"Our active array compares favourably with the latest American technology. When combined with the functionalities developed for the French Forces, it will bring a unique efficiency to the Rafale's already impressive capabilities," stresses Philippe Ramstein. The RBE2's open architecture will facilitate upgrading, and the new array is totally 'plug and play'. It can be readily adapted to standard RBE2 radar sets without any changes to the processing equipment as only a patch for the new computer programme has to be injected, and the wiring has to be slightly modified. This represents a crucial advantage for customers in need of an urgent radar update as this can be achieved within a very short time. From 2006 onwards, the active array will be available and, although no decision has been announced yet, it is widely anticipated that the French Armed Forces will switch to active radar technology at some stage.

Uprated engine

The Rafale was conceived as a swing-role fighter, and the M88 engine was designed to excel at low and high levels, and to instantly respond to pilot throttle movements. However, a few potential customers have clearly expressed some concern as they felt that the engine was not really powerful enough for certain roles, typically for the air-defence/air superiority missions. For air arms requiring more power for enhanced combat agility and improved performance, Snecma has launched the development of a growth variant of the M88, the M88-3, which will be rated at 90 kN (20,225 lb) thrust with afterburner, a 20 percent increase over the original M88-2. Improvements are not limited to power output, and durability is also expected to

be considerably ameliorated. To boost even further engine TBO, the customer can select a 75-kN (16,854-lb) peacetime rating. This change can be performed in a few minutes on the ground, but a cockpit switch is under consideration to allow the pilot to select either rating in flight.

"Every effort has been made to retain a high degree of commonality between the M88-2 and the M88-3, and the two variants have about 40 percent parts in common," claims Jacques Desclaux. "The M88-3 features a redesigned low-pressure compressor for a higher airflow, 72 kg (159 lb) per second instead of 65 kg (143 lb), a new high-pressure turbine, a new stator vane stage, a modified afterburner, and an adapted nozzle." Engine weight will grow from 915 kg (2,017 lb) for a M88-2 Stage 1 to 985 kg (2,171 lb) for a M88-3. "The M88-2 and M88-3 are interchangeable, but the introduction of the M88-3 will impose the adoption of slightly enlarged air intakes to allow for the higher airflow," stresses Jean Camus. "These fixed intakes, which can be easily retrofitted to existing airframes, have

Above and left: With the French Chief of the Air Staff, Général Job, in the front seat, Rafale B302 taxis in at Mont-de-Marsan. The aircraft is in a typical air defence loadout, with three supersonic tanks and four MICA EMs. Magic 2 acquisition training rounds are carried on the wingtips.

By mid-2002 all 13 Standard F1 Rafales will be flying, the batch including 10 operational machines for the Aéronavale. The remaining three Rafales – two Bs and a C – although built to full production standards, are essentially trials machines for the following Standard F2 batch. The Bs (B301 above and B302 left) are used for both development work and for air force familiarisation/ tactical evaluation. In the latter roles they frequently operate with CEAM at Mont-de-Marsan.

Below: Although much of the export effort focuses on the air-to-surface capability of the Rafale, the type remains a formidable air superiority fighter. Here a Rafale M demonstrates a typical short-range air defence loadout, with one supersonic tank, four MICA EMs and two MICA IRs. Using outer underwing pylons and dual launchers on the centre wing pylons, the maximum loadout is 10 MICAs. For the time being, operational Standard F1 Rafale Ms are restricted to carrying Magic 2s instead of the MICA IR.

been carefully shaped so that they will not have any impact on drag or radar cross-section."

Thanks to the new engine, Rafale take-off distance, climb rate and sustained turn rate will all be improved. Although the M88-3 is much more powerful, it will have the same specific fuel consumption as the M88-2. Full-scale development of the M88-3 was launched in 2001, and qualification is planned for 2005. The flight test programme will comprise 200 hours using Rafale C01, and the first deliveries are anticipated in 2006. Looking further ahead, other variants could give even more power.

CFTs for export

From the start the Rafale was designed to carry a large fuel load, and the internal tanks of a single-seater contain 5750 litres (1,519 US gal). Additionally, the fighter is equipped with no fewer than five wet points, and two types of external tanks are available: 1250-litre (330-US gal) supersonic tanks may be carried on any of the five wet pylons, and 2000-litre (528-US gal) drop tanks can be mounted on the centreline and inner wing stations. Finally, the Rafale is equipped with a fixed inflight refuelling probe located to the right of the nose, ahead of the windscreen.

For air forces requiring an extremely long range, Dassault Aviation has conceived two 1150-litre (304-US gal) detachable conformal fuel tanks (CFTs), which can be mounted on the upper surface of the wing/fuselage blend, causing less drag than traditional tanks, and freeing underwing stations for armament. CFTs bring the Rafale's maximum external fuel load to 10800 litres (2,853 US gal), and they can be mounted or removed in less than two hours. All Rafales have a built-in CFT capability, and the CFTs can be adapted to any variant of the fighter, including naval and two-seat versions.

CFTs first took to the air at the Dassault Test Centre in Istres on 18 April 2001, fitted to the two-seat prototype B01 and with test pilot Eric Gérard at the controls. Supersonic speeds have already been demonstrated, the fighter reaching Mach 1.4, and various configurations have been successfully tested: air-to-air with MICA missiles, and long-range strike with three 2000-litre (528-US gal) drop tanks, two Scalp stand-off missiles and four MICAs. It has been determined that the CFTs had negligible impact on aircraft handling. Rafale B01 was exhibited with CFTs at the Paris Air Show at Le Bourget in June 2001, and attracted significant interest.

Now in full operational service, the flexible, powerful, swing-role Rafale is one of the most advanced fighters in service anywhere. With the development of the RBE2 active array and of the M88-3 engine now strongly backed by the French Ministry of Defence, the Rafale appears as a very serious contender on the combat aircraft market, and numerous foreign countries have expressed a strong interest. The ongoing update programme will help ensure that the aircraft remains at the forefront of Dassault export effort, and further developments are seriously considered.

For instance, Dassault is pushing ahead for a new system to shroud external weapons to reduce radar cross-section. The stealthy shapes would be ejected before weapon release. Alternatively, tube-launched missiles could be adopted. Recent competitions have highlighted the issue of exportability of high-tech fighters, and Dassault has appeared as a natural choice for many countries: whereas the US has often been reluctant to supply, even to their closest allies, the software source-codes of their weapon systems, the French manufacturer has always agreed to do so, building strong ties with its customers.

Henri-Pierre Grolleau

The Armée de l'Air and the Aéronavale are respectively busy preparing and introducing the Rafale to supplant ageing fighters. The equipment budget of the French Armed Forces diminished from 100 billion Francs in 1991 to 80 billion Francs in 2001, with obvious results. The severe budget constraints have had a considerable impact on the Rafale programme. When the project was launched, it was planned that the first production aircraft would be delivered in 1996, and that 23 fighters would have been in service by 31 December 1999. Things have not progressed as planned, and the Rafale has only just entered operational service with the Aéronavale.

AÉRONAVALE

In preparation for the arrival of the Rafale, the aircraft-carrier *Foch* had received, among other improvements, modified catapult shuttles for the Messier-Dowty undercarriage launch bar of the aircraft. However, *Foch* was sold to Brazil in 2000, and was supplanted by the nuclear-powered aircraft-carrier *Charles de Gaulle* which will be able to carry up to 30 Rafales.

The arrival, on 4 December 2000, of Rafales M2 and M3 at Base Aéronavale Landivisiau, Brittany, home of the French Navy combat aircraft community, officially marked the beginning of the renewal of the French carrierborne naval aviation. In May 2001, Flottille 12F was re-established at Landivisiau, and it will be fully operational in mid-2002 in the air defence role with a complement of 10 Standard F1 Rafales (M1 to M10). This prestigious 'Duck' squadron, the last to operate the classic Vought Crusader anywhere in the world, disbanded in December 1999 when the final F-8P retired.

To accommodate the new type, an extensive rework programme has been launched at Landivisiau, including a new squadron building, new engine maintenance facility, state-of-the-art engine test bench, new workshop for painting and composite material maintenance, and special turnaround hangars. By November 2001, 10 pilots had converted to the type: three of them are test pilots (including two Air Force exchange officers), and most had flown Mirage 2000s while on secondment to the Armée de l'Air to gain experience of fly-by-wire aircraft. It should be noted that the first Rafales produced were delivered in the Standard LF1, which featured only limited SPECTRA and RBE2 modes.

However, at the time of writing, all Rafales were undergoing an upgrade programme that will bring them to full Standard F1 specifications. "In 2004, the first Standard F2 swing-role Rafale will be delivered to Landivisiau, and Flottille 11F will start its conversion to the new variant the following year," said Captain Brûlez, Landivisiau Station Commander, at a press conference in December 2000. "By 2010, Flottille 17F will have converted to Standard F3 Rafale, and the last aircraft is due to be delivered to the Navy in 2012."

Flottille 12F's Rafale Ms wear on the fin a toned-down representation of the unit's well-known badge, which depicts Donald Duck carrying a blunderbuss.

However, a study published in November 2001 by the Defence Committee of the French National Assembly revealed that the programme might be delayed by another year, and that the Rafale Ms due to be delivered in 2004 (two aircraft) and 2005 (five) might not be built until 2006 due to budget reductions.

ARMÉE DE L'AIR

In the past few years, the Armée de l'Air has accepted into service 55 upgraded Mirage F1CTs, 37 modernised Mirage 2000-5Fs and 86 Mirage 2000Ds. The Air Force decision-makers were not interested in introducing Standard F1 Rafales for air defence/air superiority missions as they already had large numbers of Mirage 2000C/B/-5Fs at their disposal, and it was decided instead to postpone acceptance into service until Standard

B301's public debut was at the 1999 Paris air show. Notable in the impressive array of weaponry are the AM 39 Exocet, follow-on ANF and the Damoclés FLIR/laser designation pod.

F2 aircraft were available in 2005, to supplant the outdated Jaguar.

Today, the first two Armée de l'Air production Rafales, B301 and B302, regularly operate from the Centre des Expérimentations Aériennes Militaires (CEAM, or Airborne Military Experimentation Centre) at Mont-de-Marsan, helping Air Force personnel gain first-hand experience of the performance of the new fighter. Plans for entry into service have now been fully endorsed by the French Air Force headquarters: the Rafale B/Cs will first replace the old Jaguars, and will enter operational service at Saint-Dizier in 2005 with Escadron de Chasse 1/7 'Provence'. Two Jaguar units, Escadron de Chasse 2/7 'Argonne' and Escadron de Chasse 3/7 'Languedoc', disbanded in June 2001, and an extensive rework programme has started at Saint-Dizier. However, the French government has indicated that further budget constraint could lead to another postponement of the Rafale's entry into service, and the programme could slip by a year from 2005 to 2006, 10 years later than initially anticipated when the Rafale programme was launched.

Although no decision has yet been announced, it is widely accepted that the replacement of the Mirage F1CT and Mirage F1CR fleets will be next on the agenda as the Mirage 2000s are much more recent. Consequently, EC 1/30 'Alsace' and EC 2/30 'Normandie-Niemen' at Colmar could well be the third and fourth Rafale squadrons.

Planned evolution of the AdA operational fighter force

	2002	2005	2010	2015
Jaguar	25	–	–	–
Mirage F1C/B	20	20	10	–
Mirage F1CT	40	40	20	–
Mirage F1CR	40	30	20	–
Mirage 2000N	60	60	40	40
Mirage 2000D	60	60	60	60
Mirage 2000 RDI	80	80	60	40
Mirage 2000-5F	30	30	30	20
Mirage IVP	5	–	–	–
Rafale B/C	–	20	80	140
Total	**360**	**340**	**320**	**300**

DASSAULT/SNECMA/CEV

Istres, in the south of France, has long been the main airfield from where have Rafales operated. Today, Dassault, Snecma, Thales and the Centre d'Essais en Vol operate as an integrated test team with test pilots, flight test engineers, test engineers and equipment shared and pooled for maximum efficiency and cost reduction. Rafale prototypes C01, B01 and M02 are still very active, with B01 and M02 participating – along with Rafale production aircraft B301, B302 and M1 – in the Standard F2 development programme. Rafale C01 is not fitted with a representative avionics fit, and is only used for engine trials. It will take part in the M88-3 trial programme. Naval prototype M01 has been now withdrawn from use, and is in storage at Istres.

For separation trials and weapons testing, Rafale prototypes deploy to Cazaux, in the southwest of France, where the Centre d'Essais en Vol and Dassault have dedicated facilities. Trials at Cazaux have mainly concentrated on air-to-air weapons (MICA, Magic and 30-mm cannon) so far, but the qualification of the GBU-12 and Scalp separation trials have also been completed.

Despite talk of its retirement, the matt black first prototype, C01, remains a valuable test vehicle at Istres, as well as a popular participant at selected air shows.

Air Combat Command

Headquartered at Langley AFB, Virginia, Air Combat Command (ACC) was established on 1 June 1992, tasked with combining many of the roles of the old Strategic and Tactical Air Commands in the wake of the Cold War.

Air Combat Command, United States Air Force

More than 10 years have passed since Operation Desert Storm and the end of the Cold War, during which time the United States Air Force has completed a major restructuring. It has become a much leaner force even though its mission tasking has remained fairly constant and the service continues to support operations in Europe and Southwest Asia. In addition, it has to deal with contingency and humanitarian missions around the world.

To cope with the workload now carried by a reduced number of active-duty units, the service implemented the Expeditionary Aerospace Force (EAF) concept during Fiscal Year 2000 (1 October 1999 – 30 September 2000).

The primary objective is to enhance operational capabilities available to commanders in chief (CinC), and enable the USAF to respond effectively to crises as and when they arise.

Each AEF is a cross-section of the total force, representing a responsive 'package' that can deal with a variety of contingencies. Besides combat aircraft that include fighters and bombers, the AEF is assigned reconnaissance; command and control (C²); command, control and communications countermeasures (C³CM); and combat rescue forces. Supporting forces include tankers and tactical (intra-theatre) transports.

In addition to active-component forces, units from each category serving the Air Force Reserve Command (AFRC) and the Air National Guard (ANG) are assigned to the EAF. Although the concept continues to be refined, it has already demonstrated its value, enabling deployments to be scheduled with sufficient advance notice.

The second rotation of the EAF was well underway by mid-2001, having technically begun in October 2000 when units attached to Aerospace Expeditionary Forces (AEF) 1 and 2 began their deployment training cycle. While assigned units will vary somewhat between deployment cycles, the command structure for individual AEFs will generally remain the same.

Air Combat Command

Operating all USAF bombers and continental United States (CONUS)-based combat-coded fighter, attack, reconnaissance and combat rescue aircraft, ACC is responsible for organising, training and equipping air combat-ready forces. It also provides nuclear forces for US Strategic Command (USSTRATCOM), plus theatre air forces for US Joint Forces Command

F-15C Eagles of Langley AFB's 1st Fighter Wing patrol during Operation Noble Eagle in late 2001.

Direct Reporting Units

UNIT	BASE/AIRCRAFT	TAILCODE
Air Warfare Center – Nellis AFB, Nevada		
53d Wing	**Eglin AFB, Florida**	
53d TEG	**Nellis AFB, Nevada**	
Det. 1	Holloman AFB, N.M.	
	F-117A	OT
Det. 2*	Ellsworth AFB, S.D.	
	(B-1B)	EL
85th TES 'Skulls'	Eglin AFB, Florida	
	F-15C/E,	OT
	F-16C/D (Block 42/50)	
72d TES*	Whiteman AFB, Missouri	
	(B-2A)	WM
49th TESTS*	Barksdale AFB, Louisiana	
	(B-52H)	LA
422d TES 'Green Bats'	Nellis AFB, Nevada	
	A-10A,	OT
	F-16C/D (Block 42/50/52)	
	F-15C/D/E, HH-60G	
53d WEG	**Tyndall AFB, Florida**	
82d ATRS	Tyndall AFB, Florida	
	E-9A, QF-4E/G	TD
82d ATRS, Det.1	Holloman AFB, N.M.	
	QF-4E/G	HD

UNIT	BASE/AIRCRAFT	TAILCODE
57th Wing	**Nellis AFB, Nevada**	
USAFADS 'Thunderbirds'	Nellis AFB, Nevada	
	F-16C/D (Block 32)	
57th OG	**Indian Springs AFAF, Nevada**	
11th RS	Indian Springs AFAF, Nevada	
	RQ-1A/B	WA
15th RS	Indian Springs AFAF, Nevada	
	RQ-1A/B	WA
66th RQS	Nellis AFB, Nevada	
	HH-60G	WA
414th CTS	Nellis AFB, Nevada	
	F-16C/D (Block 32/52)	WA
USAFWS	Nellis AFB, Nevada	
(Det. 1) B-1 Div.*	Ellsworth AFB, S.D.	
	(B-1B)	EL
(Det. 2) B-52 Div.*	Barksdale AFB, Louisiana	
	(B-52H)	LA
F-16 Div.	Nellis AFB, Nevada	
	F-16C/D (Block 42/52)	WA
F-15 Div.	Nellis AFB, Nevada	
	F-15C/D	WA
F-15E Div.	Nellis AFB, Nevada	
	F-15E	WA
A-10 Div.	Nellis AFB, Nevada	
	A-10A	WA
HH-60 Div.	Nellis AFB, Nevada	
	HH-60G	WA

* Unit uses aircraft borrowed from host wing

First Air Force (ANG)

Tyndall AFB, Florida
(The First Air Force is composed of gained ANG units)

UNIT	BASE/AIRCRAFT	TAILCODE
102d FW (102d OG)	**Otis ANGB, Massachusetts**	**MA**
101st FS 'Eagle Keepers'	F-15A/B	
119th FW (119th OG)	**Hector IAP, Fargo, N.D.**	**ND**
178th FS 'Happy Hooligans'	F-16A/B (ADF)	
Det. 1	Langley AFB, Virginia	
	F-16A (ADF)	
120th FW* (120th OG)	**Great Falls IAP, Montana**	
186th FS 'Vigilantes'	F-16C/D (Block 30)	
125th FW (125th OG)	**Jacksonville IAP, Florida**	**FL**
159th FS 'Jaguars'	C-26B, F-15A/B	
Det. 1	Homestead ARS, Florida	
	F-15A	
142d FW (142d OG)	**Portland IAP/ANGB, Oregon**	
123d FS 'Red Hawks'	F-15A/B	
144th FW (144th OG)	**Fresno-Yosemite IAP/ANGB, Ca.**	
194th FS 'Griffins'	C-26B, F-16C/D (Block 25)	
Det. 1	March ARB, California	
	F-16C	
147th FW* (147th OG)	**Ellington Field/ANGB, Texas**	**EF**
111th FS 'Ace in the Hole'	C-26B, F-16C/D (Block 25)	
148th FW (148th OG)	**Duluth IAP/ANGB, Minn.**	**MN**
179th FS 'Bulldogs'	F-16A/B (ADF)	
Det. 1	Tyndall AFB, Florida	
	F-16A (ADF)	
158th FW* (158th OG)	**Burlington IAP, Vermont**	**(VT)**
134th FS 'Green Mountain Boys'	F-16C/D (Block 25)	
177th FW* (177th OG)	**Atlantic City IAP/ANGB, N.J.**	**AC**
119th FS 'Jersey Devils'	F-16C/D (Block 25)	

* Unit tasked with a general purpose mission

Above: Numerically ACC's most important type, the Lockheed Martin F-16 equips ANG and AFRC units as well as numerous regular units. One of the latter is the 20th FW at Shaw AFB, which fields four squadrons, including the 79th FS 'Tigers', equipped with the F-16CJ/DJ (F-16C/D Block 50Ds equipped with AGM-88 HARM for SEAD duties).

Below: Hill AFB's long association with the F-16 began in 1979 when the resident 388th TFW became the first unit equipped with the type. In 2002, the 388th FW (as it is now known) flies F-16C/D Block 40s. A 34th FS 'Rude Rams' aircraft is pictured.

Above: Among the ANG units of First Air Force equipped with F-16s is the Fargo, North Dakota-based 178th FS' Happy Hooligans' with F-16A/B ADFs. These examples are pictured over the Pentagon during Operation Noble Eagle.

Below: The 27th FW at Cannon AFB, New Mexico flies a mixture of F-16C/Ds from Blocks 30, 40 and 52. 86-0347 is a F-16C Block 30 of the 523d FS 'Crusaders'.

Below: Tenth Air Force is made up entirely of Air Force Reserve Command units, including the Florida-based 93d FS 'Makos', 482d FW, which flies F-16C/D Block 30s.

Below: St. Louis-based 110th FS 'Lindberg's Own' is one of the ANG units gained by Eighth Air Force. In the aftermath of the 11 September 2001 attacks, the squadron has joined other ANG units mounting armed patrols under Operation Noble Eagle.

First Air Force include three squadrons equipped with F-15A/Bs; the Massachusetts ANG's 101st FS 'Eagle Keepers' (above) are based at Otis ANGB, while the Oregon ANG's 123d FS 'Red Hawks' are at Portland IAP.

Eighth Air Force
Barksdale AFB, Louisiana

UNIT	BASE/AIRCRAFT	TAILCODE
2d BW (2d OG)	**Barksdale AFB, Louisiana**	**LA**
11th BS(FTU) 'Mr Jiggs'	B-52H	
20th BS 'Buccaneers'	B-52H	
96th BS 'Red Devils'	B-52H	
5th BW (5th OG)	**Minot AFB, North Dakota**	**MT**
23d BS 'Bomber Barons'	B-52H	
7th BW (7th OG)	**Dyess AFB, Texas**	**DY**
9th BS 'Bats'	B-1B	
28th BS(FTU) 'Mohawk Warriors'	B-1B	
13th BS 'Grim Reapers'	B-1B	
27th FW (27th OG)	**Cannon AFB, New Mexico**	**CC**
428th FS 'Buccaneers'*	F-16C/D (Block 52)	
522d FS 'Fireballs'	F-16C/D (Block 30)	
523d FS 'Crusaders'	F-16C/D (Block 30)	
524th FS 'Hounds'	F-16C/D (Block 40)	
28th BW (28th OG)	**Ellsworth AFB, South Dakota**	**EL**
37th BS 'Tigers'	B-1B	
77th BS 'War Eagles'	B-1B	
65th ABW	**Lajes AB, Azores**	
	no aircraft assigned	
85th Group	**NAS Keflavik, Iceland**	**IS**
56th RQS 'Friends in Low Places'	HH-60G	
85th OS	TDY F-15, HC-130P, KC-135	
509th BW (509th OG)	**Whiteman AFB, Missouri**	**WM**
325th BS 'Alley Oop'	B-2A	
393d BS 'Tigers'	B-2A	
394th CTS 'Panthers'	T-38A	

Eighth Air Force – gained ANG units

110th FW (110th OG)	**W.H. Kellogg AP/**	
	Battle Creek ANGB, Mich.	**BC**
172d FS 'Mad Ducks'	A-10A, OA-10A	
115th FW (115th OG)	**Dane County RAP/**	
	Truax Field, Madison, Wisc.	**WI**
176th FS 'Badgers'	C-26B, F-16C/D (Block 30)	
122d FW (122d OG)	**Fort Wayne IAP, Indiana**	**FW**
163d FS 'Marksmen'	F-16C/D (Block 25)	
127th Wing (127th OG)**	**Selfridge ANGB, Mount**	
	Clemens, Michigan	**MI**
107th FS 'Wolves'	F-16C/D (Block 30)	
131st FW (131st OG)	**Lambert-St Louis IAP, Mo**	**SL**
110th FS 'Lindberg's Own'	F-15A/B	
159th FW (159th OG)	**NAS New Orleans JRB, Louis.**	**JZ**
122d FS 'Bayou Militia'	C-130H, F-15A/B	
181st FW (181st OG)	**Terre Haute IAP/Hulman Field,**	
	Indiana	**TH**
113th FS 'Racers'	F-16C/D (Block 30)	
183d FW (183d OG)	**Capital Airport, Springfield, Ill.**	**SI**
170th FS 'Fly'N Illini'	F-16C/D (Block 30)	
184th BW (184th OG)	**McConnell AFB, Kansas**	
127th BS 'Jayhawks'	B-1B	
187th FW (187th OG)	**Montgomery RAP-Dannelly Field,**	
	Alabama	**AL**
160th FS 'Snakes'	C-26B, F-16C/D (Block 30)	
188th FW (188th OG)	**Fort Smith RAP/Ebbing ANGB,**	
	Arkansas	**FS**
184th FS 'Flying Razorbacks'	F-16C/D (Block 32)	

* Squadron provides training for the Republic of Singapore Air Force
** Wing controls squadrons assigned to ACC and AMC

(USJFCOM), US Central Command (USCENTCOM), US Southern Command (USSOUTHCOM), US European Command (USEUCOM) and US Pacific Command (USPACOM). It also supplies air defence forces for North American Air Defense Command (NORAD).

Reporting directly is the USAF's Air Warfare Center (AWC), which comprises the Operational Test and Evaluation (OT&E) organisation (the 53d Wing) at Eglin AFB, Florida, plus the 57th Wing at Nellis AFB, Nevada, which focuses on advanced training and operational missions. The USAF Air Demonstration Squadron, better known as the 'Thunderbirds', is part of the latter.

Four numbered air forces are assigned to ACC. These include the First Air Force (composed entirely of ANG wings gained in operational situations), plus the Eighth, Ninth and Twelfth Air Forces, which are made up of gained ANG and active-duty units. In addition, some elements of the Tenth Air Force are gained from the AFRC.

ACC has approximately 85,000 active-duty personnel and employs 10,000 civilians. Additionally, 58,000 airmen from the AFRC and ANG come under its operational control.

In all, ACC controls some 1,700 aircraft, of which about 775 are operated by the AFRC and ANG. The command continues to phase out its oldest F-16As and less than 40 ADF variants now remain in service. They fly with two ANG wings. Meanwhile, older AFRC/ANG F-16Cs are gaining the capability to deliver precision-guided munitions through incorporation of Litening II laser targeting pods.

As examples of recent changes to ACC's force structure in support of the EAF concept, a fifth B-1B squadron has been activated, along with a dedicated combat search and rescue (CSAR) wing. The latter is being augmented through transfer of assets from the AFRC.

Individual ANG wings, previously coded as combat units and assigned to ACC's Ninth and Twelfth Air Forces, were transferred to AETC in 2000. Both are now tasked as formal training units (FTU) for F-16C/D crews. As is clear, ACC relies heavily on operational support from reserve and Guard units across all of its air forces.

A component of USJFCOM, the Eighth Air Force is composed of seven active-duty wings plus a further 11 gained from the ANG. In the case of the Ninth Air Force, which supplies assets to USCENTCOM, seven active-duty wings (including one expeditionary element) are supported by 11 ANG wings.

In the case of the Twelfth Air Force, a component of USSOUTHCOM, the composition is seven active-duty and eight ANG wings. Eight wings and one group from AFRC's Tenth Air Force are also gained by ACC.

When it was activated on 1 June 1992 Air Combat Command (ACC) controlled six Numbered Air Forces, 41 wings and one group. Subsequently, a number of the operational wings were inactivated, the strategic missile force was transferred to Air Force Space Command (AFSPC) and numerous training units were reassigned to Air Education and Training Command (AETC). Today the command is responsible for 19 active duty wings and one group. In addition eight Air Force Reserve Command (AFRC) wings and one group and 40 Air National Guard (ANG) wings are assigned. During the early 1990s the USAF operated the equivalent of 36 fighter wings, however today when reserve component units are factored into the command structure includes just 20.2 fighter wing equivalents (FWE) and these number 12.6 in the active component and 7.6 in the reserve component. Although ACC is responsible for a total of nine combat-coded fighter and composite wings, not all of these are FWEs. For record-keeping purposes an FWE is comprised of 72 combat

Ninth Air Force
Shaw AFB, South Carolina

COMPONENT	BASE/AIRCRAFT	TAIL CODE
1st FW (1st OG)	**Langley AFB, Virginia**	**FF**
27th FS 'Fighting Eagles'	F-15C/D	
71st FS 'Ironmen'	F-15C/D	
94th FS 'Hat-in-the-Ring'	F-15C/D	
4th FW (4th OG)	**Seymour Johnson AFB, N.C.**	**SJ**
333d FS(FTU) 'Lancers'	F-15E	
334th FS(FTU) 'Eagles'	F-15E	
335th FS 'Chiefs'	F-15E	
336th FS 'Rocketeers'	F-15E	
23d FG	**Pope AFB, North Carolina**	**FT**
74th FS 'Flying Tigers'	A-10A, OA-10A	
75th FS 'Sharks'	A-10A, OA-10A	
20th FW (20th OG)	**Shaw AFB, South Carolina**	**SW**
55th FS 'Fighting 55th'	F-16C/D (Block 50)	
77th FS 'Gamblers'	F-16C/D (Block 50)	
78th FS 'Bushmasters'	F-16C/D (Block 50)	
79th FS 'Tigers'	F-16C/D (Block 50)	
33d FW (33d OG)	**Eglin AFB, Florida**	**EG**
58th FS 'Gorillas'	F-15C/D	
60th FS 'Fighting Crows'	F-15C/D	
93d ACW (93d OG)	**Robins AFB, Georgia**	**WR**
12th ACCS	E-8C	
16th ACCS	E-8C	
93d TRS	TE-8A	
347th RQW (347th OG)	**Moody AFB, Georgia**	**MY**
41st RQS 'Jolly Green'	HH-60G	
71st RQS 'Kings'	HC-130P, C-130E	
363d AEW (363d EOG)	**Prince Sultan AB, Al Kharj, Saudi Arabia**	
(rotational ACC Units)		
363d ERS	Prince Sultan AB	U-2S
363d EAACS	Prince Sultan AB	E-3B/C
363d EARS	Prince Sultan AB	KC-135E/R
363d EAS	Prince Sultan AB	C-130E/H
363d EFS	Prince Sultan AB	F-15C or F-16C
763d ERS	Prince Sultan AB	RC-135V/W
763d EARS	Al Dhafra AB, UAE	KC-135E/R
763d EAS	Al Seeb AB, Oman	C-130E/H
332d AEG	**Ahmed Al Jaber AB, Kuwait**	
332d EFS	A-10A, OA-10A, F-16C	
332d ERQS	HH-60G	
9th EOG	**Ali Al Salem AB, Kuwait**	
9th ERQS	HC-130P	
9th EAS	C-130E/H	

Ninth Air Force–gained ANG units

103d FW (103d OG)	**Bradley IAP/ANGB, Windsor Locks, Connecticut**	**CT**
118th FS 'Flying Yankees'	A-10A, OA-10A	
104th FW (104th OG)	**Barnes MAP/ANGB, Westfield, Mass.**	**MA**
131st FS 'Death Vipers'	A-10A, OA-10A	
106th RQW (106th OG)	**Francis S. Gabreski AP/ANGB, New York**	**LI**
102d RQS 'ANG's Oldest Unit'	HC-130N/P, HH-60G	
111th FW (111th OG)	**NAS Willow Grove JRB, Pennsylvania**	**PA**
103d FS 'Black Hogs'	A-10A, OA-10A	
113th Wing* (113th OG)	**Andrews AFB, Maryland**	**DC**
121st FS 'Capital Guardians'	F-16C/D (Block 30)	
116th BW (116th OG)	**Robins AFB, Georgia**	**GA**
128th BS 'Georgia Bones'	B-1B	
169th FW (169th OG)	**Columbia MAP/McEntire ANGS, South Carolina**	(**SC**)
157th FS 'Swamp Foxes'	C-130H, F-16C/D (Block 52D)	
174th FW (174th OG)	**Syracuse-Hancock IAP, N.Y.**	**NY**
138th FS 'Cobras'	F-16C/D (Block 25)	
175th Wing* (175th OG)	**Martin State AP/Warfield ANGB, Baltimore, MD**	**MD**
104th FS	A-10A, OA-10A	
180th FW (180th OG)	**Toledo Express AP, Ohio**	**OH**
112th FS 'Stingers'	F-16C/D (Block 42)	
192d FW (192d OG)	**Richmond IAP-Byrd Field, Va.**	**VA**
149th FS 'Rebel Riders'	F-16C/D (Block 30)	

* Wing controls squadrons assigned to ACC and AMC

With the 4th FW, the 366th Wing at Mountain Home AFB has an 'on call' expeditionary role and has not been assigned to an AEF during the current cycle. This flexible approach sees the wing operate B-1Bs, F-15C/Ds, F-16C/D Block 52s and KC-135Rs, as well as F-15E Strike Eagles (above). The other autonomous expeditionary wing is the 4th FW at Seymour Johnson AFB, North Carolina, which fields four Strike Eagle units, including the 336th FS 'Rocketeers' (above right).

Both the B-2 and F-117 wings operate a squadron of T-38A Talons for qualification training on their respective types. These are the 394th Combat Training Sqn 'Panthers' (above) and the 7th CTS 'Screamin' Demons' (below), respectively.

Air Combat Command's stealth assets comprise the 21 B-2As (below) operated by the 509th BW at Whiteman AFB, Missouri, and the F-117As (above) of the 49th FW at Holloman AFB, New Mexico.

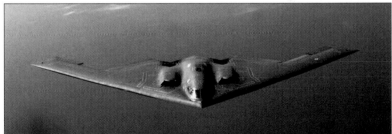

Despite their age Air Combat Command's B-52Hs continue to play a vital role as a 'bomb truck' for not only iron bombs, but also precision-guided weapons such as JDAM and ALCM, most recently in the skies over Afghanistan. The 'Buff' fleet is divided between Barksdale AFB (2d BW, below) and Minot AFB (5th BW, above).

The B-1B Lancer force includes the two squadrons of the 28th BW (above) at Ellsworth AFB and the 34th BS 'Mules' of the 366th Wing at Mountain Home AFB (below).

Twelfth Air Force

Davis-Monthan AFB, Arizona

UNIT	BASE/AIRCRAFT	TAIL CODE
Coronet Oak		
612th EAS	Luiz Munoz Marin IAP, P.R.	
	TDY ANG/AFRC C-130 Units	
Coronet Nighthawk*		
12th EFS	Hato IAP, Curaçao,	
	Netherlands Antilles	
	TDY ANG F-15/F-16 Units	
9th RW	**Beale AFB, California**	**BB**
Det.1	Akrotiri AB, Cyprus U-2S	
9th OG		
1st RS(FTU)	Beale AFB, California	
	T-38A, U-2S, TU-2S	
5th RS 'Black Cats'	Osan AB, Korea U-2S	
99th RS	Beale AFB, California T-38A, U-2S	
49th FW (49th OG) Holloman AFB, New Mexico		**HO**
7th CTS 'Screamin' Demons' T-38A		
8th FS 'Black Sheep'	F-117A	
9th FS 'Flying Knights'	F-117A	
20th FS 'Silver Lobos'**	F-4F	
55th Wing (55th OG)	**Offutt AFB, Nebraska**	**OF**
1st ACCS 'First Axe'	E-4B	
38th RS 'Fighting Hellcats'	RC-135U/V/W, TC-135W	
45th RS 'Sylvester'	OC-135B, WC-135C/W,	
	RC-135S, TC-135B/S	
Det. 1	Eielson AFB, Alaska RC-135S	
82d RS 'Hog Heaven'	Kadena AB, Okinawa, Japan	
	RC-135U/V/W	
95th RS 'Kickin Ass'	RAF Mildenhall, England RC-135V/W	
Det. 1	Souda Bay, Crete RC-135V/W	
338th CTS (FTU)	Offutt AFB, Nebraska	
	no aircraft assigned	
355th Wing (355th OG)	**Davis-Monthan AFB, Arizona DM**	
41st ECS 'Scorpions'	EC-130H	
42d ACCS 'Axe'	EC-130E	
43d ECS 'Bats'	EC-130H	
354th FS 'Bulldogs'	A-10A, OA-10A	
357th FS(FTU) 'Dragons'	A-10A, OA-10A	
358th FS(FTU) 'Lobos'	A-10A, OA-10A	

366th Wing (366th OG)	Mountain Home AFB, Idaho	MO
22d ARS 'Mules'	KC-135R	
34th BS 'Thunderbirds'	B-1B	
389th FS 'Thunderbolts'	F-16C/D (Block 52)	
390th FS 'Wild Boars'	F-15C/D	
391st FS 'Bold Tigers'	F-15E	
388th FW (388th OG)	**Hill AFB, Utah**	**HL**
4th FS 'Fightin Fuugins'	F-16C/D (Block 40)	
34th FS 'Rude Rams'	F-16C/D (Block 40)	
421st FS 'Black Widows'	F-16C/D (Block 40)	
552d ACW (552d OG)	**Tinker AFB, Oklahoma**	**OK**
960th AACS 'Viking Warriors' E-3B/C		
963d AACS 'Blue Knights' E-3B/C		
964th AACS 'Phoenix'	E-3B/C	
965th AACS 'Falcons'	E-3B/C	
966th AACS(FTU) 'Ravens' E-3B/C, TC-18E		

* FOL at Eloy Alfaro IAP/Manta AB, Ecuador
** Luftwaffe training squadron

Twelfth Air Force – gained ANG units

114th FW (114th OG)	Joe Foss Field, Sioux Falls, S.D.	
175th FS 'Lobos'	F-16C/D (Block 30)	
124th Wing (124th OG)	**Boise AT-Gowen Field, Idaho**	**ID**
189th AS	C-130E	
190th FS	A-10A, OA-10A	
129th RQW (129th OG)	**Moffett Federal Airport/ANGS,**	
	California	**CA**
129th RQS	HH-60G, MC/HC-130P	
132d FW (132d OG)	**Des Moines IAP, Iowa**	**IA**
124th FS 'Hawkeyes'	F-16C/D (Block 42)	
138th FW (138th OG)	**Tulsa IAP, Oklahoma**	**OK**
125th FS 'Tulsa Vipers'	F-16C/D (Block 42)	
140th Wing* (140th OG)	**Buckley AFB, Aurora, Col.**	**CO**
120th FS 'Cougars'	F-16C/D (Block 30)	
150th FW (150th OG)	**Kirtland AFB, New Mexico**	**NM**
188th FS 'Tacos'	C-26B, F-16C/D (Block 30/40)	
185th FW (185th OG)	**Sioux Gateway Airport, Sioux**	
	City, Iowa	**(HA)**
174th FS 'Bats'	F-16C/D (Block 30)	

* Wing controls squadrons assigned to ACC and AMC

Tenth Air Force, AFRC

NAS Fort Worth JRB/Carswell Field, Texas
(Tenth Air Force, AFRC units gained by Air Combat Command)

UNIT	BASE/AIRCRAFT	TAIL CODE
301st FW (301st OG)	**NAS Fort Worth JRB/Carswell Field,**	
	Texas	**TX**
457th FS 'Spads'	F-16C/D (Block 30)	
419th FW (419th OG) Hill AFB, Utah		**HI**
466th FS 'Diamondbacks' F-16C/D (Block 30)		
442d FW (442d OG)	**Whiteman AFB, Missouri**	**KC**
303d FS 'KC Hawgs'	A-10A, OA-10A	
482d FW (482d OG)	**Dade County Homestead RAP/ARS,**	
	Florida	**FM**
93d FS 'Makos'	F-16C/D (Block 30)	
513th ACG*	**Tinker AFB, Oklahoma**	**OK**
970th AACS	E-3B/C	
917th Wing (917th OG)	**Barksdale AFB, Louisiana**	**BD**
47th FS 'Termites'	A-10A, OA-10A	
93d BS 'Indian Outlaws' B-52H		
926th FW (926th OG) NAS New Orleans JRB, Louis.		**NO**
706th FS 'Cajuns'	A-10A, OA-10A	
939th RQW** (939th OG) Portland IAP/ANGB, Oregon		
303d RQS	Portland IAP, Oregon C-130E, HC-130P	PD
304th RQS	Portland IAP, Oregon HH-60G	PD
305th RQS	Davis-Monthan AFB, Ariz. HH-60G	DR
920th RQG		
39th RQS	Patrick AFB, Fl. C-130E, HC-130N/P	FL
301st RQS 'Guardian Wings'		
	Patrick AFB, Florida HH-60G	FL
944th FW*** (944th OG) Luke AFB, Arizona		**LR**
302d FS 'Sun Devils'	F-16C/D (Block 32)	

* Associate unit shares aircraft with active-component wing
** Wing to transition to KC-135R in 2002
*** Wing also controls AETC-aligned associate squadron

coded fighter/attack primary mission aircraft. It should also be noted that OA-10As assigned to the forward air control role and Air National Guard fighters assigned to the 1st Air Force are not included in this total.

ACC is led by General Hal M. Hornburg, who assumed command on 14 November 2001. He is also the component commander of US Air Forces Joint Forces Command and US Strategic Command (USSTRATCOM). General Hornburg serves as the executive agent for search and rescue matters throughout the continental United States (CONUS) and reports to the USAF Chief of Staff.

New aircraft

Coupled with a smaller force, the USAF now operates on a much reduced budget and the administration of President Bush is reviewing all current and future defence programmes. It is entirely possible the USAF's plan to develop two new fighters: the F-22A and the Joint Strike Fighter (JSF) could fall victim.

The FY 2001 budget authorised 81 new aircraft for the USAF, including 27 for Air Combat Command. Ten of the latter were low-rate initial production (LRIP) models of the F-22A. Pending the outcome of the Department of Defense review, however, funding is being released only on an incremental basis. Under the FY01 budget, the service continued to receive a limited number of F-15Es and F-16Cs, and support for modifications to the B-2A fleet is there. The administration was also rumoured to be considering the procurement of additional Spirits at the expense of the B-1B fleet.

Manned and unmanned reconnaissance systems were addressed with authorisation for a single E-8C Joint STARS and another RQ-1B Predator unmanned air vehicle (UAV).

Although the fiscal year 2002 Defense Appropriation has not yet been approved the portion dedicated to aircraft procurement will provide ACC with just 14 new aircraft, however it also provides the command with eight unmanned aerial vehicles. These are broken down as follows:

Northrop Grumman E-8C Joint STARS	1
Lockheed Martin F-22A Raptor	13
General Atomics RQ-1B Predator	6
Northrop Grumman RQ-4A Global Hawk	2

The budget also includes funding for upgrades for the B-2A and RC-135 and research and development for a number of programmes, including the F-35A Joint Strike Fighter.

Tom Kaminski

Air Combat Command air bases (active duty)

Above: Two regular ACC wings – the 355th Wing and the 4th FW – include A/OA-10As in their inventories. Here a 355th Wing aircraft practices formation flying with a P-51 Mustang warbird prior to an airshow appearance.

Above: Assigned to the 4th FW, the 23d FG at Pope AFB has two squadrons of A/OA-10As. Both the 74th FS 'Flying Tigers' and the 75th FS 'Sharks' adorn their aircraft with sharkmouths reminiscent of the famous wartime 'Flying Tigers' unit and use an appropriate 'FT' tailcode.

Unmanned aerial vehicles, (UAVs) like the RQ-1B (left) and long-range RQ-4A (below) seem likely to play an increasingly important role in the ACC. Funding for both types is included in the FY02 defence budget.

Right: The sole CONUS-based Air Combat Command KC-135R tankers are those of the 22d ARS 'Mules', assigned to the 366th Wing at Mountain Home AFB, Arizona. The 85th OS at NAS Keflavik, Iceland, also operates a handful of KC-135s, alongside HC-130P Hercules and F-15C/Ds on detachment.

Rescue units in ACC operate a mix of HC-130s and HH-60s. The 71st RQS, 347th RQW at Moody AFB, Georgia, is equipped with HC-130Ps (above), while the 129th RQS at Moffett Federal Airport, California, has HH-60Gs (below) and MC/HC-130Ps.

Above: In recent months the U-2S fleet has been subject to a cockpit upgrade programme in which digital instrumentation was installed.

Below: U-2 training is carried out by the 1st RS at the 9th RW's base at Beale AFB, California. The squadron flies the T-38A Talon alongside the two-seat TU-2S.

Expeditionary Aerospace Force (EAF)
Aerospace Expeditionary Force (AEF)
ACC Unit Alignments – Cycle 2

The second rotation of the EAF got underway in 1 December 2000 and was due to end on the last day of February 2002. Note that in addition to the AEFs, the 366th Wing and 4th Fighter Wing (whose squadrons were not assigned AEF duty) had an 'on call' expeditionary role. Information on wing and squadron assignments for Cycle 3 was not available at the time of writing, though it seemed likely that the Cycle 2 assignments, detailed here, would be repeated.

AEF 1: 1 Dec 2000 – 28 Feb 2001 (Note 1)
Missions: Operation Northern Watch, Counter-drug operations (Caribbean, South America), North Sea Operations (Iceland)
Lead Wing: 388th FW

149th FS	F-16C	ANG
170th FS	F-16C	ANG
176th FS	F-16C	ANG
188th FS	F-16C	ANG
421st FS	F-16C (Note 2)	
79th FS	F-16C (Note 3)	
74th FS	A-10A	
303d FS	A-10A	AFRC
706th FS	A-10A	AFRC
96th BS	B-52H	
11th RS	RQ-1B	
38th RS	RC-135V/W	
99th RS	U-2S	
43d ECS	EC-130H	
71st RQS	HC-130P	
66th RQS	HH-60G	

AEF 2: 1 Dec 2000 – 28 Feb 2001 (Note 1)
Mission: Operation Southern Watch
Lead wing: 7th BW

58th FS	F-15C
9th BS	B-1B
15th RS	RQ-1B
42d ACCS	EC-130E
41st ECS	EC-130H
41st RQS	HH-60G

AEF 3: 1 March 2001 - 31 May 2001 (Note 1)
Missions: Operation Northern Watch, Counter-drug operations (Caribbean, South America), North Sea Operations (Iceland)
Lead wing: 3d Wing (PACAF)

4th FS	F-16C (Note 2)
77th FS	F-16C (Note 3)
75th FS	A-10A
96th BS	B-52H
38th RS	RC-135V/W
99th RS	U-2S
11th RS	RQ-1B
43d ECS	EC-130H
71st RQS	HC-130P
66th RQS	HH-60G

AEF 4: 1 March 2001 - 31 May 2001 (Note 1)
Mission: Operation Southern Watch
Lead wing: 48th FW (USAFE)

60th FS	F-15C	
157th FS	F-16C (Note 3)	ANG
9th BS	B-1B	

15th RS	RQ-1B	
42d ACCS	EC-130E	
41st ECS	EC-130H	
41st RQS	HH-60G	

AEF 5: 1 June 2001 - 31 August 2001 (Note 1)
Missions: Operation Northern Watch, Counter-drug operations (Caribbean, South America), North Sea Operations (Iceland)
Lead wing: 355th Wing

138th FS	F-16C	ANG
163d FS	F-16C	ANG
194th FS	F-16C	ANG
354th FS	A-10A	
23d BS	B-52H	
38th RS	RC-135V/W	
99th RS	U-2S	
11th RS	RQ-1B	
43d ECS	EC-130H	
41st RQS	HH-60G	
71st RQS	HC-130P	

AEF 6: 1 June 2001 - 31 August 2001 (Note 1)
Mission: Operation Southern Watch
Lead wing: 20th FW

27th FS	F-15C	
78th FS	F-16C (Note 3)	
112th FS	F-16C (Note 2)	ANG
124th FS	F-16C (Note 2)	ANG
125th FS	F-16C (Note 2)	ANG
178th FS	F-16A(ADF)	ANG
179th FS	F-16A(ADF)	ANG
184th FS	F-16C	ANG
186th FS	F-16C	ANG
128th BS	B-1B	ANG
127th BS	B-1B	ANG
15th RS	RQ-1B	
42d ACCS	EC-130E	
41st ECS	EC-130H	
301st RQS	HH-60G	AFRC (Note 5)
304th RQS	HH-60G	AFRC (Note 5)
305th RQS	HH-60G	AFRC (Note 5)

AEF 7: 1 Sep – 30 Nov 2001 (Note 1)
Missions: Operation Northern Watch, Counter-drug operations (Caribbean, South America), North Sea Operations (Iceland)
Lead wing: 27th FW

524th FS	F-16C	
103d FS	A-10A	ANG
104th FS	A-10A	ANG
118th FS	A-10A	ANG
131st FS	A-10A	ANG
172d FS	A-10A	ANG
190th FS	A-10A	ANG
20th BS	B-52H	
93d BS	B-52H	AFRC
38th RS	RC-135V/W	
99th RS	U-2S	
11th RS	RQ-1B	
43d ECS	EC-130H	
41st RQS	HH-60G	
71st RQS	HC-130P	

AEF 8: 1 Sep - 30 Nov 2001 (Note 1)
Mission: Operation Southern Watch
Lead wing: 28th BW

94th FS	F-15C
55th FS	F-16C

107th FS	F-16C	ANG
175th FS	F-16C	ANG
174th FS	F-16C	ANG
93d FS	F-16C (Note 4)	AFRC
302d FS	F-16C (Note 4)	AFRC
457th FS	F-16C (Note 4)	AFRC
466th FS	F-16C (Note 4)	AFRC
37th BS	B-1B	
15th RS	RQ-1B	
42d ACCS	EC-130E	
41st ECS	EC-130H	
102d RQS	HH-60G	ANG (Note 5)
129th RQS	HH-60G	ANG (Note 5)
210th RQS	HH-60G	ANG (Note 5)

AEF 9: 1 Dec 2001 – 28 Feb 2002 (Note 1)
Missions: Operation Northern Watch, Counter-drug operations (Caribbean, South America), North Sea Operations (Iceland)
Lead wing: 2d BW

101st FS	F-15A	ANG
110th FS	F-15A	ANG
122d FS	F-15A	ANG
123d FS	F-15A	ANG
159th FS	F-15A	ANG
199th FS	F-15A	ANG
111th FS	F-16C	ANG
113th FS	F-16C (Note 4)	ANG
119th FS	F-16C	ANG
120th FS	F-16C (Note 4)	ANG
121st FS	F-16C (Note 4)	ANG
134th FS	F-16C	ANG
160th FS	F-16C (Note 4)	ANG
20th BS	B-52H	
38th RS	RC-135V/W	
99th RS	U-2S	
11th RS	RQ-1B	
43d ECS	EC-130H	
41st RQS	HH-60G	
71st RQS	HC-130P	

AEF 10: 1 Dec 2001 - 28 Feb 2002 (Note 1)
Mission: Operation Southern Watch
Lead wing: 1st FW

71st FS	F-15C	
34th FS	F-16C (Note 2)	
523d FS	F-16C	
77th BS	B-1B	
15th RS	RQ-1B	
41st ECS	EC-130H	
42d ACCS	EC-130E	
66th RQS	HH-60G	
303d RQS	HC-130P	AFRC

Notes:
1. E-3B/C and E-8C squadrons will deploy as part of each AEF. However, as of 5 September 2000 there was no approved AEF alignment. Additionally, JSTARS has not reached Full Operational Capability (FOC) status and is therefore not capable of steady-state deployment during Cycle 2
2. LANTIRN-equipped F-16CG squadron
3. HARM-equipped F-16CJ squadron
4. Litening II-equipped F-16C squadron capable of delivering precision-guided munitions (PGM)
5. Squadrons will combine assets to fulfil deployment requirement

The venerable F-4 Phantom continues to serve as an aerial target. This QF-4G drone (below) is marked as an aircraft of the 53d WEG at Tyndall AFB, Florida, its 'TD' codes indicating service with the 82d ATRS, a detachment of which also operates from Holloman AFB. The squadron also flies a handful of radar-equipped E-9As (left) for range support. The only other F-4s in USAF 'service' are the Luftwaffe's F-4F training aircraft (right) which make up the 20th FS 'Silver Lobos' at Holloman AFB and carry full USAF markings.

Above: Apart from a small number of aircraft assigned to PACAF, most E-3 AWACS (including this E-3C) are stationed at Tinker AFB, Oklahoma with the 552d ACW.

Below: The E-8C Joint STARS fleet of the 93d ACW at Robins AFB (including 94-284, pictured) is divided between the 12th and 14th ACCS.

Based at Davis-Monthan AFB, Arizona, the 355th Wing operates EC-130E ABCCC command post aircraft (right) and EC-130H Compass Call stand-off jammers (above).

Above: Under the umbrella of the 55th Wing, the 1st ACCS 'First Axe' at Offutt AFB operates four E-4B National Emergency Airborne Command Post aircraft.

Below: The bulk of the 55th Wing's operations from Offutt AFB are those carried out by its RC-135 Elint platforms, which equip two squadrons (the 38th and 45th RS) at Offutt and two overseas-based units. The former each have training aircraft in their inventories; a TC-135W of the 38th RS is pictured.

Below: The 57th Wing at Nellis AFB is responsible for the USAF Air Demonstration Squadron ('The Thunderbirds'), which flies F-16C Block 32s.

The Air Warfare Center includes the 57th Wing at Nellis and the 53d Wing at Eglin. Nellis-based aircraft include F-16Cs (above) and F-15Cs (below) with 'WA' tailcodes.

The Eglin AFB-based 53d Wing operates various types; the 85th TES 'Skulls' flies late-model F-16C/Ds (above) and F-15Cs (below) and F-15Es.

Ängelholm

Scheduled to close in 2002, Ängelholm is home to one of Sweden's two Gripen wings, as well as the basic flying training school, national aerobatic team and Flygvapnet's historic flight.

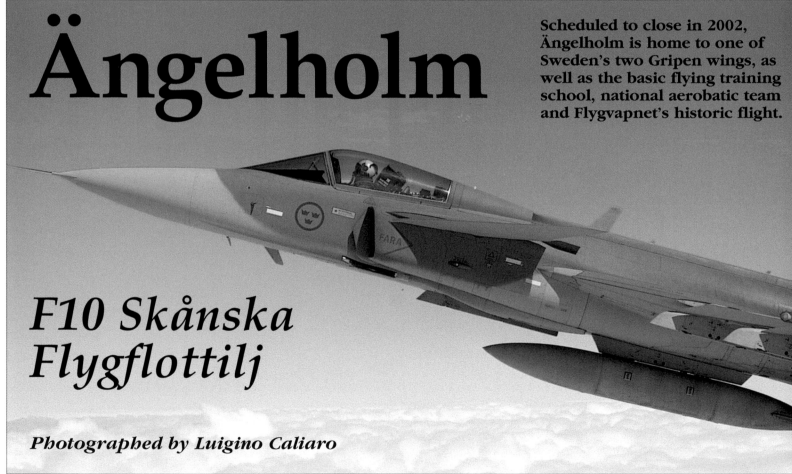

F10 Skånska Flygflottilj

Photographed by Luigino Caliaro

F10 Flygflottilj (Wing), named for the local region (Skåne), was established at Bulltofta, near Malmö, on 28 October 1940 to provide air defence for southern Sweden. With German-occupied Danish territory plainly visible across the narrow Öresund, this task was of utmost importance during World War II. In 1945 F10 moved a few miles north to its present location at Ängelholm-Barkåkra. In the post-war period F10 was the only user of the Saab J 21R jet fighter, which it operated briefly from 1949. The J 29 Tunnen was flown between 1953 and 1964, and the J 35 Draken from 1964 to 1999. AJS 37 Viggens were operated by one squadron from 1993 to 2000.

As part of a major cost-cutting review of Flygvapnet (the Swedish air force), Ängelholm is to close on 31 December 2002, despite having been upgraded at some considerable cost for the operation of Gripens. Its aircraft will be dispersed to other units, the Gripens going to F17 Blekinge Flygflottilj at Ronneby-Kallinge.

Three generations of Saab fighter fly in formation near one of southern Sweden's best-known landmarks – the bridge over the Öresund linking Malmö with Copenhagen in Denmark. F10 operated Drakens for many years, although one squadron briefly converted to Viggens before both of F10's constituent fighter units upgraded to Gripens.

Having traditionally applied the wing number to the forward fuselages of its aircraft, Flygvapnet briefly adopted a new system for the Gripen, using the corresponding letter of the alphabet small on the fin. This practice has been dropped, although one of these F10 Gripens continues to wear the 'G' which signifies a previous assignment to F7 at Såtenäs. A lack of unit insignia simplifies central maintenance and eases the transfer of aircraft between wings. Gripen tailcodes are the last two or three digits of the serial number.

JAS 39A Gripen – 1. Division *'Johan Röd'*

In 2000 1./F10 began its conversion from the AJS 37 Viggen to the JAS 39A Gripen, a process completed by the end of 2001. Sweden's Gripen purchases total 204, comprising 175 single-seaters and 29 two-seaters. Batch 1 consisted of 30 JAS 39A single-seaters, while Batch 2 consists of 96 JAS 39As and 14 JAS 39B two-seaters. Batch 3, to be delivered from 2003, comprises 50 JAS 39C single-seaters and 14 JAS 39D two-seaters.

The force will now equip eight squadrons within four wings (originally scheduled to be 12 squadrons in six wings, F10 and F16 Upplands Flygflottilj at Uppsala being the unlucky units). F7 Skaraborgs Flygflottilj is well established at Såtenäs and includes the TIS 39 training unit which flies the JAS 39Bs. F10's two squadrons at Ängelholm are fully equipped but will move to F17 Blekinge Flygflottilj at Ronneby-Kallinge, displacing JA 37Ds, on the closure of Ängelholm at the end of 2002. F21 Norrbottens Flygflottilj at Luleå-Kallax is due to begin conversion from the JA 37D to JAS 39 in 2002, with F4 Jämtlands Flygflottilj at Östersund-Frösön following. The Batch 3 JAS 39C/Ds, incorporating many improvements such as full-colour cockpits, inflight refuelling capability and new central computers, are slated for delivery to the four squadrons of F4 and F21.

Swedish squadron names/callsigns are derived from the unit's designation. The male name is derived from the wing number (in this case 10) – J being the 10th letter in the alphabet, and 'Johan' being the Swedish phonetic name for J. Other current wings use 'David' (F4), 'Gustav' (F7), 'Petter' (F16), 'Quintus' (F17) and 'Urban' (F21). The colour suffix represents the number of the squadron within the wing – Röd (red) represents 1. Division, Blå (blue) for 2. Division, Gul (yellow) for 3. Division and Svart (black) for a fourth unit (usually designated 5. Division).

JAS 39A Gripen – 2. Division *'Johan Blå'*

Having been the last front-line Swedish operator of the Draken, F10's 2. Division was the third squadron to receive Gripens, following the two units at F7 Såtenäs. The first two JAS 39As for 2./F10 arrived at Ängelholm on 30 September 1999. Gripen is a true multi-role aircraft, and it has a variety of weapons to match. A Mauser BK 27 27-mm cannon is semi-recessed into the port underside of the fuselage, next to the air intake, firing through a shallow trough in the lower port nose. Under the starboard intake is a pylon for podded sensor carriage. For the air defence role Gripen employs Rb 74 (AIM-9L) Sidewinders and Rb 99 (AIM-120B) AMRAAMs, with the IRIS-T expected to replace the Rb 74 and the Meteor to eventually replace the Rb 99. For anti-ship missions the giant Rbs 15F missile is available, while Rb 75 (AGM-65A/B) Mavericks provide precision attack in both anti-ship and air-to-ground roles. The BK 90 (DWS 39) glide munitions dispenser provides long-range attack against area targets. Weapons slated for future Gripen use include laser-guided bombs and the Taurus stand-off missile, while the adoption of a reconnaissance pod will allow the type to fulfil the 'S' (Spanings = reconnaissance) in its JAS 39 designation and take over the role from AJSF 37 Viggens.

J 35J Draken (2. Division)

Ängelholm will forever be associated with the J 29 Tunnen and the J 35 Draken, and examples of both types are maintained at the base in flyable condition by Flygvapnet's historic flight. Shown above is the airworthy J 35J, the last version in Swedish service. Following a trial conversion, 66 J 35F2s were raised to 'Johan' status (the variant suffix was chosen over the logical J 35G to match F10's callsign), primarily to bridge the gap in air defence capability until the JAS 39 Gripen entered service. The J 35Js were redelivered between 1987 and 1991, but in 1993 1. Division began operating AJS 37s, and many of the J 35Js were placed in storage at an underground facility at F9 Säve.

Another consideration in modifying J 35Js for F10 was the desire to keep a Draken unit, and its training group, operational to support foreign users (Austria, Denmark and Finland) until contracts expired in 1995. F10 had trained Draken pilots since 1986, when the TIS 35 (Typ Inflygnings Skede 35 = type conversion unit for J 35) was relocated from F16 Uppsala. The last student went through the course in 1996 and the Draken officially retired on 12 December 1998.

AJS 37 Viggen (1. Division)

In 1993 1./F10 gave up its Drakens to operate AJS 37 Viggens made surplus by the disbandment of F6 and F13. The last of these aircraft in service was given this special scheme which highlighted the squadron's colour (Röd = red) and the historic ghost badge. This is said to date back to the war years, when 1. Division's fighters were always ready on alert – often with engines running – before dawn broke. The motto 'The Show Must Go On' dates from the transition from J 29 to J 35. The 'Ghost' Viggen was finally retired on 11 June 2001.

SK 60 – Krigsflygskolan

In 1998 the Krigsflygskolan (Military Flying School) relocated to Ängelholm from its historic base at F5 Ljungbyhed. The unit uses Saab Sk 60s to provide the GFU (Grundläggande Flygutbildning = basic flying training) course, which lasts for 12 months and encompasses 125 flying hours. The fate of the Krigsflygskolan after the closure of Ängelholm has not yet been announced. The SK 60A (above) is the main Saab 105 variant in use, augmented by the SK 60B light attack platform and SK 60C recce/attack aircraft. The dedicated four-seat SK 60D transport and SK 60E navigation trainer were not re-engined and have been retired. However, the basic two-seat SK 60A can be modified to four-seat transport configuration in about 90 minutes.

Saab SK 60A – *Team 60*

Manned by instructors from the Krigsflygskolan, *Team 60* flies six Saab SK 60As. They are specially marked with the air force crest on the fin, 'Flygvapnet' across the top of the wings and with a blue/yellow sunburst pattern underneath. The team is not full-time, and performs only a limited number of displays within Sweden and occasional forays to other European air shows. The SK 60s carry a smoke-generating pod on the port wing. Around nine aircraft have been given the Team 60 scheme, and are fully employed for day-to-day Krigsflygskolan training purposes. SK 60As in use with the school (and the team) have been re-engined with the RM15 engine (Williams/Rolls-Royce FJ44 turbofan) which offers much better economy and reliability, reduced noise and higher performance with minimal airframe changes and reduced weight.

Beriev Be-12 'Mail'

Developed from Beriev's Be-6 flying-boat, the Be-12 'Mail' was the first Soviet aircraft to be designed from the outset as an anti-submarine platform. Search and rescue, firefighting and transport variants have evolved during the aircraft's 40-year career, and the aircraft remains in both military and governmental service today.

In the mid-1950s, anti-submarine aviation in the USSR was formed as a new branch of the air forces, focused specifically on anti-submarine warfare. The intensive development of the American submarine fleet, including nuclear-powered vessels, was the main cause for concern in the Soviet navy. Another worry was the considerable augmentation of submarine performance by the use of long-range, self-homing torpedoes and nuclear weapons.

During the development of submarine-hunting aviation, the most complicated task was to create the means of underwater detection. The mission equipment of Soviet ASW aircraft differed little from that used in common reconnaissance aircraft. In 1953 the first hydro-radio-acoustic system was developed, which included a set of 18 RGB-H buoys and a SPARU-55 receiver. In 1955 the system was tested on a Be-6 flying-boat and, after successful completion of the tests, it was accepted by the military and put into the service under the designation Baku. The system allowed detection of a diesel submarine travelling between periscope and 50 m (165 ft) depth at a speed of 9 to 11 km/h (5 to 6 kt) at a range of 1.5 to 6 km (0.8 to 3.2 nm) from the buoy. An AMP-56 magnetic detector was designed almost simultaneously with the Baku system.

The Be-6 flying-boat was the first Soviet anti-submarine aircraft modified for installation of this detection equipment and formed the basis of the first anti-submarine regiments of Soviet naval aviation. However, they did not meet all the requirements for submarine hunting. The main problem was that the Be-6 had been designed as a long-range reconnaissance aircraft and could not perform the search and destroy tasks in one mission; it could either search or destroy, but not both. Despite these limitations, the Be-6 successfully performed its role as an interim aircraft.

Under an order of the Central Committee of the Communist Party of the Soviet Union and the Council of Ministers of the USSR, dated 28 March 1956, the design of a new anti-submarine amphibian aircraft was initiated. The aircraft's main purpose was ASW, operating at short to medium range from the coastline. The design was also to include a search and rescue variant. The powerplant for the new aircraft was intended to be the Kuznetsov Design Bureau's NK-4F turboprop.

Beriev had created draft designs for the aircraft long before the official order was issued, under the leadership of chief of the preliminary design

Right: From the first prototype onwards the Initsiativa-2 radar was mounted in a nose radome. The repositioning of the radar restricted the previous 180° forward-hemisphere field-of-view from the navigator's position enjoyed by the first prototype.

Below: With its engines mounted beneath the wings and the radar positioned in the forward fuselage, the first prototype looked markedly different from its successors.

An early production Be-12 commences its take-off run from an unpaved runway during the military trials programme. Note the photo-calibration markings on the forward fuselage beneath the cockpit.

Beriev's Be-12 was the largest amphibious aircraft in the world at the time of its introduction. For operations from water the tail section was positioned to be clear of the jets of water created by the first planing step during take-off. This aircraft was the first converted to the Be-12P firefighting configuration and is seen here during the flight test programme in 1992.

department Aleksey K. Konstantinov. After numerous calculations of the results of wind tunnel testing and water-towing testing at TsAGI (the Central Aero-Hydro Dynamics Institute), as well as results of the testing of a remotely-controlled model of the amphibian, the chief designer approved the Be-12 design, which was given the production designation *izdeliye* 'Ye'. Work on the aircraft proceeded rather slowly, as the OKB was busy with other projects such as the jet-powered Be-10, so a full-scale model of the *izdeliye* 'Ye' was manufactured and submitted only in November 1957.

The choice of turboprop engines instead of jet ones was accepted without question; turboprops were the most effective for an aircraft that needed long range and long endurance at flight speeds up to 550 km/h (297 kt).

Although the Kuznetsov NK-4 powerplant was specified in the project's documents, the AI-20 engine designed by Zaporozhye Engine Design Bureau, headed by A.I. Ivchenko, appeared as a competitor. The engines' characteristics differed little, and they had a common deficiency: they needed plenty of electric power to start. This electricity was not available from onboard storage batteries, thereby necessitating a ground power source. This was unsatisfactory, since the aircraft would lose its autonomy as an amphibian working independently at sea.

It was decided that a small auxiliary power unit (APU) would be designed and installed on the aircraft, both to provide multiple engine starts and to supply power to onboard equipment

Leaving a characteristic trail of black smoke in its wake, a Ukrainian navy Be-12 departs its Saki base in the Crimea for an ASW sortie over the Black Sea in August 1996. The Ukraine maintains a single unit of Be-12s for ASW and SAR work and, with airframe life remaining, is considering an upgrade programme.

while based at sea or at a remote airfield. Specialists from Kuznetsov's and Ivchenko's design bureaux visited the Beriev OKB. As the task's importance was realised, the Zaporozhye Design Bureau allocated an increasing amount of time to the task, resulting in the AI-8 turbine-generator power unit.

At this time, Ilyushin's OKB was testing the AI-20 and NK-4 engines for one of its projects, and the results were presented to Beriev. The AI-20 engines showed greater reliability than the NK-4, and AI-20D engines, uprated to 5,180 hp (3864 kW), were eventually accepted as the powerplant for the Be-12.

Flight testing

Construction of the first prototype was completed on 30 June 1960 with factory tests commencing shortly afterwards. Flight tests began on 18 October 1960, and after early problems had been eliminated, test flights proved that the aircraft's aero/hydrodynamic design, stability and controllability met the requirements.

The Be-12 first prototype had the antenna for the Initsiativa radar located in the hull in front of the first planing step. It was retracted prior to landing and the port was closed with a cover driven by complex-drive kinematics. If the actuator failed landing on water was impossible. Beriev decided to move the radar to the aircraft's nose and install it above the navigator's cockpit windows. From the second prototype, all Be-12s featured this 'duck' nose.

Testing of the first prototype continued for more than a year. However, on 24 November 1961 during a test flight over the Azov Sea, near the city of Zhdanov (now Mariupol, Ukraine), the Be-12 prototype crashed and sank. The news reached Beriev at his office and he left for Zhdanov, going to the hospital where the two

surviving crew members – co-pilot Pankin and radio operator Perebaylov – had been taken. As soon as Beriev opened the door of hospital ward, he heard the co-pilot's voice "Georgiy, it was our fault, the aircraft is not guilty..."

During single-engine performance trials, co-pilot Pankin had shut down the starboard engine and, after completion of interim tasks, began the procedure to restart it. The turboprop reached start-up rpm, but no fuel combustion occurred. Pankin repeatedly tried to start the engine, but without results. Seeing the co-pilot's fruitless efforts, the commander, Bobro, decided to start the engine himself. He transferred controls to Pankin, and pushed the feathering button – of the port (wrong) engine! Bobro initiated a glide, while the starboard engine began autorotation, significantly increasing the drag. The navigator's call for a 'dead-stick' landing was ignored, and Bobro attempted to restart the port engine. The Be-12's speed degraded to the point where the aircraft entered a stall at low altitude. It hit the water hard, broke into two pieces and sank.

Pankin managed to open the hatch cover and get to the surface. Perebailov was already there. Holding on to aircraft parts sticking out from the water, they were in water for some time until picked up by the crew of a small fishing boat. Test pilot Bobro, navigator Antonov and leading designer Petrochenkov were killed in the crash.

The second prototype Be-12 was not built until September 1962 and incorporated refinements based on tests of the first prototype. The state tests of the second prototype were completed without incident by 20 April 1965.

Service history

The first pair of Be-12s was supplied to the 33rd Navy Aviation Training Centre in Nikolayev in the spring of 1964, where crew training took

Flying the 'Chaika'

At higher weights the Be-12 sat characteristically low in the water, as this example undergoing trials shows.

Lt Col Artemyev was the chief test pilot of the flight research department of the 555th Anti-submarine Composite Aviation Regiment based at Ochakov. He flew a number of Be-12s including the aircraft depicted in the 3-view artwork (page 103). Here he recounts his opinions on the flying qualities of the aircraft.

Take-off in Be-12 from a land-based runway, while not very complicated, is quite unpleasant, especially with wind from the starboard side. The clockwise rotating propellers cause a tendency to roll to the right. The subsequent increase in loading on the starboard wheel exacerbates the situation. With flaps deployed, the axis of propeller wash passes below the fins and increases the tendency to swing to the right. Additionally, the large side area of the fuselage works as a sail and the aircraft tries to 'weathercock' into the wind. As a result, in order to maintain direction during take-off with a crosswind from the right it is necessary to use the brakes while fully depressing the rudder and turn the control column wheel to the left, while not raising the tail before a speed of 130-140 km/h (70 to 75.5 kt) has been reached. The combination of inputs was difficult to master, and almost impossible if the pilot was shorter than 170 cm (5 ft 7 in) in height. To compensate, shorter pilots often had to to have a rigid pillow placed under their backs. As a speed of 180 km/h (97 kt) is reached and the pilot prepares for rotation, the ailerons need to be deflected against the wind, although care has to be taken as too much bank may lead to the float touching the ground. The flight restriction for starboard crosswind take-off is set at 8 m/s (18 mph). During take off from an unpaved field it is not recommended to let the aircraft rotate after hitting an undulation, but to continue the run until a take-off speed of 200-220 km/h (108-119 kt), depending on take-off weight and conditions, is reached.

Landing gear retraction causes a tendency to turn to the left accompanied with a slight drop in altitude. This is compensated by reducing the pressure on the left pedal by 25-40 kg (55-88 lb). As more power is applied so is pressure on the left pedal. When the flaps are retracted, the propeller wash axis is raised and its influence on the upper part of the fins decreases. The aircraft becomes more balanced and the left rudder pedal is unloaded accordingly.

In flight the Be-12 is well balanced, and has similar characteristics to other aircraft with similar powerplants, exception being rather heavy control of ailerons. However, the level of noise and vibration are high enough for the crew to suffer discomfort on longer flights.

Descent can be made at various vertical speeds, but rapid descent is not recommended as it causes an unpleasant reaction in the ears, due to the unpressurised cockpit. When the landing gear is released a small but sharp descent is experienced. Deployment of the flaps results in an increase in lift and simultaneously the airstream over the tailplane increases, causing the aircraft not only to gain altitude but also the nose to pitch up. In order to counter this, it is necessary to move the control column forward. Landing approach, after full deployment of the flaps, is conducted at 240-250 km/h (130-135 kt) and touchdown is at 180-190 km/h (97-102.5 kt) . At the moment of impact, the fuel handle-indicator is moved to zero, and the propeller safety pitch control (which is deployed in flight to prevent pitch reduction and 'negative thrust') is de-activated at 145-150 km/h (78-81 kt). To avoid skidding the brakes must be applied smoothly (subsequently anti-skid devices were added).

Taxiing on water with wind speeds of up to 10 m/s (22 mph) is completed by changing the heading by 5°-6 ° using the water-rudder and asymmetric engine settings. On turns and at the crest of waves water spray from beneath the hull can impact the propellers, and cockpit glazing. The turn radii on water are different depending on the nature of the turn, due to the 'same-handedness' of the propellers: left it is larger than right. Taxiing long distances is carried out by 'water gliding' on the first planing step at 120-140 km/h (65-75.5 kt) if the height of a wave does not exceed 0.4-0.5 m (1.3-1.6 ft). The Be-12 lifts onto the first planing step at around 100 km/h (54 kt). At this speed the hydrodynamic lift force 'squeezes' the aircraft onto the surface of water.

Take off from water is usually made into wind. The direction at the start of the run is maintained with the help of the water-rudder. Banking to the right at the beginning of the take-off run is eliminated by full deflection of the ailerons. With the increase of speed there are appreciable impacts of water on the hull, which decrease as the aircraft lifts up onto first planing step. During landing on water the Be-12 is quickly 'seized' by the water.

place. The amphibians were operated from an unpaved airfield of the 555th Anti-submarine Composite Aviation Regiment, Ochakov.

The 318th Separate Long-range Aviation Regiment of the Black Sea Fleet was the first operational unit to receive Be-12s, in July-August 1965 (with unit crew training completed in April 1968). A separate anti-submarine squadron of the Pacific Ocean Fleet received the aircraft in 1967, and the 403rd Separate Anti-submarine Aviation Regiment of the North Fleet in 1968. The 49th Separate Anti-submarine Aviation Squadron of the Baltic Fleet began training in March 1970, the last unit to do so.

Radar problems

As training began, so did the troubles with the Initsiativa radar. The radar waveguides were cracking and bursting, and most of the time the radar was inoperative. Initsiativa blamed the OKB, insisting that the aircraft's design was the cause of the problems. After an investigation, it was revealed that the harsh vibrations caused by operations from unpaved runways was damaging the radar and all blame was removed from the airframe. To solve the problem the support frame for the electronics was reinforced and shock absorbers for the radar were modified.

In early operations, sub-hunting was performed mainly with radar or MAD, as the quantity of radio sonobuoys was limited due to financial constraints. The use of MAD varied with the theatre of operations. It was effective over the deep waters of the Pacific Ocean or the Barents Sea, but in the shallow Baltic Sea – the bottom of which is littered with metal debris from two world wars – the MAD was largely useless.

Once a submarine's approximate location had been established, RGB passive sonobuoys were used to accurately 'fix' a submarine's location course and speed. Once the data had been correlated, AT-1 self-orienting anti-submarine torpe-does were used to attack the submerged vessel.

In service from 1962, the AT-1 torpedo was dropped from an altitude of 200-400 m (656-1,312 ft) and employed two parachutes, one for braking and one for stabilising. After entering the water and reaching its operating depth, the torpedo switched to search mode. Once a submarine had been detected, the torpedo was

Be-12 (*izdeliye* 'Ye')
The first production Be-12 left Plant No. 86 at Taganrog on 12 December 1963, but the type was not officially commissioned into service with the Soviet navy until 29 November 1968. Production ended in June 1973 after 143 examples.

Be-14 (*izdeliye* '2Ye')
Designed to be the primary naval search-and-rescue aircraft for the Soviet military, this was the only Be-14 to be completed. Despite showing considerable ability in the role, the less-radical Be-12PS was ordered due to financial constraints. After successful trials the Be-14 entered operational service with the 49th Separate Anti-submarine Aviation Squadron of the Baltic Fleet.

Be-12 incidents and accidents

Of the 143 production Be-12s nine (or 6.3 per cent) were lost between 1964-1992. This compares favourably with the only contemporary amphibious ASW type, Japan's PS-1, of which six of 21 (28.6 per cent) were lost between 1972-1989.

The first crash of a Be-12 (s/n 6600503) operated by the military occurred on 26 September 1969. The Pacific fleet aircraft, while on approach to a ground bombing range, smashed into the the side of the Avachinsk volcano. At night on 1 October 1970, while taking off from Donuzlav air base in the Crimea, Be-12 (s/n 6600503) of the 318th Separate Long-range Anti-submarine Aviation Regiment crashed after multiple bird strikes. On 3 June 1971, while taxiing at Leonidovo airfield, the unit commander of the 289th Separate Anti-submarine Aviation Regiment failed to turn on the braking system. As a result, the aircraft turned 180° and smashed into a nearby aircraft. Both aircraft (s/ns 9601401 and 0601903) were destroyed by fire. Be-12 (s/n 7600803) of the 318th Separate Long-range Aviation Regiment of the Black Sea Fleet, flooded and sank on 20 July 1972 after hitting a submerged object during a water landing. A probable port engine failure was thought to be the cause of the crash of a 318th Separate Long-range Anti-submarine Aviation Regiment Be-12 on 9 August 1974 during a night ASW exercise. On 4 September 1975, after take-off from Berdyansk air base, the crew of a Voroshilovgrad Higher Air Force Navigators College Be-12 found the left aileron to be jammed, the aircraft fell into a right bank from which it did not recover. On 14 June 1982 Be-12 (s/n 7600805) of the 130th Training Aviation Regiment, was returning to its Zhdanov base after a training flight carrying live ordnance when there was a failure of the right engine air bleed system. Mistakes made by the crew led to both engines being shut down and the aircraft landed hard on water, broke into three parts and sank. The last crash involving a military Be-12 happened on 17 June 1987, when a 318th Separate Long-range Anti-submarine Aviation Regiment Be-12 (s/n 8601202) flipped over during a landing run over water at Donuzlav hydro-airdrome. Apart from the loss of the first prototype, the only other loss was of the first Be-12P (s/n 9601404), which aquaplaned onto the shore after an engine suffered damage from a foreign object (above right). The aircraft was airlifted by Mi-26 to Taganrog (right) but was destined never to fly again.

navigated towards it. At 5-6 m (16-19 ft) from the target, the proximity fuse exploded; the contact-sensitive fuse was actuated if the torpedo hit the target. If the target was lost, the search was repeated. If the submarine was not detected in nine minutes, a self-destruction mechanism was activated. The first use of an AT-1 torpedo from a Be-12 was performed on 14 May 1966 by a crew of the 318th Separate Long-range Aviation Regiment.

In addition, crews were trained in mine-laying and depth charge deployment. The PLAB-250-120 depth charge was equipped with contact-sensitive and hydro-acoustic proximity fuses and contained 61 kg (135 lb) of explosive, enough to damage or destroy a submarine at an effective range of 10 m (33 ft). In a non-conventional war, the Scalp SK-1 nuclear depth charge, with an effective range of up to 800 m (2,624 ft) could be deployed by Be-12SK. It was equipped with a thermostatically-controlled weapons bay, to carry one such bomb and up to 10 radio-acoustic buoys on external attachments. To prevent accidents several secret codes had to be received from Moscow in order to use the weapon.

Egyptian deployment

Changing political situations in the Middle East led to an agreement between the USSR and Egypt for the temporary deployment of six Northern Fleet Tu-16R maritime reconnaissance aircraft under the unit designation 90th Separate Long-range Reconnaissance Squadron of Special Purposes. Two An-12RR radio reconnaissance aircraft soon joined, and on 19 August 1968 three

A dedicated search-and-rescue variant, the Be-12PS dispensed with much of the ASW equipment and weaponry, and the cabin interior was redesigned to hold special rescue equipment, ladders, ropes, life rafts (including a motorised version) and medical supplies. The procedure on arrival at a rescue scene was to land on the water close to the incident and dispense the life rafts (which automatically inflated using compressed air) through an additional hatch positioned on the starboard side of the fuselage (right). Up to 13 survivors could be accommodated in the redesigned cabin (bottom). If the weather conditions prevented landing on water the aircraft could carry survival capsules on four underwing pylons. Outwardly similar to the Be-12, the Be-12PS retained the MAD boom to preserve the aircraft's centre-of-gravity (below).

Be-12 Technical description

The Be-12 is a high-wing monoplane, designed with a seagull-type wing. The aerodynamic quality of the aircraft is compromised by the hydrodynamic demands of the hull. Allowing take-off from water, planing steps, bilges, spray guards and underwing floats all create significant additional air resistance in flight.

The **fuselage** of the aircraft is designed with two planing steps akin to that of a powerboat. The aircraft's construction boasts 71 ribs and a set of stringers covered by a stressed skin 0.8-3 mm (0.03-0.12 in) thick.

Planing steps provide the crucial longitudinal separation of the water jets during take-off, reducing the fuselage/water contact area. The first planing step begins at the wing centre section. The second planing step is in the stern section of the aircraft's hull and is designed to lift the Be-12's tail section from the water during take-off, thus helping the aircraft onto the first planing step. For reduction of shock loads during take-off and landing, while also providing greater rigidity and reliability, the bottom of the hull has a variable deadrise angle (an angle of rise from the keel to the boards). At the first planing step it is 27° and at the bow section of the boat it is around 60°. To deflect water spray downwards, the base of the hull, in the area of the bilges, has negative inclination. Buoyancy is provided by waterproofing the bottom of the hull from the first planing step to a height of 3.3 m (10 ft 10 in).

The hull consists of 10 compartments, eight of which are waterproof. This allows the aircraft to maintain buoyancy and stability in the case of damage to any two compartments. For movement

between the compartments, the partitions have strengthened hatches with waterproofed doors.

There are two **crew entrance doors** on the port side of the front and rear fuselage. The doors, as is the convention on ships, are designed to open inward, facilitating the abandoning of the aircraft in the case of flooding.

The tapered shape **wing design** is based around a two-spar torsion box. The centre-section has 20° dihedral, which in a combination with the high wing, is intended to lift the engines and propellers as far away from the water as possible. The outer section of the wing has 1.5° anhedral. Wing-mounted high-lift devices consist of single-slotted flaps and ailerons, the latter with electrically-controlled trim tabs and balance tabs.

There are eight compartments for **fuel tanks** (in the wing centre-section) and two pressurised tank compartments in the mid-wing torsion box.

Non-retracting single-planing step **support floats** are attached to the outer wing on pylons to ensure the dynamic and static stability of the plane

while afloat. Each float is divided into five waterproof compartments.

The **tail unit** is attached to rib Nos 56 and 60 and comprises a set of surfaces to provide longitudinal and directional stability of the aircraft. The tailplane/elevator, set at an angle of 5.5° dihedral, incorporates vertical endplate fins and rudders. To cope with the reactive moment from the 'same-handed' propellers, the fins are canted 2° to starboard. The placement of the tail unit allows propeller wash to enhance their efficiency. Anchor and 'flood in a compartment' lights are installed on top of the rudders.

The main **landing gear** retracts into wells in the fuselage between rib Nos 22 and 26. The tail wheel retracts into the fuselage aft of the second planing step. Undercarriage legs, struts and other support units of the main undercarriage are constructed from steel, however, much of the tailwheel, due to its proximity to the MAD, is constructed from low-magnetic titanium alloys. All landing gear has hydro-pneumatic-type shock absorption.

Up to s/n 9601504 Be-12s were powered by two Zaporozhye AI-20D series 3 (later series 4) **turboprops**. The engine's turbine works at a constant 12,300 rpm, and only on the ground at idle are they lowered to 10,400 rpm. The constant-speed propellers rotate at 1,075 rpm. To meet the requirements for cold engine starts it was necessary to reduce, as far as possible, the pitch of the propeller blades. If spontaneous transition of the propeller blades to fine pitch in flight occurs, there is significant frontal resistance called 'negative thrust'. After landing, transferral of the blades to autorotation mode causes

creation of the 'negative thrust', which in turn reduces the landing distance.

The aircraft is provided with main and back-up **hydraulic systems**. The main system is driven by two 435F pumps and is used for the extension/retraction of the landing gear, flaps and cargo bay doors, as well as driving the water-rudder and tailwheel. The back-up system is intended for normal and emergency braking and, if necessary, can operate the landing gear, flaps and water rudder using a 465K pump. Working pressure in the system is 150 kg/cm² (2,133 lb/sq in).

The **electrical-supply system** issued for independent start of engines and various onboard equipment and sensors.

Specification

Be-12 (izdeliye 'Ye')
Powerplant: Two 5,180 ehp (3863-ekW) Progress/Zaporozhye AI-20D turboprops
Span: 97 ft 11 in (29.84 m)
Length: 98 ft 9 in (30.11 m)
Height: 26 ft 1 in (7.94 m)
Wing area: 1,066 sq ft (99.00 m²)
Fuel load: 19,842 lb (9000 kg)
Armament: AT-1 (AT-1M torpedoes, PLAB-250-120 and PLAB-50-65 depth charges, AMD-2-500 mines, RGB-N, RGB-NM, RGB-1 and RGB-2 sonobuoys up to a maximum load of 6,614 lb (3,000 kg)
Weights: empty (operating) 54,013 lb (24,500 kg); maximum take-off 79,366-lb (36000 kg)
Performance: maximum level speed 329 mph (530 km/h); range with full payload 932 miles (1500 km); service ceiling 26,246 ft (8000 m)

Above: A pair of Be-12s from the Black Sea Fleet formate over Sebastopol in 1999.

Left: The 20th Independent Aviation Repair Plant is responsible for all Be-12 modifications and repairs.

Above left: The Be-12's cockpit incorporated dual controls with a number of flight instrumentation dials replicated for the co-pilot. This is the Be-12P-200's cockpit and incorporates extra controls for firefighting purposes including water release levers.

Bottom left: The Be-12's tail boom contained the APM-60Ye (later APM-73S) magnetic anomaly detection (MAD) gear, and was magnetically-isolated from the rest of the airframe.

Left: The navigator is positioned in the glazed nose section beneath the radome.

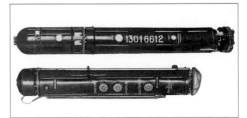

Above: A retractable underwater wing was tested in an attempt to improve water taxiing characteristics, but was not introduced on service aircraft.

Left: Two types of sonobuoys employed by the Be-12 are the RGB-N (top) and more capable RGB-NM (bottom).

Wing tip floats
Each wing tip float is divided into five watertight compartments. At empty weight and even keel, both floats sit just above the water surface. The floats are angled at a 5° bow-high attitude to prevent the front of the float burying itself into the water.

Anti-icing system
The anti-ice system comprises both hot air and electro-thermal devices. Hot air protects the engine intakes, wing leading edges, tailplane and fins. The electrical system de-ices the cockpit, propellers, spinners and pitot head.

Powerplant
The AI-20D turboprops are installed in nacelles and attached to the front spar of the wing. Lateral, front and rear cowlings provide easy access for maintenance, while also serving as platforms for engineers both on the ground and at sea.

Fire prevention
Inert gas is used to fill the empty space in the fuel tanks to exclude a possibility of an explosion due to battle-damage. Fire-prevention equipment also includes a fire-warning system, firewalls installed in the engine nacelles and four extinguishing canisters filled with freon gas.

Wing design
The tapered shape wing design is based around a two-spar torsion box. The wing is of relatively thick profile to aid low-speed flight performance, giving a stalling speed of 200 km/h (124 mph) at a weight of 30.5 ton (31000 kg). The wing centro section has 20° dihedral and the outer wing 1.5° anhedral. Landing lights are positioned on the lower surface of the wing, with navigation lights on the wing tips. A marker bomb container can be installed on the underside of the port wing.

Beriev Be-12 'Mail'
5600302, Bort yellow 25
555th Anti-submarine Composite Aviation Regiment, 33rd Navy Aviation Training Centre, Ochakov, winter 1973-74

MAD boom
The rear hull, tailwheel and empennage are constructed of non-magnetic materials so as not to effect the performance of the MAD gear. APM-60Ye MAD was originally fitted although some aircraft later received the improved APM-70S.

Flight crew
The Be-12's crew of four consisted of two pilots, navigator and radio/weapons operator. The navigator's position in the nose of the aircraft was glazed in the forward lateral and downwards hemisphere giving excellent visibility. The pilots were housed in an unpressurised cockpit with the radio/weapons officer stationed in the middle section of the fuselage. He was provided with an observation blister.

Flight controls
The pilot's are provided with dual steering columns and rudder pedals. A pneumatic system automatically tilts the control columns forward in the case of the pilots abandoning the aircraft in an emergency.

Water rudder
The water rudder, used for directional control, is attached under the rear hull, aft of the second planing step.

Be-12s of the 318th Separate Long-range Aviation Regiment of the Black Sea Fleet became part of the squadron. Egyptian markings were applied to the aircraft and they were ferried to Cairo West air base through Hungary and Yugoslavia. Later, they were relocated closer to the shore, at Mersa-Martrukch air base.

After initial training in the area, the Be-12s participated in joint search exercises conducted in the Mediterranean Sea, for the first time.

In March/April 1970 the group participated in operation Okean (Ocean) the largest military exercise in the history of the Soviet Naval Fleet. The Be-12s conducted extensive ASW search operations together with the *Moskva* and *Leningrad* anti-submarine cruisers/helicopter carriers (carrying 28 Ka-25PL ASW helicopters), 20 ships and 10 submarines. The Be-12s performed 10 flights from Mersa-Martrukch air base and dropped 360 buoys. A target submarine was detected and monitored for 12 minutes, before being transferred to the ship-based helicopters. In June 1971 the Be-12s in Egypt were replaced by Ilyushin Il-38s.

Be-12s operating with other fleets had their share of submarine detections. In 1968, Northern

Fleet aircraft made their first contact shortly after entering into service, and in the same year the Be-12s of the Pacific Ocean Fleet were successful in the Japan Sea. The total Naval Aviation flight hours on ASW operation between 1965-1981 totalled 81,124, of which 37,205 was by Be-12s.

In addition to navy service, the Be-12 was operated by the air force from 1971-1983, initially with the 163rd Training Aviation Regiment and then the 130th Training Aviation regiment of the Voroshilovgrad Higher Air Force Navigators College, for training in ASW techniques.

Declining numbers

The most successful year of ASW operations was 1989, the crews of Be-12s detecting some 29 foreign submarines. However, the 1991 collapse of the USSR triggered a dramatic drop in operations and many Be-12s were withdrawn from use and placed in reserve. By 1992, the 555th Instructor-research Anti-submarine Aviation Regiment and the 316th Separate Anti-submarine Aviation Squadron of the 33rd Combat Training Centre, 403rd Separate Anti-submarine Aviation Regiment of the Northern Fleet aviation, 318th Separate Long-range Anti-submarine Aviation

regiment of Black Sea Fleet aviation, 49th Separate Anti-submarine Aviation Squadron of Baltic Fleet aviation, 289th Separate Anti-submarine Aviation Regiment of Pacific Ocean Fleet aviation and 317th Separate Composite Aviation Regiment of Pacific Ocean Fleet aviation were the surviving Be-12 operators.

Following the break-up of the USSR, four Be-12PSs, based at Kala air base were left in Azerbaijan and three Be-12s of the 316th Separate Anti-submarine Aviation Squadron became the property of Ukraine. During the division of the Black Sea Fleet, nine Be-12s, one Be-12N and one Be-12PS of the 318th Separate Long-range Anti-submarine Aviation Regiment were also transferred to Ukraine.

The number of Be-12s in the Russian navy gradually declined in the 1990s, with their tasks being transferred to the Il-38. In 1993 there were 55 Be-12s in service, and in 1996 little more than 40 remained. By 2001 the 318th Separate Long-range Anti-submarine Regiment of the Black Sea Fleet was the sole remaining operational Russian Be-12 unit and an upgrade programme for the remaining aircraft is under consideration.

Ukraine also continues to operate the type in small numbers and the headquarters of the Naval Forces of Ukraine is considering equipping the youngest airframes with new searching-targeting systems, sonobuoys and weapons.

The Be-12 has also been developed for governmental and civilian purposes within the last decade including the development of the Be-12NKh transport and the Be-12P and 12P-200 firefighting variants.

Aleksandr Zablotskiy and Andrey Salnikov;
additional material from Col Anatoily Artemyev

Be-12P-200
Used as a test and demonstration airframe for Beriev's jet-powered Be-200 firefighting amphibian, the sole Be-12P-200 is a capable firefighter in its own right and has been used for this purpose in Russia. The aircraft has additional air/drainage holes on the forward fuselage aft of the cockpit, expelling the water clear of the engines during aquaplaning runs (above).

Be-12 'Mail' first prototype

Designed to replace the Be-6 flying-boat, the Be-12 emerged as a twin-turboprop, cantilever monoplane with a high-mounted, 'seagull'-type wing and was equipped with two underwing floats and a twin-fin tail.

The Kuznetsov NK-4 engine was specified in the Request for Proposal, but a competitor surfaced immediately -- the AI-20, designed by Zaporozhye Engine Design Bureau headed by A.I. Ivchenko, and it was variants of this engine that were to power all subsequent versions.

The new aircraft exceeded the previous Be-6 in weight and dimensions, and was the largest amphibian aircraft in the world at the time of its first flight. Range was 3300 km (1,782 nm) and its patrol time at 600 km (324 nm) from base was three hours. Maximum mission load was 3 tons, including depth charge bombs, AT-1 (AT-1M) torpedoes, mines, and the RGB-N Iva, RGB-NM Chinara and RGB-1 radar hydro-acoustic buoys. Mission loads varied with the task: the search version of the Be-12 could be loaded with 90 buoys; the attack version was equipped with three AT-1 torpedoes; and the search-attack version could be loaded with 24 RGB-NM buoys and one torpedo. Self-defence armament consisted of a dorsal gun turret.

The Be-12's search and targeting equipment included the Baku system of radio sonobuoys, APM-60E search magnetometer (Magnetic Anomaly Detector, or MAD), Initsiativa-2B radar, ANP-1V-1 automatic navigation instrument and PVU-S Syren-2 targeting computer. The radar antenna extended through a hatch in the hull and was located behind the cockpit, being retracted prior to landing. The stern hull, tail unit and tail undercarriage were made from non-magnetic alloys to provide proper operational conditions for the magnetic detecting equipment.

Completion of the first Be-12 prototype came on 30 June 1960 and it was transferred to the LII's test centre. After all systems were tested and checked on the ground, the maiden flight took place on 18 October 1960. The crew that flew the aircraft from the manufacturer's unpaved runway comprised commander P.P. Bobro, co-pilot V.G. Pankin, navigator V.P. Antonov, and radio operator V.P. Perebailov.

Tests revealed structural defects, some quite serious. The critical flutter speed of the wings was found to be low, so anti-flutter weights were added. A number of structural improvements were incorporated in December, and the first substantial improvement to the seaplane's hull was made in May 1961 when the design of the first planing step was changed. This greatly decreased pitch vibrations, which arose at hydroplaning speeds ranging from 60 to 120 km/h (32 to 64 kt).

A government decision soon halted the tests and the aircraft was ferried to Moscow for the Tushino air show on 9 July 1961. By this time, the Be-12 had become known as 'Chaika' (Seagull). The tests resumed as soon as the aircraft returned to Taganrog. From 19 July 1961, state tests were made with the participation of the Military Representative Committee.

One serious problem that came to light was a result of taking off from choppy water. The propeller blades were suffering damage from water spray – the blade tips were being deformed and eroded. The aircraft was transferred for modifications in September. New splash deflectors 20 cm (7.9 in) wide were installed along the port and starboard sides of the nose.

After more than 12 months of flight tests on 24 November 1961, during a test flight over the Azov Sea near Zhdanov, the Be-12 prototype crashed and sank.

Above: Seen during initial water taxiing tests in the autumn of 1960, the Be-12 prototype lacks the anti-flutter weights, which compensated for a lack of wing stiffness.

With the wing tip anti-flutter weights fitted, the Be-12 prototype was ferried to Moscow in July 1961, completing this flypast at the Tushino air show on 9 July.

Be-12 'Mail' first prototype

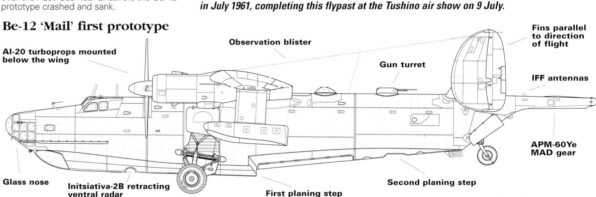

AI-20 turboprops mounted below the wing

Observation blister

Gun turret

Fins parallel to direction of flight

IFF antennas

APM-60Ye MAD gear

Glass nose

Initsiativa-2B retracting ventral radar

First planing step

Second planing step

Be-12 'Mail' second prototype

The second Be-12 amphibious prototype was manufactured at the OKB's experimental plant in September 1962. All the problems discovered during the flight tests of the first aircraft were addressed in the design of the second. The engines were located farther from the water by being set on top of the wings, which, in turn, were freed of anti-flutter weights as their stiffness was increased. Additional splash deflectors were installed on the nose section. The antenna of the Initsiativa radar was relocated to the extreme nose and installed above the navigator's cockpit, a position which was standardised on all following models.

The dorsal gun turret was deleted with the agreement of the military authorities, as the anti-submarine aircraft were supposed to operate beyond enemy fighter reach. The radio-electronics and other equipment were modified and expanded.

During testing of the second prototype improvements were made to increase reliability and serviceability. The tail wheel was controlled by rudder pedals and the electric drive of the windshield wipers was replaced by a more efficient hydraulic model. To trim the propellers' torque reaction, the end-plate fins were canted 2° to starboard.

Following the successful conclusion of the state tests on 20 April 1965, the Be-12 was accepted by the Soviet navy under an Order of the Minister of Defence of the USSR, on 29 November 1968.

Above: The raising of the engines to the overwing position did much to cure the problems of water ingestion during take-off and landing on water. In addition, further splash guards were added, stretching back beyond the wheel wells.

Above right: The nose section was extensively modified, the most notable change being the relocation of the Initsiativa radar from a forward ventral position to a newly-designed radome (with IFF on the upper surface) above the navigator's position. These changes forced a re-design of the nose glazing.

Be-12 'Mail' second prototype

Fins canted 2° to starboard

Entrance door

AI-20 turboprops mounted above wing

Nose glazing modified

Entrance door

IFF antenna

Reprofiled MAD boom

Cargo bay

New main gear doors

Additional splash guards

Initsiativa radar moved to nose position

Be-12 'Mail' (*izdeliye* 'Ye')

Serial production of the Be-12 was organised at the Taganrog Aviation Plant named after Georgiy Dimitrov (now JSC Taganrog Aviation) and lasted from 1963 to 1973. Manufacture of the first serial aircraft (c/n 4600201) was completed on 12 December 1963 and of the last (c/n 3602802) in June 1973. A total of 143 aircraft of all versions was produced.

Improvements were implemented during production to increase tactical performance. From the first production aircraft, all Be-12s were equipped with an 'infloat' refuelling system, located on the lower forward starboard part of the fuselage, in addition to a pressure ground refuelling system. Such a system had previously been tested on a Be-6. Refuelling at sea was performed as the aircraft moved directly behind the tanker. Via a cable passing from aircraft to tanker, the special towing hose was pulled to the aircraft by means of a hoist and then attached to the refuelling unit. After refuelling, the towing hose was detached from the cable and the refuelling unit. Power came from two 5,180-shp (3863 kW) AI-20DK turboprops, replaced halfway through the production run by improved AI-20DM engines.

For operations from unpaved runways, the support frame for the electronic equipment was reinforced and shock absorbers for the radar were introduced.

As the aircraft became operational in military service, tyres began to break during the landing roll. Anti-skid devices were installed in an effort to combat this.

From 1973, Be-12s were equipped with more efficient RGB-NM-1 Zheton sonobuoys. These were able to detect lower frequency sound, which transmits farther in water, thereby increasing the maximum detection range of submarines.

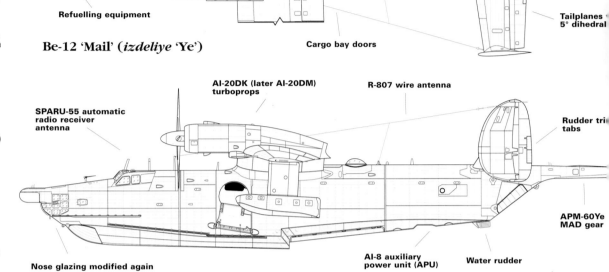

Navigator's escape hatch

Refuelling equipment

ARK-11/ARK-U2 radio compass antenna

SPARU-55 antenna

Fins canted to starboard

Cargo bay doors

Tailplanes 5° dihedral

Be-12 'Mail' (*izdeliye* 'Ye')

SPARU-55 automatic radio receiver antenna

AI-20DK (later AI-20DM) turboprops

R-807 wire antenna

Rudder trim tabs

APM-60Ye MAD gear

Nose glazing modified again

AI-8 auxiliary power unit (APU)

Water rudder

Right: Both the main and tailwheels of the Be-12 are retractable, the latter being largely constructed of titanium due to its proximity to the MAD tail boom. As part of the design effort to lift the engines and propellers as far from the water as possible, the wing centre section has 20° dihedral. The outer section has 1.5° anhedral to help prevent the aircraft entering a sharp turn in the case of a single engine failure. The tailplane is also of sharp dihedral with large endplates canted 2° to starboard to counteract the effects of the 'same-handed' propellers.

Left: The main production version of the 'Mail' was intended primarily for ASW, and it was this role that production 'Ye' aircraft would perform for the majority of their careers. With aircraft such as the Il-38 performing the long-range maritime patrol role, the Be-12s would provide 'inner-ring' coverage, detecting NATO submarines on numerous occasions.

Be-12/M-12 record breaker

A modified production Be-12 'Ye' entered the world record books in some 42 categories under the designation M-12. Allocated the Bort number 02, the aircraft (below) had all unnecessary equipment removed, the engines were tuned to give optimum performance and the crew reduced to two for the record-breaking attempts – consisting for the majority of flights of test pilots M. Mikhailov and Yu. Kupriyanov.

After the record-setting flight programme had been completed the aircraft was returned to standard production configuration.

Be-12SK 'Mail' (*izdeliye* 'YeSK')

This modification allowed the Be-12 to carry SK-1 Skalp nuclear depth charges. The charge, or bomb, was designed to be dropped from altitudes between 2000 and 8000 m (6,562 and 26,247 ft) and exploded at depths of 200 to 400 m (656 to 1,312 ft). If the sea was shallow in the drop area, a delay fuse was installed to let the aircraft get as far as possible from the explosion.

The 'special item' weighed 1600 kg (3,527 lb) and the destruction radius of its underwater explosion was 800 m (2,624 ft). The Be-12SK could carry one such bomb internally in a thermostatically controlled weapons bay, and, externally, up to 10 radio-acoustic buoys for confirmation of received contact. Electric heaters were used to warm the weapons bay.

In 1961 the aircraft was trialled to carry the nuclear anti-submarine weapon. The tests – which did not include a live nuclear explosion – were conducted at the 71st test base of the air force, located in Bagerovo, the Crimea.

An SK-1 Skalp nuclear depth charge rests on a loading cart. This was the principle weapon of the Be-12SK.

Be-14 (izdeliye '2Ye')

From the beginning of the project, the Beriev OKB planned to modify the Be-12 to create a search and rescue version for operations at sea. Work began according to Order No. 28-110 of the Central Committee of the Communist Party of the Soviet Union and the Council of Ministers of the USSR, dated 2 March 1962. The new aircraft, designated Be-14, was intended to find and retrieve the crews of aircraft and ships lost at sea. It was designed to be an all-weather, day/night SAR platform. In 1965 the prototype aircraft was manufactured at the design bureau's plant with the participation of Taganrog Aviation Plant No. 86.

The aircraft could carry up to 29 people when overloaded and up to 15 normally, and was equipped with extra entrance hatches, rafts, floating ropes, life-belts, medicine and other rescue and medical equipment. Extensive modifications to the fuselage were required to accommodate these changes. The load compartment was converted to carry passengers and the deck and bottom hatches were eliminated. The radio operator workstation and blister were moved forward and the APU was relocated to the tail section of the fuselage. The lifeboat container was moved from the starboard fuselage side to

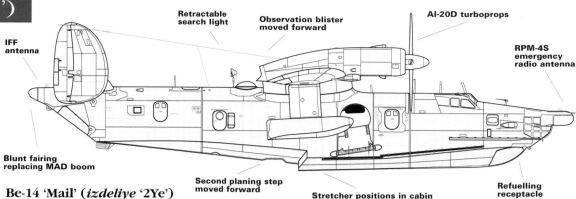

Be-14 'Mail' (izdeliye '2Ye')

Labels: Retractable search light · Observation blister moved forward · AI-20D turboprops · IFF antenna · RPM-4S emergency radio antenna · Blunt fairing replacing MAD boom · Second planing step moved forward · Stretcher positions in cabin · Refuelling receptacle

the cabin floor. Entry doors were included on both port and starboard sides.

The radio electrical equipment included the RPM-S beacon radio receiver, which was linked with airborne radar, ARK-U2 VHF auto direction finder and RSIU-5B VHF radio station. The flight crew increased by two – a doctor and flight technician. To improve operating conditions at night, red lighting of the consoles and instrument panels was

provided. A retractable searchlight was installed behind the dorsal blister.

The Be-14 successfully passed its tests but was not put into serial production due

to economic considerations. The only example manufactured was operated by the 49th Separate Anti-submarine Aviation Squadron of the Baltic Fleet.

With an appropriate Bort (code) number, the sole Be-14 is seen during land and water tests in 1969. The aircraft could carry rescue capsules beneath each wing (above), along with further life rafts stored in the rear cabin. The removal of the MAD boom, the addition of rear cabin doors and various SAR equipment necessitated moving the second hull planing step forward due to a change in the centre-of-gravity.

Be-12PS 'Mail' (izdeliye '3Ye')

Although the SAR-optimised Be-14 was not put in production, the military could not abandon the development of search and rescue aircraft. Following a new requirement, Beriev created a rescue aircraft based on the Be-12 but having more modest specifications than the Be-14. In particular, it could not operate at night.

Despite the fact that it was redundant in the SAR role, the MAD boom and its

associated equipment was retained to prevent redesigning the airframe due to a shift in the centre-of-gravity. The first '3Ye' modification was applied in the summer of 1972 to Be-12 c/n 2602503 at the OKB's facilities. This involved removing special equipment and armament, and fitting the necessary medical and rescue apparatus, a motor lifeboat, a mechanised ladder to bring casualties onboard through the

special starboard hatch, and other gear. The Be-12PS (Poiskovo-Spasatel'nyi, or search and rescue) could carry up to 13 people.

In 1972 the aircraft underwent manufacturer's tests and was built at the Taganrog plant in a small production run. The first serial Be-12PS aircraft (c/n 3602901) was accepted on 20 April 1972,

and the last (c/n 3603002) was taken from the assembly shop on 25 November 1973.

Ten aircraft were produced, in addition to two (c/n 2602503 and c/n 2602603) that had been modified earlier at the experimental plant. Two (c/n 0601905 and c/n 3602801) were modified as rescue aircraft by operational units, using modification kits supplied by Taganrog.

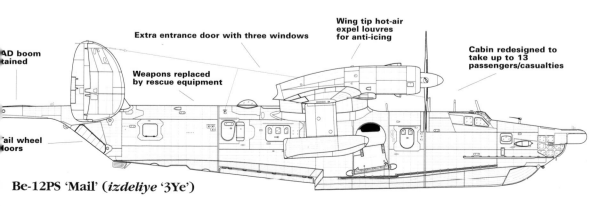

Be-12PS 'Mail' (izdeliye '3Ye')

Labels: Extra entrance door with three windows · Wing tip hot-air expel louvres for anti-icing · Cabin redesigned to take up to 13 passengers/casualties · Weapons replaced by rescue equipment · MAD boom retained · Tail wheel doors

Although only 10 Be-12PSs were built as new by Beriev at Taganrog, two examples were modified from existing ASW airframes by Beriev OKB and a further two modified by the Soviet navy. An additional rescue/entry door with three small windows was added beneath the trailing edge of the wing as seen on Bort 32 (above). Converted examples varied in their fit. Seen during sea trials, this Be-12PS (above left) has only one window in the additional doors giving away its status as one of the modified aircraft. Note the waterborne refuelling receptacle aft of the navigator's station.

Be-12 infra-red testbeds

In 1968/69 there was an attempt to fit Be-12s with new equipment for submarine searching. Two aircraft (c/n 8601102 and c/n 8601202) received the Gagara-1 (Loon) imaging infra-red system. The infra-red searching station was supposed to detect the thermal signature of a submarine's wake.

To install the Gagara-1 station, the ARK-U2 radio direction finder was removed and the AGD-1 gyrorate sensor was replaced by the CGV-10 vertical gyroscope.

A special hatch was installed behind the load compartment for the station receiver head. In flight, the equipment was operated by the navigator or a supplementary operator, whose workstation was installed in the navigator's compartment. The aircraft were modified at a branch of the Yevpatoria repair plant located in Saki, the Crimea.

The first modified aircraft was supplied to the 403rd Separate Anti-submarine Aviation Regiment for military tests in the north. Be-12s equipped with the Gagara-1

station later flew in the Far East, over the Black Sea, and even around the Mediterranean Sea while based in Egypt in 1970.

Unfortunately, test results were poor. The system's ability to detect submarines was influenced by sea temperature. It could track a submarine in the north, but in the Black Sea the results were ambiguous, and in the Mediterranean the only thing the equipment detected clearly was crossing the coast. For these reasons, Gagara-1 was not accepted by the military, but work on imaging infra-red equipment continued.

In 1983, a further attempt to apply infra-

red imaging technology to the submarine and surface ship detection role was initiated. One aircraft (Bort 30) was fitted with the Nablyudatel-1 new-generation infra-red imager prototype, and was designed to detect the wakes of ships and submarines travelling on or under the surface. The station was located in the cargo compartment and trials of the new equipment were conducted at Taganrog, Gelendzhik and Feodosiya. The system was able to detect ships satisfactorily, but returned the same unclear results for submerged submarines as had the Gagara-1 equipment.

Be-12N 'Mail' (*izdeliye* 'YeN')

Above: The Ukrainian navy inherited 14 Be-12s, Be-12PSs and Be-12Ns following the break-up of the Soviet Union. This example is seen taxiing-out at its Saki base in the Crimea in the late 1990s.

Above right: Be-12s and Be-12Ns were regular 'customers' for Scandinavian maritime patrol and fighter aircraft during the Cold War. This example was photographed by the Swedish air force.

The Be-12N, 'N' for Nartsiss, (narcissus) entered service in April 1976. It was equipped with a search-targeting system that included AMP-73S search magnetometer (MAD), modified Initsiativa-2BN radar, Nara all-channel standardised receiver, SPARU-55 airborne receiver, Nartsiss targeting computer with target analyser and PK-025 reception and target designation equipment. The AMP-73S search magnetometer and Nartsiss targeting computer were tested on Be-12s c/n 6600602 and c/n 7600901, respectively. In addition to RGB-NM and RGB-NM-1 sonobuoys, the aircraft were equipped with 10 RGB-2 passive oriented sonobuoys as a part of a Berkut complex, hinged on the outer supports. These buoys provided the bearing of the submarine.

Some 27 aircraft were brought to Be-12N standard at the Yevpatoria aircraft repair plant while being overhauled.

Be-12 Ice research laboratory

Given the nickname 'Polar Bear', this modified Be-12 was used extensively by Beriev and the air force to test anti-ice systems for both the Be-12 and other types. The airframe was converted from the 13th production Be-12 and tested a

range of anti-ice equipment and ice accumulation sensors in the mid-late 1960s and early 1970s. The cabin was also rearranged, with non-essential equipment removed to house the project scientists and their associated equipment.

Above: The Be-12's undercarriage rotated vertically upwards to be housed in watertight wheel wells beneath the leading edge of the wing. The Be-12's distinctive 'gull-shaped' wing is readily apparent on this pair of Northern Fleet examples departing its base for an ASW patrol.

Be-12 inflight refuelling laboratory

In the 1970s tests were conducted to prove the suitability of the Be-12 to receive inflight refuelling, using a probe-and-drogue system. A single production Be-12 'Ye' was equipped with a dummy refuelling probe installed over the navigator's cabin, and an

Antonov An-12, equipped with a retractable 50-m (164-ft) hose and drogue, simulated the tanker. The tests were successfully conducted by the Flight Research Institute (LII) at Zhukovskiy, but the system was never incorporated into service aircraft.

In deference to its nickname, the single ice research Be-12 (c/n 6600501) had this amusing polar bear artwork applied on the nose section beneath the cockpit. Although retaining the MAD gear and associated systems, other ASW equipment was removed. The aircraft is seen here at Beriev's Taganrog plant during the test programme.

Be-12 export

In 1981 four anti-submarine aircraft were exported to Vietnam (c/n 9601701, 0601801-0601803). Their airframes, engines and equipment were modified to operate in the tropics. The aircraft originally were shipped from Odessa to Cam Ranh Bay. It is believed that these aircraft are no longer in airworthy condition.

Be-12LL 'Moskit' flying laboratory

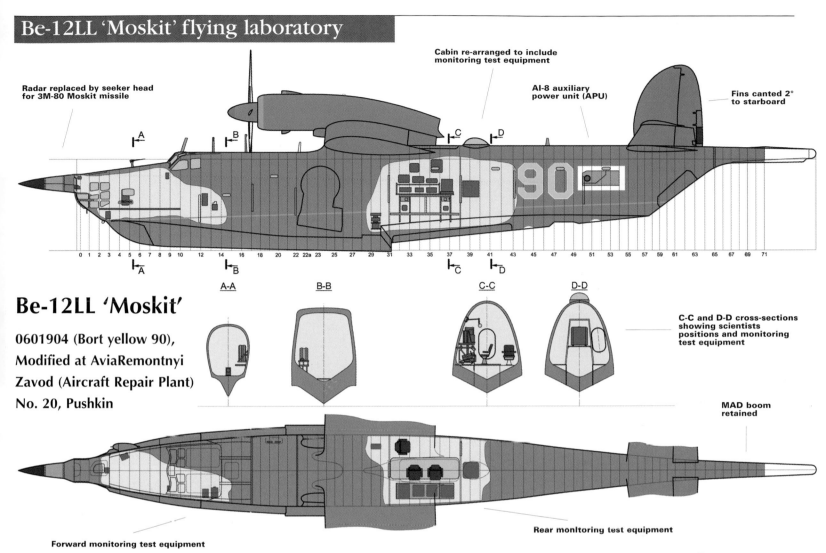

Radar replaced by seeker head for 3M-80 Moskit missile

Cabin re-arranged to include monitoring test equipment

AI-8 auxiliary power unit (APU)

Fins canted 2° to starboard

Be-12LL 'Moskit'

0601904 (Bort yellow 90),
Modified at AviaRemontnyi
Zavod (Aircraft Repair Plant)
No. 20, Pushkin

A-A B-B C-C D-D

C-C and D-D cross-sections showing scientists positions and monitoring test equipment

MAD boom retained

Forward monitoring test equipment

Rear monitoring test equipment

Be-12 constructor's number 0601904 was modified as a testbed at the 20th Aircraft Repair Plant in Pushkin to test the homing head of the 3M-80 Moskit (SS-N-22 'Sunburn') anti-shipping missile. The designation 3M-80 apparently refers to the Mach 3 speed of the 1980 weapons. The head unit of the missile was installed above the navigator's position, replacing the Initsiativa radar, and giving the aircraft its distinctive nose profile.

All other equipment and the workstations of the operators were arranged in the re-equipped load compartment.

Below: By the late 1980s the Be-12 'Moskit' was in a derelict state at the 20th Aircraft Repair Plant's airfield. A number of the aircraft's components (including the engines and rudders) have been removed as spares for other Be-12s.

Above: By August 2000 the 'Moskit' was still extant, although is destined never to fly again. After the trials programme the missile was introduced into service on 'Sovremennyy'-class destroyers.

Be-12P (*izdeliye* 'YeP')

In the early 1990s the Beriev Aircraft Company received from the navy a number of Be-12s whose length of service exceeded navy limits but still had airframe life remaining. One aircraft (c/n 9601404, bort 40 yellow) was modified to become a Be-12P (P, Pozharnyi, or fire-fighter) water bomber. Sub-hunting mission equipment was removed, although the Initsiativa radar remained. Water tanks with a 6-m³ (212-cu ft) total capacity were installed in the load compartment and behind the cockpit, and a water tank drain system, water scooping system and water dropping system were fitted, as was a system for water filling on the ground. The aircraft also received equipment required to operate in Russian civil air traffic zones.

The Be-12P can fill its tanks with water in 25 to 30 seconds while aquaplaning and can operate efficiently 60 to 70 km (37 to 43 miles) from a water reservoir. Aircraft

The second Be-12P, RA-00041, taxies from the water at Taganrog. The aircraft carries a 'Taganrog seal' badge on the nose, celebrating 300 years of the city.

performance has not changed. The Be-12P can operate from airfields with Class II classification as well as from water in a sea state up to 3.

Three additional amphibians were later brought to this standard (c/n 8601004, 0601704, 2602505) and registered as RA-00041, RA-00049 and RA-00073. The fire-fighting system of the first (RA-00041) differs from that of the other two.

The conversions were financed by Irkutsk regional authorities and by the federal forestry service of Russia, which ordered the modification.

The maiden flight of the Be-12P (c/n 9601404) was made on 27 April 1992 from the factory airfield in Taganrog, and the first fire-fighting mission was completed in July 1992 near Migulinskaya Stanitsa.

In seven years of operations, these aircraft have extinguished over 140 forest fires and prevented over 350. Some 2,798 water drops were made on fires, with the total weight of dropped water about 17,188 tons. The record output by a Be-12P aircraft is 132 tons of water dropped on a fire on a single sortie. In 11 weeks in 1997, one aircraft made 242 water drops comprising 1,452 tons of water.

There has been only one accident involving a Be-12P (c/n 9601404), which occurred on 14 July 1992. The aircraft was scooping water from the Don river at Veshenskaya Stanitsa when the engine was hit by a foreign object, and the aircraft aquaplaned to shore (no crew members were injured). The aircraft was later transferred to Taganrog by Mi-26 helicopter on an external cable, and, due to the extent of the damage, written off.

Based on operational tests, a number of adjustments and modifications were made to the aircraft. A new intercom was installed (SPGS-1 instead of SPU-7), the water scoop device and pipelines were reinforced, the flap retracting-extending system was modified, and rear-view mirrors were installed on RA-00041, so pilots could monitor the filling of the water tanks.

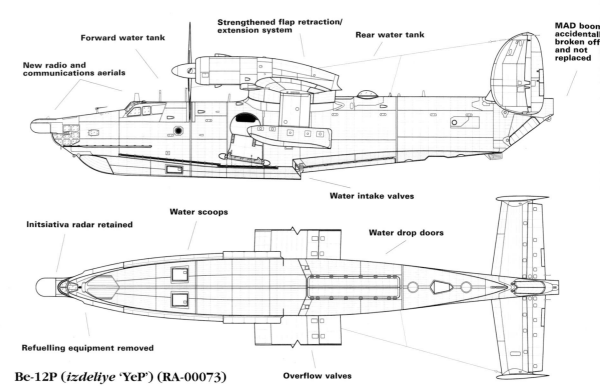
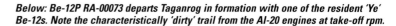

Be-12P (*izdeliye* 'YeP') (RA-00073)

Below: The only one of the four Be-12Ps to be finished in a military scheme, c/n 9601404 was painted as Bort yellow 40. The aircraft was written off shortly after entering service after hitting a foreign object during waterborne operations.

Below: Be-12P RA-00073 departs Taganrog in formation with one of the resident 'Ye' Be-12s. Note the characteristically 'dirty' trail from the AI-20 engines at take-off rpm.

Be-12NKh

Simultaneous with the construction of the first Be-12P, two serial amphibians (c/n 9601403 and c/n 9601702) were modified to the Be-12NKh (NarodnoKhozyaistvennyi) civilian passenger transport version. The passenger compartment was arranged in the re-equipped weapons compartment.

Be-12NKh aircraft were operated over the Kuril Islands (in the Sea of Okhotsk, near Japan) by SAKTOAR, the Sakhalin aviation transport company. During 1992/93 regular flights were performed both by Beriev crews and SAKTOAR crews on routes from Sakhalin to Kounashir Islands, incorporating a water landing on South Kuril Bay after which the aircraft taxied to shore.

These amphibians made 260 flights lasting 628 hours, and carried 2,055 passengers and 401.5 tons of cargo. Sixty-eight flights were made searching for fish runs, and 20 for ecological monitoring and free economic zone inspection. Another 60 flights came under a programme of training civilian aviation personnel to operate the Be-12NKhs. For the first five days after an

Two of the three Be-12NKhs were destroyed at Kounashir Island within a few months of each other during supply operations to the island. The aircraft seen here (c/n 9601403, Bort 65) was damaged on 30 March 1993 and began to take on water in South Kuril Bay (below). The aircraft would almost certainly have been repairable but, as the aircraft was towed from the water, the salvage crew inadvertently failed to raise the landing gear, which sank into the mud. The increasing strain ripped the tail section from the airframe (right) destroying the aircraft, which was subsequently scrapped.

earthquake in October 1993, the only transport able to deliver cargo to Kounashir Island was a Be-12NKh operated by a crew from Taganrog.

Unfortunately, these aircraft did not serve long: both crashed on South Kuril Bay, Kounashir Island, one on 30 March 1993 and the other on 30 October 1993.

Another amphibian modified to the Be-12NKh version (c/n 9601505, bort 96) is kept at the aircraft parking area of the aviation plant air base in Taganrog.

The sole surviving Be-12NKh remains in airworthy condition at Taganrog. Note the additional passenger cabin windows.

Be-12 EKO

Revealed in 1991, the Be-12 EKO was an unbuilt ecological reconnaissance and research variant, its design coinciding with the resurgence of the Be-12 in the early 1990s, as the firefighting variant entered service. It was intended that the aircraft would be fitted with specialised water monitoring equipment and scientist work stations for the monitoring of water quality. Alighting onto remote rivers or lakes the aircraft would take water samples for onboard (or later laboratory) analysis.

Be-12P-200

In creating the new multi-purpose Be-200 amphibious aircraft – whose basic configuration is intended for fire-fighting – the Beriev OKB faced a number of technical decisions. Of particular importance was the design of a special water-scooping system, which would place minimum strain on pipes during water scooping while aquaplaning and would fill the tanks quickly, yet present the least stress to the aircraft's hull. It would have low fuel consumption and stabilise the service life of the engine and airframe. The amount of time needed to fill the tanks by water scooping had to be lowered to between 12 and 20 seconds.

This system was tested on another modification of the Be-12 – the Be-12P-200 testbed (c/n 8601301, Bort 46, later registered as RA-00046). Beriev made the modifications between August 1994 and June 1996. All the special anti-submarine equipment was removed, including the Initsiativa-2B radar, Syron 2 targeting computer, APM 60 MAD, SPARU 55 receiver, PP-1 receiver-processor unit, and SPO-10 radar illumination warning station. Part of the radio electronic equipment was replaced.

As a testbed, the aircraft was equipped with the same fire extinguishing system as that used in the Be-200. The Be-12P-200 has two water tanks with a total capacity of 6.0 m³ (212 cu ft). They are divided into two equal sections and provided with doors for water dropping, with drain lines routed to port and starboard. Water scooping is performed at an aquaplaning speed that is 90 to 97 per cent that required for take-off, from any body of water – sea, lake or river. The tanks can also be filled with water from a ground source at base.

Tests of the Be-12P-200 were conducted between August and October 1996 at Taganrog Bay and Gelendzhik Bay on the Black Sea. During testing, 37 modes of water scooping and water dropping were examined. After that, during the 1997-1998 summer fire seasons, the aircraft successfully operated against forest fires in the Irkutsk and Khabarovsk regions. The aircraft has been demonstrated for potential B-200 customers at a number of international aviation trade exhibitions and air shows, such as Gelendzhik-96, MAKS-97, Gelendzhik-98, the Berlin Air Bridge event of 1998, MAKS-99, Gidroaviasalon-2000 and MAKS-2001.

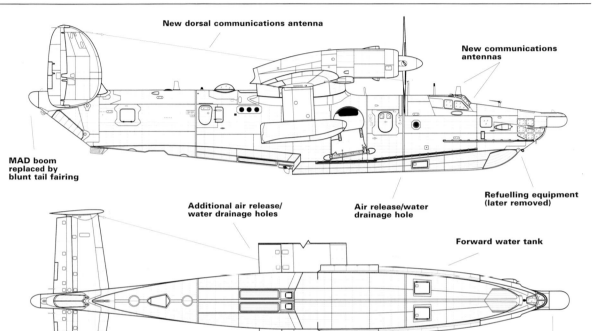

New dorsal communications antenna

New communications antennas

MAD boom replaced by blunt tail fairing

Refuelling equipment (later removed)

Additional air release/ water drainage holes

Air release/water drainage hole

Forward water tank

Additional water intake/release equipment (as on Be-200)

Water scoops

Initsiativa radar removed, but radome retained

Be-12P-200 (RA-00046)

Used as a testbed for the Be-200 firefighting amphibian, RA-00046 (c/n 8601301) incorporates a new rapid water scoop system with air release/water drainage holes. The aircraft retains the nose radome and navigator's glazing and has a more prominent observation blister on the starboard side (above left), but has dispensed with the MAD boom and its associated equipment. Operated by a four-man crew, the Be-12-200 is cleared for operations from water (above) of a minimum depth of 2 m (6 ft 7 in). For release of water or fire retardant (left) normal procedure is to have the trailing-edge flaps deployed, reducing airspeed and increasing accuracy of the drop. Deploying the flaps also gives a characteristic 'nose down' flight attitude as the load is released.

'Dragonships'

USAF gunships in Vietnam

Part One: Gunship I – the FC/AC-47D Spooky

In 2001 the fixed-wing gunship platform continues to be an important type in the USAF Air Force Special Operations Command (AFSOC) inventory. The most modern of these aircraft are the Lockheed Martin AC-130U Hercules of the 4th SOS, a unit that can trace its origins to the first operational application of the concept in Vietnam in the mid-1960s. Then the platform was Gunship I – a conversion of the venerable C-47 transport, designated AC-47D and named Spooky.

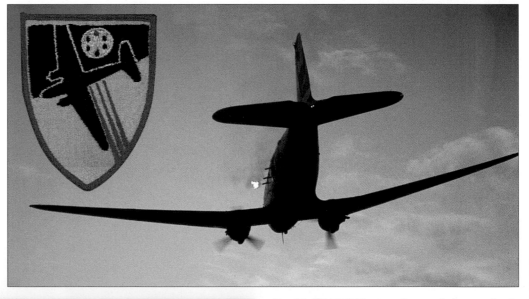

Top: AC-47D 44-76593 returns from a mission over South Vietnam in late 1967. Tail codes were adopted by gunship units from mid-1967 – 'EN' was that of the 4th ACS, 14th ACW. Note the webbing straps in the aircraft's cargo door, used as a restraint by the flare handler when the aircraft was heavily banked during an attack.

Above and left: A 'Dragonship breathes fire' from a 7.62-mm Minigun, during an attack in 1966. Note the 12° angle of the gun; initially SUU-11s were mounted 'flat', necessitating a 30-33° bank by the aircraft.

Modern fixed-wing side-firing airborne weapons systems, commonly referred to as 'gunships', are generally credited in their conception to USAF Captain Ron Terry. Captain Terry's contribution actually came some 35 years after the original concept was tested, for it was he who developed the idea into an operational weapons system. The concept of a side-firing weapon goes back to

C-131B 53-7820 was the test bed for the first installation of the SUU-11A/A gun pod firing laterally from an aircraft. Capt Ron Terry test flew the aircraft from Eglin AFB, Florida, in August 1964, with some success.

about 1927, when US Army Air Corps pilot First Lieutenant Fred Nelson mounted a single 0.3-in machine-gun in a de Havilland DH.4 trainer and fixed it to fire 90° to the forward line of flight. However, despite conducting several successful test firings Nelson was unable to interest the Army Air Corps in his idea.

Enter Army Lt Col G. C. McDonald, a weapons specialist with the fledgling Army Air Force. During the early part of World War II, McDonald proposed installing in an aircraft a single 0.5-in machine-gun, and later a 2.5-in Bazooka, mounted so that the weapon fired laterally, i.e., to the side, rather than forward or aft as in normal gun installations. He proposed using the side-firing aircraft against ground troops, armoured vehicles and submarines.

'Pylon turns'

McDonald's proposal was developed around the aircraft flying what is known as a tight 'pylon turn'. The pilot put the aircraft into a tight turn to the side on which the weapon was mounted, usually the left side, holding it steady while lining up the target using the wingtip. This brought the weapon to bear on the target, and the pilot simply pressed the trigger. Although McDonald's idea proved itself in subsequent testing, the Army Air Force once again showed scant interest. World War II was winding down and there seemed little need for such a novel proposal.

Although a side-firing weapon system – such as multiple machine-guns in a C-47 – would have been ideal against North Korean and Chinese troop concentrations during the

Korean War, the Air Force appears never even to have considered it. Ten years later, the USAF found itself embroiled in another war, this time in Vietnam.

The Army of the Republic of Vietnam (ARVN) had troops defending villages, hamlets and cities throughout South Vietnam. These troops were being trained and 'advised' by US Army Special Forces personnel. Air support initially came from Vietnamese Air Force (VNAF) A-1 fighter-bombers and Douglas B-26B light bombers – often flown by USAF 'advisors'.

One of the biggest problems faced by air force pilots, be they USAF or VNAF, was that

often the Viet Cong (VC) would be in such close proximity to the defending ARVN forces that pilots could not bring their weapons to bear without fear of hitting friendly troops. And when USAF jet fighters entered combat in mid-1964, the problem of close air support was compounded due to the jets' much higher speeds over the target areas.

In late 1961, the Air Force held a conference at Eglin AFB, Florida, in the hope that it could come up with ideas that could alleviate the close air support problem being encountered in South Vietnam. Attending the conference was (now) Air Force Reserve Lt Col G. C. McDonald,

Project Tailchaser/Gunship I

The first Tailchaser C-47 was Air Systems Division C-47D, 43-48462 (above). Named *Terry & The Pirates*, the aircraft was fitted with a single 7.62-mm Gatling-type machine-gun (far right) by Capt Terry and his 'pirates' and tested at Eglin during September 1964. Tailchaser aircraft initially trialled a modified Mk 20 Mod 4 unit gunsight of the type fitted to the A-1E Skyraider. The FC-47 installation is pictured (right).

FC-47D 43-48579

The aircraft for which the 'Puff, the Magic Dragon' nickname was first coined was 43-48579 (right), one of the trio of original FC-47Ds operated by the 1st ACS. Built during World War II as a C-47B-5-DK, the aircraft was later modified to C-47D standard (its superchargers having been removed). Before conversion to FC-47D standard, 43-48579 was assigned to the 1st ACS as a mail courier aircraft in early 1964 (below). By late 1965 *Puff* had been resprayed in SEA camouflage (below right), over which its nose art was reapplied.

and Mr Ralph Flexman, an assistant chief engineer with Bell Aerosystems.

On 13 December 1961, Flexman and McDonald met for the first time, and discussed the problems. The first thing McDonald did was bring up his old proposal for the side-firing weapon system. On 27 December 1961, following his discussion with McDonald, and using McDonald's old test results from World War II as an example, Flexman wrote a proposal that eventually evolved into a family of laterally-firing aircraft weapons systems.

Flexman pointed out that the pilot of a forward-firing aircraft could completely lose sight of a target between initial sighting and firing his weapons, then lose the target again

during pull-out and turn-around for a possible second pass. However, the pilot of an aircraft with a side-firing system could line up its weapons on the target while maintaining the pylon turn, and never lose sight of the target while continuously bringing fire to bear.

Flexman found an ally in USAF Captain John Simons, who was working in Air Force Systems Division (ASD). However, as with all previous laterally-firing proposals, the Pentagon rejected the idea.

Simons and Flexman persisted, and resubmitted the proposal in May 1963, this time to the Flight Test Section of ASD. Going beyond his immediate superiors' authority, Simons obtained approval for initial tests, and Project

Tailchaser was borne. Simons flew some of the early tests himself, using a North American T-28D Trojan and a Convair C-131 twin-engined cargo aircraft. With grease pencil markings on the canopy acting as a rudimentary gun sight, and flying a tight pylon turn, Simons held the target within the marks with extreme ease. Once again, a lack of funding held up further development of the idea for another full year. By the time funding was reacquired, Simons was forced to give up Project Tailchaser for other duties.

This is when Captain Ron Terry became involved. Terry was a fighter pilot who had recently returned from South Vietnam, where he had been studying the many problems of a 'limited war', the weapons systems that were being used, and which weapons he felt would be needed. Captain Terry took over Project Tailchaser in the summer of 1964, just prior to the Tonkin Gulf Incident, which put the United States squarely in the middle of the war in Southeast Asia.

Project Gunship I

Captain Terry drew up a proposal based on his recent experience in South Vietnam, which called for a laterally-firing weapon system to be used in the defence of the small hamlet or village, as found in that country. He forwarded his proposal to the newly organised Limited War Office within ASD, which promptly examined and approved it.

Flight tests were approved by ASD beginning in August 1964, using a modified cargo aircraft as the prototype. Captain Terry took a standard Convair C-131B (serial 53-7820) to Eglin AFB

7.62-mm Miniguns are shown installed in one of the first FC-47Ds at Bien Hoa in 1965. Two of the guns were fitted in the modified rear window openings with a third in the open cargo door, so as to clear the trailing edge of the wing. Later installations had all three Miniguns installed in window openings.

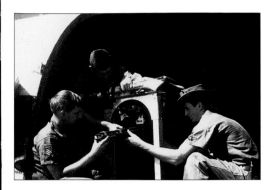

Above: Gunship I team members (from left) A1Cs Ron Snyder and James Schmeisser and A3C Allan Sims assemble an SUU-11A/A pod in FC-47D 43-48471 at Tan Son Nhut in 1965. The SUU-11 pods were shipped to South Vietnam in pieces and reassembled before being installed in the FC-47Ds. Most of the early aircraft had only two guns until production caught up with demand.

Right: The Gunship I test team are seen installing one of the SUU-11 pods in FC-47D 43-48471 at Tan Son Nhut in the autumn of 1965. The first FC-47Ds were assigned to the 1st Air Commando Squadron until the arrival of the 4th ACS in November 1965. Note the yellow colour in the national insignia, suggesting that the aircraft may have been short-listed for transfer to the VNAF.

and began installation of the weapons system. The weapon of choice was the new General Electric GAU-2/A 7.62-mm Gatling gun. However, the only weapons available were those mounted in SUU-11A/A pods, designed for carriage under the wings of standard aircraft like the Douglas A-1 Skyraider.

An SUU-11A/A pod was procured and installed in the cabin of the C-131B, the cargo door having been removed. The SUU-11A/A was an electrically-driven, six-barrelled Gatling gun that fired the 7.62-mm projectile. Its two-speed motor varied the firing rate at either 3,000 or 6,000 rounds per minute (rpm). A makeshift sight, of the type fitted to the A-1, was installed on the left side of the cockpit.

Initial test results astounded even Captain Terry. Firing from various altitudes between 500 and 3,000 ft (152 and 914 m), with a slant range of up to 9,000 ft (2743 m), a one-second burst of fire riddled a 10-ft (3-m) rubber raft with 25 hits! A second test was flown against a 50-ft (15-m) raft floating in the Gulf of Mexico – 75 hits, again with only a single burst of fire lasting only one second. Setting up target mannequins on the Eglin AFB land range gave even more astonishing results: 25 mannequins were set in positions over an area of approximately one acre (4047 m²), and in three seconds, the C-131 gun crew hit 19 of them, with 10 being considered 'kill shots'.

ASD promptly took over complete control of the project to give it full support and funding, and Project Gunship I was borne. With the war in South Vietnam suddenly going 'hot', the project was given a 'Top Priority' classification. In September 1964, Captain Terry flew more tests, this time using a C-47D aircraft. A single SUU-11A/A gun pod was installed in the open cargo door area of C-47D 43-48462. On the nose of the aircraft was the inscription *Terry &*

This FC-47D at Sewart AFB, Tennessee, in 1965 was assigned to Training Detachment 8 of the 1st Air Commando Wing, the gunship training unit. Det 8 became the 4th Air Commando Squadron on 15 August 1965. Note the aircraft has only two guns due to SUU-11 production and delivery problems.

The Pirates, an appropriate name since almost every part of the weapons system had been 'pirated' from other units at Eglin.

Throughout September and early October, tests of the C-47 gunship prototype were flown. The results were the same as they had been with the C-131 – astounding. As the gun mounts were improved (vibration caused the firing arc to jump around until the mounts were sufficiently buffered), and the sighting was stabilised and perfected, the killing zone became devastatingly deadly. Each SUU-11 gun pod had a capacity of 1,500 rounds on internal storage before it needed to be reloaded. A one-minute pass with the single gun-equipped C-47D could put one bullet in every square foot of a US football field, some 22,500 sq ft (2090 m²).

Testing in combat

ASD then authorised combat testing of the new weapons system, and a designation for the new aircraft – FC-47D. The FC-47D was also standardised with not one, but three SUU-11A/A gun pods per aircraft; one in the open door of the C-47, and two mounted in modified window openings just forward of the open door. All three guns were bore-sighted to fire just aft of the trailing edge of the wingtip. The gun sight chosen for the FC-47D was the

FC-47D 43-48471 rests on the ramp at Bien Hoa in 1965, armed with 10 0.3-in M2 machine-guns prior to SUU-11 installation later that year.

proven Mk 20 Mod 4 sight as used in the Douglas A-1 Skyraider. The trigger button, also borrowed from the A-1, was mounted on the C-47 pilot's control wheel.

Combat tests, of course, would be conducted in the environment for which the weapon system was designed – the jungle of South Vietnam. In October 1964, Captain Terry arrived at Bien Hoa AB, home of the 1st Air Commando Squadron, with his Gunship I Test Team – First Lieutenant Edwin Sasaki, Master Sergeant Tom Ritter, Staff Sergeant Estelle Bunch, First Lieutenant Ralph Kimberlin, Airman First Class James Schmeisser and

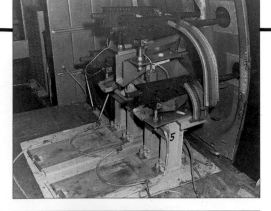

The '0.30-caliber kit'

Captain Ron Terry designed several 'kits' for installation of the M2 0.3-in machine-guns in the FC-47D aircraft. The four-gun mount (right) was fired through ports cut in the rear half of the cargo door. Another four-gun mount was installed in the other open doorway, and two guns were installed in each of the last two windows on the left side. Three hundred M2 machine-guns awaiting disposal in a warehouse at McClellan AFB were soon requisitioned by Captain Terry for his gunship programme. *Get-Em Bullit*, one of the three FC-47Ds armed with 10 0.3-in machine-guns and assigned to the 1st ACS, was pictured on the PSP ramp at Bien Hoa AB in 1965.

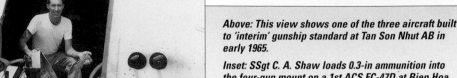

Above: This view shows one of the three aircraft built to 'interim' gunship standard at Tan Son Nhut AB in early 1965.

Inset: SSgt C. A. Shaw loads 0.3-in ammunition into the four-gun mount on a 1st ACS FC-47D at Bien Hoa AB in 1965; each mount carried 20,000 rounds – 5,000 rounds per gun.

Airman Third Class Allan Sims. Terry immediately obtained the flight line maintenance records of the C-47Ds delivering mail to troops throughout Southeast Asia, and picked out three of the lowest-time aircraft in the theatre (43-48471, 43-48491 and 43-48579).

Captain Terry obtained authorisation to pull them from their normal mail run duties, and had them brought up to the latest modification standards. The engines were checked and tuned, the fuel systems were checked, the flight control systems were tightened up, and all hydraulic pumps were replaced and reservoirs were topped with fluid. Anything broken or worn was replaced with a new part, using Project Gunship I as the authorisation. None of the maintenance crews was informed why these three C-47Ds were getting the royal treat-

ment; they were just told to make them the best C-47Ds in the fleet.

On 2 December 1964, Captain Terry arrived back at Bien Hoa, bringing with him enough SUU-11A/A gun pods, mounts and Mk 20 Mod 0 gunsights to build three FC-47D aircraft. He began his modifications on the lowest-time C-47D of the group (43-48471). The folding metal seats were removed from the interior, except for five on the forward right side of the cargo compartment. Three aluminium gun mounts were installed in the left rear area of the cargo bay – two facing the last two windows, and a third in the open cargo door.

FC-47D modifications

The forward area of the cargo bay was modified to hold 24,000 rounds of 7.62-mm ammu-

nition, on belts and in cans similar to those used with the M60 machine-gun used by the infantry. A large box was installed opposite the open cargo door, to hold 24 Mk 6 200,000-candlepower flares. Additional radio equipment included both a VHF and UHF radio, an FM command radio so that the FC-47D could communicate with friendly ground troops, and the normal TACAN and IFF transponders. The FC-47D carried a crew of seven – pilot, co-pilot, navigator, three gun mechanics, and a South Vietnamese interpreter.

Captain Terry and his Gunship I team, along with armourers and mechanics from the 1st ACS, finished the first FC-47D on 11 December 1964. The second FC-47D was finished four days later. The SUU-11A/A pods installed in the first two aircraft had suffered so many failures during early tests at Bien Hoa that the guns slated for the third aircraft were pressed into use with the first two FC-47Ds.

Captain Terry then put together the combat flight test teams, using the crews who had formerly been flying mail in the C-47s. Captain Jack Harvey and Captain Lee Johnson were the pilots of the two FC-47Ds at Bien Hoa. Both aircraft were still assigned to the 1st Air Commando Squadron, so the 1st ACS became the first FC-47D operational unit, even though the aircraft technically were under the control

This seemingly 'new' AC-47D with the 4th ACS at Sewart AFB in late 1965 sports the new Southeast Asia camouflage, but with a 'daylight' light grey underside. The 4th ACS deployed to Southeast Asia on 14 November 1965.

Sitting on the ramp at Tan Son Nhut AB in late 1965, this 4th ACS AC-47D wears a coat of the new 'daylight' SEA camouflage. At this time AC-47s flew daylight missions as convoy escorts, as well as fire support sorties.

of Project Gunship I and Captain Ron Terry.

Captain Terry briefed Captains Harvey and Johnson about laterally-firing weapons and the problems associated with computing the firing arc and slant ranges, how to operate the sight and what to expect. Captain Harvey was elated with the concept, stating that he was tired of being fired on and not being able to fire back. On 15 December 1964, Captain Harvey fired the weapons in anger for the first time, during a daylight mission looking for any targets of opportunity.

The first results were apparent on a 21 December mission. An O-1 Bird Dog called on Captain Terry and his FC-47 to fire on a large building into which he had observed 14 Viet Cong enter. Captain Terry put the FC-47D into a 20° bank to the left, lined up on the building, and pressed the trigger. Using only one gun, the FC-47D left the building full of holes – "It looked like a sieve!" – and 21 bodies were counted inside.

Captain Terry's test team then began flying night missions. The VC used the night to their advantage, attacking hamlets and outposts throughout South Vietnam with relative impunity. Most US and VNAF air power was of the clear air, daylight type, rather then the all-weather or night interdiction variety. The C-47D, flying low and slow, was equally adept at flying in daylight or at night. C-47B psywar aircraft, callsign GABBY and known throughout the theatre as 'Bullshit Bombers', had been flying night missions since 1961.

Puff, the Magic Dragon

On the night of 23/24 December 1964, an airborne FC-47D was sent to Thanh Yend, a small outpost west of Can Tho in the Mekong Delta. The outpost was under heavy attack from the VC and was defended by ARVN troops with US Army Special Forces advisors. Captain Harvey contacted the troops on the ground, commonly known as 'TICs' or 'Troops In Contact', and asked them what they wanted. Not knowing the FC-47D carried guns, they called for flares. Captain Harvey dropped 17 flares, which made the entire situation clearer.

The gunsight used in the FC/AC-47D gunships was a modified version of the Mk 20 Mod 4 unit used in the A-1 Skyraider and tested in the Project Tailchaser aircraft. As both the C-47 and A-1 were Douglas products, their electrical systems used the same voltage, easing the adaptation of the sight for use in the Spooky. The Mk 20 Mod 4 sight was readily available as the Navy had begun phasing the A-1 out of service.

Above: One of the production mounts for the SUU-11A/A pods is seen here, as mounted in the open rear doorway of an FC-47D at Bien Hoa AB in 1965.

Right: This interior view of a 4th ACS FC-47D shows the original 'flat' gun pod installation, angled at 12° downwards from 1966. Ballistic cloth covers the cabin windows to the right, to prevent AA fire from hitting the ammunition supply. The flare box behind the guns, held 48 Mk 24 flares.

This 'natural metal' 4th ACS AC-47D, seen on the ramp at Tan Son Nhut in 1965, seems ready for a coat of camouflage paint. C-47s modified for gunship operations were drawn from low-time airframes; this aircraft had been used as a VIP transport prior to conversion.

attacking VC; they were soon seen to retreat.

The VC didn't know what it was. Nor did the ARVN troops or South Vietnamese civilians, but it was loud. A reporter from *Stars And Stripes* had been within the walls of Trung Hung and witnessed the effects. He reported that the sight of the tracer and the growl of the Gatling gun gave the impression to the Vietnamese, both defender and VC, of a large dragon sweeping the area clear with its breath. Upon reading the account of the action, the commander of the 1st ACS noted jokingly, "Well, I'll be damned! *Puff, The Magic Dragon!*" (a reference to a children's song popularised in the US at the time by the singing trio Peter, Paul, and Mary).

Wing requested

The press soon picked up on the name and began referring to the FC-47Ds as 'Puff' or 'dragonship' whenever they operated. The crew of FC-47D 43-48579 painted the name and a dragon outline on the nose of its aircraft.

Forward air controllers, both airborne and on the ground, began using the term so often that 2nd Air Division authorised PUFF as the call-sign for all FC-47 missions.

So impressed were the commanders in the field throughout South Vietnam with the success of the FC-47Ds (there still were only two aircraft in the theatre) that they demanded that Air Force deploy an entire wing of the dragonships, some 75 aircraft, to the theatre. The Air Force did not have enough aircraft for even a flight of three. On 18 June 1965, Pacific Air Force Headquarters made an official request for a full squadron of 16 FC-47Ds.

ASD met the request by putting the FC-47D into production. Air International, based at Miami Airport, would be the primary contractor to modify 20 C-47D aircraft into FC-47D gunships, bringing all the C-47D airframes and flight systems up to full accordance with the latest Air Force TCTOs (Time Compliance Tech Orders), and building and installing the mounts for the Gatling guns. Wiring harnesses were standardised and installed.

It was the intention that the SUU-11A/A pods used in the first Gunship I aircraft be replaced with GAU-2B/A gun 'modules'. However, General Electric was having difficulty just keeping up with demand for the SUU-11A pods, much less making new modules, so the first

Left: Another 4th ACS AC-47D, this time on the ramp at Nha Trang in late 1965, shows the production version of the gun installation. Note the smoke extractor extending from one of the forward windows. The 4th ACS was the only gunship squadron in Southeast Asia until the 7th Air Force activated the 14th ACS on 25 October 1967, using assets taken from the 4th ACS.

Upon locating the main thrust of the VC attack, Captain Harvey asked the defenders if they wanted him to fire.

Although somewhat puzzled, the Army advisors said, "Yes!" Captain Harvey obliged and the 7.62-mm Gatling guns growled. Using first one gun, then another, he expended 4,500 rounds on Thanh Yend and the VC attack ceased. With the

A pair of 14th ACS AC-47Ds sits on the ramp at Nha Trang in October 1967. Tail codes were not applied to gunships until 1968, by which time the 14th ACS had become the 3rd ACS. Note the replacement left elevator on the near aircraft, possibly the result of repairs to damage from anti-aircraft ground fire.

attack on Thanh Yend abruptly stopped, Captain Harvey went back on station over the Mekong Delta. Quickly, another cry for help came over the radio, this time from Trung Hung, 20 miles (32 km) west of Thanh Yend.

The defenders of Trung Hung had been under attack all night. A VNAF C-47 had already dropped over 70 flares to illuminate the area for VNAF B-26B Invaders and US Army helicopter gunships. Captain Harvey arrived over the outpost, dropped seven flares to locate the attacking VC, and the guns growled again. Again, using just one gun at a time, Captain Harvey fired another 4,500 rounds against the

Top: An AC-47D drops four 2 million-candlepower Mk 24 flares over Saigon in March 1967. Forty-eight flares were normally carried inside the AC-47's rear cargo door.

Above: An AC-47D opens fire with all three SUU-11 Minigun pods on a target in the Mekong Delta, tracer rounds forming a cone of light in this time-lapse photograph. Tracers hitting the ground have set the elephant grass alight.

Above: A 4th ACS AC-47D opens fire on a target near Saigon during the early dusk hours of March 1967. Normal operations called for the pilot to use just one gun at a time as the ammunition magazine held only 1,500 rounds, or about a 15-second burst, before having to be reloaded.

Below: C-47 psywar 'Bullshit Bomber' 42-92111 of the 5th ACS was photographed at Da Nang in December 1966. Vietnamese personnel used the large speaker to broadcast to the enemy on the ground, flying in the same orbit as the AC-47 gunship. If the enemy fired upon the psywar aircraft, the 'dragonship' would return fire.

aircraft delivered to the new gunship unit still had the SUU-11A pods. Even they were in short demand due to their increased use by Air Force, Army and Navy aircraft in the Southeast Asia theatre.

Captain Terry sought to relieve the problem by installing an alternative weapon in place of the 7.62-mm Gatling gun, now called a 'Minigun' in deference to the M61 20-mm rotary cannon. His proposal centred around the 0.3-in and 0.5-in machine-guns, because of their availability: the 0.5-in gun had been the main armament of US aircraft since World War II, and the M2 0.3-in was the primary ground weapon of all US ground forces. Captain Terry found a supply of these weapons at McClellan AFB, California, in the late spring of 1965.

Using the Gunship I authorisation, Terry took control of the 300 0.3-in machine-guns he found at McClellan, and he and the test team designed an installation which put 10 of the old weapons in the cargo bay of a C-47. Small holes were cut in the cargo doors and six of the 0.3-in guns protruded through; two more were mounted in each of the last two window openings, for a total of 10 0.3-in guns per FC-47 kit.

Terry had the Gunship I team build six FC-47Ds using the 0.3-in kit at Eglin. Three aircraft would be assigned to the new unit forming at Forbes AFB, Kansas, and the rest would stay at Eglin for tests. Terry's team then returned to Bien Hoa, and installed the 0.3-in kit in four more C-47Ds assigned to the 1st ACS. That made a total of six 'dragonships' operational in South Vietnam with the 1st ACS – two Minigun-equipped aircraft and four with the 10 0.3-in

gun installation. The breakdown rate of both the SUU-11A/A pods and the much older and worn out 0.3-in M2 machine-guns meant that all six were rarely operational at the same time.

Once the aircraft were introduced into combat, very few modifications were made to the basic aircraft weapons system. As SUU-11A/A pods became available in Southeast Asia, the 0.3-in machine-gun-armed aircraft were re-armed with Miniguns. Later, when the GAU-2/A gun module became available from General Electric, the SUU-11A/A pods were replaced. The GAU-2B/A module used the same components as the older SUU-11, but rather than being in a pod originally designed for underwing carriage, it was designed from the start to be mounted inside the gunships. The flares were replaced with Mk 24 flares which gave 2 million candlepower.

Cost-effective modifications

The cost of modifying the standard C-47D cargo aircraft to FC-47D dragonship specifications totalled US$4,288,975 for all 20 aircraft modified by Air International and equipped with the GAU-2/A weapons. Other changes to the FC-47D included ballistic cloth over the windows on the left side to stop small arms fire

from penetrating the aircraft and striking the crew or, worse, the ammunition and flares; racks to hold the maximum ammunition load of 21,000 rounds; new wiring for the gun systems and gun sight; and the new radio and navigation equipment.

One major change came in September 1965. When the first gunships were in operation, the Air Force designated them FC-47Ds, implying a 'fighter' aircraft, albeit one derived from a transport type. An outcry went up from fighter pilots throughout the Air Force. Captain Jack Harvey, pilot of the first FC-47D in Vietnam, said, "You could hear the fighter pilots' teeth grinding at 100 yards whenever someone mentioned the FC-47. A rusty old 'Gooney Bird' was not a fighter aircraft, no matter how many guns it carried!" In September 1965, the Air Force redesignated the aircraft AC-47D.

In the field, the armourers and pilots worked on perfecting the system. Eventually, the guns were declined 12°, allowing the pilot to decrease the angle of the bank of the aircraft, making it easier for him to 'roll in' onto the target; it also stabilised the aircraft, and permitted easier handling of both the ammunition boxes and flares. The decreased slant range brought an increase in aircraft altitude, which

Above: This 4th ACS AC-47D on the ramp at Bien Hoa in early 1966 has the daylight camouflage applied after the aircraft was painted at Clark AB. Production AC-47Ds were modified by Air International at Miami Airport and then flown to the combat arena in Southeast Asia. Some were painted at Clark before going on to Vietnam.

Below: An armed AC-47D with C Flight, 14th ACS is seen on the alert aircraft ramp at Bien Hoa in 1967. 7th Air Force activated the 14th Air Commando Squadron at Nha Trang on 25 October 1967, with aircraft and crews taken from the 4th ACS. The 14th ACS became the 3rd ACS on 1 May 1968.

Below: Seen on the ramp at Binh Thuy AB in 1968 is a 4th Special Operations Squadron AC-47D. The 4th ACS had been redesignated the 4th SOS on 1 August 1968. By then its normal mission was night fire support/interdiction and its aircraft had been resprayed with black undersides.

reduced the danger to the crew from small arms fire, and raised the impact velocity of the 7.62-mm bullet. Night operations were enhanced in early 1966 when gunship pilots began using a 'starlight scope', or Night Observation Device, to spot targets.

In 1967, the new GE MXU-470/A Minigun module became available, marking a leap in reliability for the 'dragonship'. It featured electric ammunition loading from a vertical drum that held 500 more rounds than the old SUU-11 or GAU-2 system. The electric loading all but eliminated gun jamming, which could be traced to manual loading of the belted ammunition.

The unit was smaller in size and easier to work on during flight. The new modules were rushed into production and were available for combat operations in the late summer of 1967. Several modules were sent to Vietnam and installed in three gunships from C Flight, 4th ACS, at Phu Cat. Other upgrades included addition of a new ceramic, armour-plated flare holder, with an integral water fire extinguisher system.

Before the end of operations in Southeast Asia, Air International and USAF crews at Clark AB, Philippines, and at in-country bases such as Tan Son Nhut, had built at least 53 FC/AC-47D gunships. This does not take into account the aircraft modified in-country to AC-47D standard using spare parts or parts from scrapped airframes, nor those built for nations like Cambodia and Thailand as 'interim' gunships, using non-standard armament such as multiple 0.5-in machine-guns.

Deployment and operations

As soon as the Air Force made the commitment for an entire unit to be equipped with gunships, a training school was formed to train pilots in laterally-firing weapons systems, slant ranges, night operations and inflight maintenance of the weapons system. On 9 August 1965, Training Detachment 8 of the 1st Air Commando Wing was established at Forbes AFB, Kansas, equipped with four FC-47Ds – one with SUU-11A/A Minigun pods, and three armed with the 10 0.3-in machine-gun kit. Six days later, on 15 August, as part of Operation Big Shoot, Det 8 was redesignated as the 4th Air Commando Squadron.

The training at Forbes AFB continued until late October 1965, when the new unit was alerted for combat operations in the Republic of Vietnam. The first of 20 FC-47Ds arrived at Tan Son Nhut AB on 14 November, where they were placed under operational control of the 6250th Combat Support Group in the 2nd Air Division. They arrived in-country without weapons, as the SUU-11A/A Minigun pods were shipped separately to save weight during the long overwater deployment of the FC-47Ds. The aluminium mounts were already installed in the aircraft, and the guns arrived within days aboard C-124 Globemaster cargo aircraft.

All missions flown during 1965 were flown with a maximum of only two guns, and many with only one gun. The official mission of the 4th ACS, according to 7th AF Operations Order 411-65, was "To respond with flares and firepower in support of hamlets under night attack, supplement strike aircraft in the defence of friendly forces, and provide long endurance escort for convoys."

The 4th ACS took control of the half-dozen FC-47Ds that had been operating in South Vietnam and had been assigned temporarily to the 1st ACS. This should have brought aircraft strength to 26 aircraft – 16 combat aircraft, four spares, and the six ex-1st ACS aircraft that were already in-country. However, between the lack of guns delivered, and a new mission with which 7th Air Force tasked the 4th ACS (i.e. out-country missions) the actual in-country inventory was reduced to about 18 aircraft.

Based at Tan Son Nhut AB, the 4th ACS deployed three-ship flights to bases in each of the four Military Regions in South Vietnam. In I Corps region bordering the Demilitarized Zone were the three AC-47Ds of A Flight at Da

An E Flight, 4th SOS AC-47D rests on the ramp at Binh Thuy AB in 1966. Note the white ballistic cloth covering the aircraft's forward windows adjacent to the Spooky's 24,000-round ammunition supply. The numerous antennas fitted were associated with equipment for communications with airborne and ground air controllers. Whip antennas atop the forward fuselage were for communication with ground troops.

Nang AB; B Flight was at Pleiku AB in II Corps; C Flight was at Tan Son Nhut with 4th ACS Headquarters in III Corps; and D Flight was based at Binh Thuy AB in the Mekong Delta region of IV Corps. In May 1966, Headquarters 4th ACS moved from Tan Son Nhut AB to Nha Trang AB, where the squadron fell under operational control of the new 14th Air Commando Wing, which had been activated at Nha Trang on 8 March 1966.

The fifth flight was assigned to Operation Cricket, and had an out-country mission, meaning its aircraft would operate outside South Vietnam. Accordingly, during the second week of December 1965, the 7th AF ordered the 4th ACS to deploy four AC-47Ds to Udorn AB, Thailand. Their mission was armed reconnaissance of the Steel Tiger area of Laos, and interdiction of the Ho Chi Minh Trail, which wound through the jungle mountains of Laos from the Mu Gia Pass into Cambodia and South Vietnam.

Crews operating over Laos found that their mission was much more difficult than those in the relatively safe airspace over South Vietnam. The missions to the Steel Tiger area were mostly during daylight, and the slow-flying AC-47D was an easy target for the North Vietnamese and Khmer Rouge gunners on the Plaine des Jarres. The AC-47Ds flying Cricket missions faced heavy anti-aircraft artillery, both 37-mm and 57-mm, some of which was radar-directed. Plus, the weather and mountainous terrain combined to make the night mission very hazardous.

The result was predictable. An AC-47D, 43-49492, was shot down on 17 December 1965 with the loss of all crew members. Another,

Above: The GE MXU-470 module was a vast improvement over the older SUU-11A/A pods installed in make-shift mounts in the gunships. The MXU-470 had adjustable mounts, allowing various angles of declination, had a 2,000-round magazine, and was reloaded electrically. The latter feature eliminated the jamming problems experienced with the earlier pods.

Right: A late production AC-47D assigned to D Flight, 4th ACS at Pleiku AB, in December 1967. Newer aircraft had another 'window' cut into the aft fuselage forward of the cargo door opening. It was here that the third MXU-470 Minigun module was installed. The 4th ACS used an 'EN' tail code from late 1967.

45-1120, went down over Laos on 24 December 1965, again with the loss of all hands. By August 1966, the missions had cost a total of six AC-47Ds, four over Laos, and 7th AF ordered the detachment to withdraw. The remaining aircraft returned to the 4th ACS at Nha Trang and were redistributed to the flights in the four South Vietnam Military Regions.

Callsigns

A word here about callsigns. The first aircraft operational in South Vietnam with the 1st ACS used the callsign PUFF, as explained previously. When the 4th ACS deployed to South Vietnam in November 1965, 7th AF wanted to know what callsign the unit was going to use. At 4th ACS Headquarters, the discussion went on between the pilots, many of whom were ex-fighter pilots. One pilot exclaimed, "What! Give that damn spooky Gooney Bird a tactical callsign? I'll kiss your ass!" The 7th AF officer on the other end of the phone line heard the words, but the emphasis in the voice had been on the word "spooky". The 7th AF officer replied, "OK, SPOOKY it is!" At least, that's how they tell it at the reunions.

Throughout 1966, both the in-country and out-country missions continued at an advanced pace. Replacement aircraft were being built rapidly and were matching the attrition count, but the pace of constant night operations with a 40-year-old aircraft began to tell. The crews and aircraft were becoming tired, and on many

nights a full complement of gunships was not available. Compounding matters, 7th AF authorised the resumption of the out-country missions which had been stopped in August 1966.

On 1 November 1966, 7th AF assigned eight new AC-47Ds to the 606th Air Commando Squadron, a composite squadron at Nakhon Phanom AB on the Thai/Laos border. The 606th ACS, 'Lucky Tigers', flew a variety of aircraft including North American AT-28D WATER PUMP attack aircraft, Douglas A-1E Skyraiders, U-10B GABBY Couriers, C-123B CANDLE flareships, HH/CH-3 'Jolly Green Giant' helicopters, and the AC-47D gunships. When the eight new AC-47Ds were operational with the 'Lucky Tigers', it raised AC-47D numbers in Thailand to approximately 10.

The increased combat in South Vietnam brought a cry from 7th AF to the Pentagon for even more gunships. The Department of Defense responded with the creation of another squadron during the summer of 1967, but it came at the cost of reduced AC-47D assets within the veteran 4th ACS. On 25 October 1967, 7th AF activated the 14th Air Commando Squadron at Nha Trang AB, RVN. Its strength was drawn from aircraft and crews within the 4th ACS, when 7th AF reduced its assets from 22 AC-47Ds to 16. Those six aircraft, plus 10 brand-new AC-47Ds, made up the 16 gunships allocated to the 14th ACS.

Like the 4th ACS, the 14th ACS was made up

This 3rd SOS AC-47D crashed at Da Nang following a mid-air collision with an OV-1 FAC aircraft on 13 December 1968. The 14th ACS became the 3rd ACS (later SOS) on 1 May 1968. This aircraft, AC-47D 43-49274, was one of the original 1st ACS FC-47Ds armed with 10 0.30-calibre guns, and served with the 4th and 14th ACSs, before being written off following this crash landing at Da Nang.

Special Operations units. Thus, the 14th ACW became the 14th SOW, and the 3rd and 4th ACSs became the 3rd and 4th SOSs.

A typical mission for the AC-47D crews began at dusk, every night and in all kinds of weather. At least three aircraft were involved in each night's operations. One aircraft would be airborne from dusk until the mid-point of the night hours; the second would relieve the first at that time and remain aloft waiting for a call for help until dawn; the third aircraft stood five-minute ground alert in case of an inflight emergency with the airborne alert aircraft, or the need for additional firepower.

"Bad things will happen..."

Gunships often worked in conjunction with other aircraft of the air commando force, such

of flights of three, four or five aircraft. Headquartered at Nha Trang, the 14th ACS placed A Flight at Nha Trang, B Flight at Phan Rang, C Flight at Bien Hoa, and D Flight at Binh Thuy. When the 14th ACS became operational, on 15 January 1968, the 4th moved its flights as follows: A Flight remained at Da Nang, B Flight was at Pleiku, C Flight at Phu Cat, D Flight was at Nha Trang, and a fifth flight was at Udorn assigned to the 606th ACS.

On 1 May 1968, the 14th ACS was redesignated the 3rd ACS (Fire Support), with the 3rd ACS and took over all the former's AC-47D assets and FOLs (Forward Operating Locations) that had been used by the 14th. On 1 August, all Air Commando units were redesignated

Medal of Honor mission

Major Bill Platt, an AC-47D navigator with the 3rd SOS, describes a mission in which his aircraft was not only hit by enemy fire, but also threatened by one of its own flares.

SPOOKY 71's crew:
Aircraft Commander: Major Ken Carpenter
Co-pilot: 1Lt Frank Slocum
Navigator: Major William Platt
Flight Engineer: SSgt Edward Fuzie
Loadmaster: A1C John Levitou
Gunners: Sgts Ellis Owen and Tom Baer
VNAF Liaison Officer: (identity not recorded)

"The '69 Tet Offensive started at about 02.00 on 23 February 1969. Every hamlet, city, outpost, and firebase in South Vietnam was attacked at the same time. One NVA division came down through War Zone C towards Saigon, while a second NVA division came through War Zone D in an attempt to take Bien Hoa, Long Binh, and eventually Saigon. The night of 23/24 February was Major Ken Carpenter's first mission as aircraft commander.

"We usually kept two AC-47 aircraft on combat air patrol from dusk to midnight, while a second pair of 'Spookies' sat ground alert, taking off just prior to midnight to relieve the airborne CAP aircraft. They would then land, refuel and reload, then take our spot on ground alert until dawn the following morning.

"In the early hours of 24 February, the NVA attack against Long Binh commenced. At midnight, it was the turn of 'Spooky 71' to fly the midnight shift.

"We took off just before midnight and relieved 'Spooky 70', taking up station near Long Binh Army Post. Things were quiet until about 4 a.m. I guess the VC and NVA were regrouping after their maximum effort the previous night. As soon as the attack on Long Binh began, they immediately put in a call for help. 'Spooky 71' answered the call. Long Binh was already under heavy rocket and mortar attack, with ground troops trying to break through the perimeter defences. The major assault was against the southeast side of Long Binh. Our job was to provide illumination and to suppress the rocket and mortar fire. Major Carpenter spotted several mortar tube flashes just southeast of the fort. He immediately put the aircraft into attack position and opened fire on the mortar positions. After two passes, we had fired about 3,000 rounds of 7.62-mm ammunition, and the mortars were silent.

"Bien Hoa Control then called for us to proceed about two 'clicks' (kilometres) due south, and begin dropping flares for some friendly units that were engaged with a large force of VC and NVA troops. As we approached, we again saw numerous mortar tube flashes, with a large mass of small arms fire covering an area about a click in diameter. The ground controller called for us to make a flare drop.

"Airman First Class John Levitou grabbed one of the Mk 24 flares and passed it to Sergeant Ellis Owen. Owen's job was to hook up the lanyard and pull the safety pin, then throw the flare out the open cargo door. Each of these flares put out 2 million candlepower in light. Normally, Sergeant Owen would hook up the lanyard, put his finger through the safety pin, stand in the door, and wait for the signal to throw it out. As the lanyard catches on the wire, it starts a 10-second fuse, which causes a small explosive charge to blow off the cap and deploy the parachute. The explosion is small but quite violent, and at least one guy was decapitated when a parachute separated while he was still holding it. After the parachute deploys, it starts a second 10-second fuse that will ignite the flare itself.

"As Levitou passed Owen the flare, Major Carpenter put the gunship into a 30° bank to bring the guns to bear on the enemy mortar teams. Owen hooked his finger through the safety, hooked up the

3rd SOS AC-47D 43-49770 was the aircraft struck by a North Vietnamese mortar round while over Long Binh Army Post on 24 February 1969.

lanyard, and prepared to throw the flare out the door. Several of our crew members could now plainly see a large number of mortar tube flashes.

"Suddenly, we heard a large explosion, and the entire aircraft was filled with a blinding flash and thousands of pieces of shrapnel. The aircraft lurched down and to the right. An NVA 82-mm anti-personnel mortar round had hit the bottom of the right wing, penetrating the skin and exploding right between the wing fuel tanks up against the wing spar. The explosion blew a hole in the upper wing about 40 in [102 cm] in diameter.

"When the mortar round exploded, it knocked everyone in the gun compartment off their feet. Shrapnel riddled the fuselage, wounding everyone. Sergeant Fuzie had at least 10 wounds, Owen was struck by several pieces, and Levitou had over 40 shrapnel wounds. When the mortar round hit the wing, Sergeant Owen was knocked off his feet, and headed directly for the open doorway. He was supposed to have a lifeline on, but it was rarely done. Even if he had fastened his lifeline, it was not the best position to be in – dangling in the airstream at about 200 mph [321 km/h].

As a CV-2 Caribou climbs out in the background, A2C Clarence Darby stands guard in front of a 4th ACS AC-47D outside Flight B's operations office, adjacent to the ramp at Pleiku. It is October 1967, by which time the 4th ACS had five flights based in South Vietnam with Headquarters at Nha Trang.

as a C-47B GABBY psywar aircraft. GABBY had a large speaker in the cargo door, and a South Vietnamese talked to the people on the ground, pleading with them to throw down their weapons and come over the the side of the good guys (the Americans). The VC had no appreciation for GABBY and often fired heavy weapons and small arms at the aircraft; however, they did not know that a gunship was in the same orbit as GABBY, about 3,000 ft (914 m) higher and a quarter orbit behind the psy-war aircraft. GABBY would inform the VC troops not to fire on it – "Bad things will happen if you do". When the VC opened fire on GABBY, the pilot of the Spooky gunship would line up his guns on the area where the fire originated, and sweep the area clean. Whereupon, GABBY would announce, "See, I told you so!"

The gunships were often used in conjunction with other night operations aircraft, like the A-1 Skyraider and B-26K Counter Invader, for which the gunship could use its flares to illuminate an area. It could then turn around and mop up any remaining enemy troop concentrations that the rapid movers missed. An experiment with a Marine helicopter gunship fitted with a Night Observation Device (NOD), code-named Night Hawk, was successful, but was never implemented as standard operating procedure on the AC-47. The use of the NOD, experimentally installed in AC-47s in 1966, did

become standard in both the AC-130 and AC-119 gunship series.

By 1969, the career of the AC-47D was coming to an end. Plans were already in the works to deactivate the squadrons and replace them with AC-119G Shadow gunships for the hamlet defence mission. Out-country missions against the Ho Chi Minh Trail and other targets in Laos

AC-47D 43-49770 (foreground) is seen at Phan Rang in November 1967, assigned to the D Flight, 4th ACS.

Above: Over 3,500 shrapnel holes were found in the right side of the fuselage of 43-49770 after the 82-mm mortar exploded inside its right wing.

Below: Here the hole torn in the underside of the starboard wing of 43-49770 by the North Vietnamese mortar round may be seen.

"Despite his own wounds, Levitou reacted instantly, and reached out for Owen. Grabbing for his parachute harness, he pulled the wounded Owen back inside the aircraft. But the explosion had also done something far more dangerous than just knocking Owen out of the door: it had knocked the flare out of his hand, which pulled the safety and the lanyard activating the 10-second fuse that separated the flare and the parachute. With everyone knocked to the floor of the gunship, the armed flare was rolling toward the front of pitch-black aircraft interior. Levitou then did something completely against all flare handling procedures.

"Procedures called for only one man to handle any inflight emergency with regards to flares. In this case it was Sergeant Owen. But Sergeant Owen was injured and in no position to grab the flare and throw it out the door.

"As Major Carpenter fought to regain control of the aircraft (the main spar was damaged, and all fuel, oil and hydraulic lines had been cut), Levitou first struggled to bring Owen to a safe position in the middle of the aircraft interior. The gunship bucked, rolled and struggled to remain aloft. Something caught Levitou's eye in the darkness. It was the armed flare. Precious seconds had already slipped by. How long before it blew the lid off the chute canister? Levitou didn't know, nor did he even consider it.

"The flare was resting between the No. 1 Minigun and the main ammunition supply – over 18,000 rounds of 7.62-mm ammunition! Levitou reached for the smoking flare – but each time he thought he had it, the aircraft would buck or jump, and the flare would roll further away. Still more

seconds went by. Finally, Levitou forced himself to a kneeling position and jumped on the smoking flare, pulling it to his chest. He then painfully crawled to the open doorway, leaving a large trail of blood behind him, and threw the flare out. The flare barely cleared the door when the first stage exploded and the parachute started to deploy. With the immediate danger of the armed flare now out of the aircraft, Levitou and I began tending to the wounded. Remember, Levitou had 40 wounds himself. Then we started to tie down the ammunition in preparation for what we all thought would be a crash landing at best, or jumping from the aircraft if there was no other choice.

"Major Carpenter finally regained control of the wounded aircraft and turned back toward Bien Hoa. Lieutenant Slocum got on the radio and called the 'May Day', telling Bien Hoa that we had many wounded and to have the ambulances and a 'dust-off' (medevac) helicopter standing by. Despite the huge hole in the upper right wing, which caused the aircraft to stall out at a much higher speed than was normal for landing, Major Carpenter made a smooth landing. The 'dust-off' air-evacuated Levitou to a waiting ambulance aircraft, which took him to Japan. He returned two months later and flew another 20 missions before rotating home.

"Sergeant Fuzie had a piece of shrapnel lodged so close to his spinal cord that the surgeons were afraid to operate. He refused even to go to the hospital until Major Carpenter gave him a direct order to do so. Sergeant Baer had the worst injuries. He'd been closest to the explosion and shrapnel had penetrated his right shoulder and cut the nerves to his right arm before puncturing his lung. He was on

the operating table at Long Binh within 10 minutes of our landing. Everyone made a full recovery, including the aircraft, which had over 3,500 holes in it, not counting the huge one in the right wing.

"On 14 May 1970, President Richard Nixon presented the Medal Of Honor to Airman First Class John L. Levitou on the White House lawn, for his "selfless heroism that saved his fellow crew members and the aircraft". I'm very proud to say that I served with John Levitou on that and many other missions.

After the AC-47D was phased out in favour of the AC-119G, most aircraft were transferred to the VNAF or RLaoAF. However, three 4th SOS aircraft were sent to Udorn and assigned to the 432nd Spooky Detachment. Carrying the 'OS' tail codes of the 432nd TRW, the Udorn-based Spookies flew base defence, Lima Site support, and Trail interdiction missions. One of the detachment's AC-47Ds, 43-49010, is pictured; its 'kill board' (above) shows 100 troops KIA, five AAA guns silenced, and three trucks destroyed.

and Cambodia would be handled by the sophisticated AC-130A Spectre and AC-119K Stinger gunships, which ultimately became the supreme truck-killers in Southeast Asia.

Although the gunship war was heating up, the impending arrival of the new-generation gunships (the AC-119 and AC-130), and 'Vietnamisation' of the war as directed by President Richard Nixon, brought an end to USAF AC-47D operations. A 3rd ACS AC-47D flew the 14th ACW's 150,000th mission on 2 March 1969. AC-119Gs assigned to the 71st SOS had deployed to Nha Trang in November 1968 and were slowly taking over the hamlet defence missions of the AC-47D crews.

On 7 August 1969, the 3rd SOS flew its last mission, and the squadron was inactivated on 1 September; its aircraft were transferred to the Vietnamese Air Force (NVAF). D Flight at Binh Thuy AB turned over its remaining four AC-47Ds to VNAF crews on 30 June. Following

the inactivation of the 3rd SOS, Headquarters 4th SOS shuffled its remaining aircraft to other FOLs for maximum coverage of the four regional combat areas. Flights remained at Da Nang, Pleiku, Nha Trang and Bien Hoa, with a detachment at Udorn.

Final USAF Spooky mission

As more of the next-generation gunships began arriving in Southeast Asia, 7th AF transferred more AC-47Ds to other friendly nations' air forces. On 30 November 1969, Lieutenant Colonel Adam Swigler, commander of the 4th SOS at Phan Rang, took off for a standard midnight shift airborne alert patrol. He landed at 07.10 on 1 December 1969. It was the final mission for the 4th SOS and and brought an end to USAF Spooky operations in South Vietnam. The 4th SOS was inactivated on 16 December 1969 and its AC-47D assets redistributed as follows: three aircraft went to the 432d

TRW Spooky detachment at the FOL Udorn, Thailand, three were turned over to the VNAF, and the remaining eight AC-47Ds went to crews of the Royal Laotian Air Force.

Over the five years in which FC/AC-47D gunships operated in defence of South Vietnam, they defended well over 6,000 hamlets and forts in South Vietnam. Their record speaks for itself – not a single hamlet or fort was lost while an AC-47 gunship was actively defending the outpost. The 4th ACS/SOS, by itself, successfully defended 3,926 hamlets and forts. Spooky crews from the 4th fired over 97 million rounds of 7.62-mm and 0.3-in ammunition, and dropped over 270,000 flares, between 14 November 1965 and 16 December 1969, accounting for more than 5,500 confirmed enemy kills. Officially, a total of 53 FC/AC-47D gunships was converted, 47 of them going to Southeast Asia.

Larry Davis

AC-47s in Indochinese service

When the AC-47Ds were phased out of service by the 7th Air Force in December 1969, aircraft were transferred to the 815th Combat Squadron, VNAF. This example (right) sits on the ramp at Tan Son Nhut AB in May 1970, waiting for dusk and another mission. The VNAF fielded single squadrons of AC-47Ds and AC-119Gs. Eight 4th Air Commando Squadron AC-47Ds, inactivated on 15 December 1969, were transferred to the Royal Laotian Air Force. Based at Nakhon Phanom RTAFB, Thailand, the Laotian AC-47Ds (below) continued night interdiction missions on the Ho Chi Minh Trail, air support of Lima sites throughout Laos and Cambodia, and airborne flare illumination of outposts under attack.

Above: Cambodian air force gunship crew stand in front of their aircraft at Nakhon Phanom RTAFB in June 1971. Note that it is armed with three 0.50-calibre machine-guns instead of Miniguns. Almost all of the Cambodian AC-47s were converted in the field; the actual number of conversions completed is unknown.

FC/AC-47D operators

1ST AIR COMMANDO WING

The 1st Fighter Wing was re-activated at Eglin AFB, Florida, on 27 April 1962, and re-designated the 1st Air Commando Wing on 1 June 1963. On 8 July 1968, the 1st ACW became the 1st Special Operations Wing. AC-47D training was assigned to the 4412th Combat Crew Training Squadron, 4410th Special Operations Training Wing at England AFB, Louisiana. Subsequently, the 4412th CCTS was re-designated the 548th Special Operations Training Squadron.

Training Detachment 8 of the 1st Air Commando Wing began operations at Forbes AFB, Kansas, on 9 August 1965 with four Douglas FC-47D aircraft. Subsequently, on 15 August, Det 8 became the foundation of the 4th Air Commando Squadron which was activated at Forbes AFB.

2ND AIR DIVISION

1ST AIR COMMANDO SQUADRON

The first FC-47D gunships were initially assigned to the 1st Air Commando Squadron, based at Tan Son Nhut AB, RVN. The FC-47 gunships within 1st ACS were under the control of 2nd Air Division from August 1965 until 8 March 1966, when the aircraft were transferred to the 14th ACW and assigned to the newly arrived 4th Air Commando Squadron. The first FC-47D was operational on 11 December 1964, flying the first mission (daylight) on 15 December. The first night mission was flown on the night of 23/24 December 1964.

14TH AIR COMMANDO WING

The 14th Air Commando Wing was activated at Nha Trang AB, Republic of Vietnam, on 28 February 1966. The 14th ACW was the host unit at Nha Trang. On 1 August 1968, the 14th ACW became the 14th Special Operations Wing, remaining in that capacity until inactivated on 30 September 1971. By mid-May 1968, the 14th ACW had flown its 100,000th combat sortie in SEA. In March 1969, the 14th SOW reached the next plateau of 150,000 combat sorties, and flew its 200,000th combat sortie in late 1969.

Units within the 14th ACW/SOW included the 1st ACS with A-1 Skyraiders, 3rd ACS/SOS with AC-47Ds, 4th ACS/SOS with FC/AC-47Ds, 5th ACS/SOS with C-47D and U-10A/B Psy-War aircraft, 6th ACS/SOS with A-1 Skyraiders, 9th ACS/SOS with C-47D and U-10A/B psywar aircraft, 14th ACS with AC-47Ds, 15th ACS/SOS with MC-130E COMBAT TALON, 17th SOS with AC-119Gs, 18th SOS with AC-119Ks, 20th ACS/SOS with UH-1F Huey helicopters, 37th ARRS with HH-3E and HH-53 helicopters, 71st SOS with AC-119Gs, 90th SOS with A-37A/Bs, 602d ACS with A-1 Skyraiders, and 604th ACS/SOS with A-37A/Bs.

The 14th ACW/SOW was based at Nha Trang AB, RVN beginning on 8 March 1966; then transferred to Phan Rang AB, RVN, on 15 October 1969. The 14th SOW was inactivated at Phan Rang on 30 September 1971.

4TH AIR COMMANDO SQUADRON
4TH SPECIAL OPERATIONS SQUADRON

Activated at Forbes AFB, Kansas on 15 August 1965, taking over the FC-47 assets of Training Detachment 8, 1st ACW, under Operation BIG SHOOT. In November 1965, the 4th ACS deployed to South Vietnam as part of Operation Sixteen Buck, arriving at Tan Son Nhut AB, RVN, on 14 November 1965, with 20 FC-47D aircraft (16 combat aircraft plus four spares). Initially headquartered at Tan Son Nhut AB, RVN, the 4th ACS was assigned to the 14th ACW on 8 March 1966, moving to Nha Trang AB, RVN, in May 1966, and to Phan Rang AB, RVN, on 15 October 1969.

Initially had three detachments: 'Det A' at Tan Son Nhut AB, RVN, 'Det B' at Da Nang AB, RVN. 'Det C' went to Udorn AB, Thailand, in December 1965 as part of Operation Cricket. 'Det C' operations at Udorn were terminated on 20 July 1966. In 1966, all detachments were re-designated as 'flights'. On 1 November 1966, eight new AC-47Ds were transferred to the control of the 606th ACS at Nakhon Phanom AB, Thailand. By September 1967, 4th ACS Headquarters was at Nha Trang AB, RVN, with five

flights; Flight A at Da Nang AB, RVN; Flight B at Pleiku AB, RVN: Flight C at Phu Cat AB, RVN; Flight D at Nha Trang AB, RVN; and Flight E at Binh Thuy AB, RVN.

The 4th ACS became the 4th Special Operations Squadron on 1 August 1968, by then having adopted the tailcode 'EN'. In late 1969, with the de-activation of the 3rd SOS, the 4th SOS took over many of the mission requirements previously flown by the 3rd SOS, and moved its flights to cover some of the 3rd SOS area of operations: Headquarters 4th SOS moved to Phan Rang AB, RVN; A Flight remained at Da Nang, B Flight remained at Pleiku, C Flight moved from Phu Cat to Nha Trang, D Flight moved from Nha Trang to Bien Hoa AB, RVN; and E Flight moved to Udorn AB, Thailand.

Crews from the 4th ACS/SOS set many records during their assignment to Southeast Asia. The 4th flew 5,584 combat sorties during 1966, accounting for at least 4,000 enemy troops KIA. C Flight at Udorn RTAFB, accounted for 204 enemy trucks destroyed in Laos. On 11 October 1966, one 4th ACS AC-47D set a record for the most 7.62-mm rounds fired, defending an outpost in Kien-Phong Province – 43,500 rounds plus 96 flares. That record was subsequently broken on the night of 27/28 February 1969, when A Flight out of Da Nang, fired a total of 219,800 rounds.

The 4th ACS/SOS lost a total of 16 AC-47D aircraft during its operational tour in SEA, including three shot down over Laos, 45-1120 on 24 December 1965, 43-49268 on 13 March 1966, and 43-49546 on 15 May 1966. Except for one aircraft destroyed by a Viet Cong sapper attack on 29 August 1968, all other aircraft were shot down over South Vietnam, usually by heavy machine-gun fire.

The last AC-47D Spooky mission was flown by Lt Col Adam Swigler (SPOOKY 41) on 30 November 1969. Following the last mission, the 14 AC-47D aircraft of the 4th SOS were passed to the Vietnamese Air Force (three aircraft) and the Royal Laotian Air Force (eight aircraft). The 4th SOS was officially inactivated on 16 December 1969.

Spooky was the 4th Special Operation Squadron's crew van at Da Nang in 1968.

14TH AIR COMMANDO SQUADRON

The 14th Air Commando Squadron was activated on 25 October 1967 at Nha Trang AB, RVN, equipped

A 3rd SOS AC-47D cruises over the central highlands of South Vietnam in 1968. Aircraft of the 3rd SOS were tailcoded 'EL' and carried a small 'ghost' emblem on their noses. Inactivated on 15 September 1969, the unit then passed its aircraft to the VNAF's 817th Combat Sqn. This aircraft has new production gun pod mountings, with all three guns positioned in cabin windows.

Left: During Colonel Frank Eaton's time as commander of the 14th Air Commando Wing, this somewhat irreverent business card was produced and distributed!

with Douglas AC-47D Spooky aircraft. The 14th ACS did not become operational until 15 January 1968. However, several AC-47Ds from the 14th ACS were assigned and operational with the 432nd Tactical Reconnaisance Wing at Udorn between 28 October 1967 and 5 November 1967, being used for air base defence and support of Lima Sites throughout Laos. The 14th ACS was inactivated and replaced by the 3rd Air Commando Squadron on 1 May 1968.

While operationally active, the 14th ACS had four flights: Headquarters and A Flight were at Nha Trang AB, RVN; B Flight at Phan Rang AB, RVN; C Flight at Bien Hoa AB, RVN; and D Flight at Binh Thuy AB, RVN. The 14th ACS lost two aircraft during the time it was operational in SEA – AC-47D 43-48925 was shot down over Laos on 3 June 1966, and 43-49330 was hit by Viet Cong rocket fire on 16 March 1968.

3RD AIR COMMANDO SQUADRON
3RD SPECIAL OPERATIONS SQUADRON

The 3rd Air Commando Squadron was activated on 1 May 1968 at Nha Trang AB, RVN, taking over the assets of the 14th ACS. The 3rd ACS was redesignated the 3rd Special Operation Squadron on 1 August 1968. The 3rd ACS had five flights: Headquarters was at Nha Trang AB. RVN; A Flight at Nha Trang AB, B Flight at Phan Rang AB, C Flight at Bien Hoa AB, and D Flight at Binh Thuy AB. The unit's tail code was 'EL'.

The 3rd ACS/SOS lost one aircraft during operations in South Vietnam, AC-47D 43-49274 on 13 December 1968. The 3rd SOS flew its last AC-47D mission on 7 August 1969, then turned over all AC-47D assets to the Republic of Vietnam Air Force The 3rd SOS was inactivated on 15 September 1969.

432ND TACTICAL RECONNAISSANCE WING

432ND SPOOKY DETACHMENT

After the 4th SOS flew its last mission in November 1969, 11 of its AC-47Ds were passed to the the VNAF and RLaoAF. The remaining three AC-47Ds were assigned to the 432nd Spooky Detachment at Udorn under control of the 432nd Tactical Reconnaisance Wing with the tail code of 'OS' between 12 October 1969 and 29 May 1970, being used for air base defence as well as out-country missions defending Lima Sites throughout Laos.

606TH AIR COMMANDO SQUADRON

Eight AC-47D aircraft taken from 4th ACS assets in South Vietnam were assigned to the 606th Air Commando Squadron at Nakhon Phanom AB, Thailand, under Operation Cricket for operations over Laos, beginning on 1 November 1966. The 606th ACS reported directly to the 13th Air Force until April 1967, when it came under the control of the 56th Air Commando Wing.

Consolidated
B-24 Liberator

Part One: Development and Allied operations, 1939-1943

Consolidated-Vultee's 'global bomber' served in greater numbers, in more theatres and with more air arms during World War II than its great rival, the Boeing B-17, thanks to its spectacular range and greater carrying capacity. It is not true, as has been said, that the Liberator was 'infinitely' better than the Flying Fortress, but staunchly loyal crews could be forgiven for thinking so. Compared with the 7,366 Lancasters and 12,731 Flying Fortresses which poured from the Western industrial machine, no fewer than 19,276 Liberators were built. At the Consolidated maintenance and overhaul facility on the high desert at Tucson, Arizona, Liberators were parked wingtip-to-wingtip as far as the eye could see, stretching off the end of the airfield and out into the desert. "We were building them faster than we could muster pilots to fly them overseas," a veteran remembers. "You could look out, and there was no end to the sight of Liberators heading toward the horizon."

The Eighth Air Force had operated its B-24s for only a matter of months before detachments were sent to North Africa. Here Joisey Bounce leads a flight of four B-24Ds of the 93rd Bomb Group in the ETO. This aircraft (41-24226, formerly Utah Man) was a survivor of Operation Tidal Wave, the famous Ploesti raid of 1 August 1943, but was reported missing on 13 November, having been attacked by Axis fighters. The aircraft on its starboard wing, The Duchess (41-24147) was another Ploesti veteran.

Above: Reuben Fleet was persuaded to test the 'Davis wing' on the Model 31 flying-boat – a private venture intended as a 52-seat commercial transport and as a possible replacement for the PBY Catalina. A prototype (NX21731) flew on 5 May 1939, greatly impressing CAC's chief test pilot William Wheatley and allaying any concerns Fleet may have had about the new wing design which, along with the Model 31's tailplane, had been adapted for use in the Model 32. (The Model 31 was tested by the USN as the XP4Y-1 Corregidor but remained a one-off. Though 200 were ordered by the Navy all were cancelled, along with development of the civil derivative.)

US Patent No. 1942688 – the 'Davis wing'

Filed on 25 May 1931 and issued on 9 January 1934, David R. Davis's US Patent covered a family of new aerofoils of high aspect ratio that, Davis felt, would offer particular benefits in long-range flying-boat design. Turned down by one major aircraft manufacturer in his attempts to develop his ideas, an undaunted Davis offered the products of his research to Consolidated in 1937. Company President Reuben H. Fleet was sceptical, as was chief designer 'Mac' Laddon, especially as extensive aerofoil research by NACA did not include anything similar to the 'Davis wing'. With a hunch that they should look further into Davis's work, Laddon convinced Fleet to agree to a wind tunnel test at the California Institute of Technology (CalTech). The results were so spectacular that CalTech personnel recalibrated their wind tunnel twice to recheck their figures. Their report confirmed the wing's properties but was unable to explain, with any certainty, why it performed as it did! A still sceptical Fleet then authorised a full-size test in the Model 31 flying-boat and, by the end of 1939, the XB-24 had taken to the air on a 'Davis wing'.

To make all this possible Davis had entered a licence agreement with Consolidated in February 1938, whereby he would receive $2,500 for each prototype built, plus a royalty per subsequent aircraft based on a sliding scale starting at $\frac{1}{2}$ of 1 per cent of the selling price of the aircraft (less Government-furnished equipment), decreasing to $\frac{1}{8}$ of 1 per cent when orders reached $10 million. When royalty payments reached $50,000, the rate reduced to $\frac{1}{16}$ of 1 percent. With almost 3,000 Liberators delivered and thousands more on order the royalty payment was renegotiated in April 1943 and the then current $92.71 was reduced to a flat $5.00 per aircraft. Another 17,383 examples were completed over the next two years!

When the XB-24 took to the air over the ridges and seashores of San Diego, California on 29 December 1939, it was a great silvery giant, glinting like a mirror in the glare of the southern California sun, much in the manner of the chrome-plated children's toy aircraft that had been so popular during the inter-war Depression years.

It was the end of one era and the beginning of another. The United States already had staked its future on the four-engined bomber – the Boeing B-17 Flying Fortress, a maturing design entering service at the time – and Britain was nurturing several long-range bombers. The other world powers – France, Germany, Japan – seemed not to be paying attention. In all the world, there was not yet anything else quite like the XB-24, being put through its paces by Consolidated Aircraft Corp.'s test pilot William B. Wheatley.

Wheatley flew the bomber wearing a necktie and a railway engineer's cap. Perhaps it was significant that he seemed a little different, because the others who spawned the B-24 Liberator seemed different, too. Reuben Fleet had founded Consolidated 17 years earlier and ran it with such a deeply-entrenched belief in 'one-man rule' that he was destined to give up his role atop the corporation just when it began to grow with the arrival of war. David R. Davis,

who designed the high aspect ratio (11.55:1) 'fluid foil' wing of the Liberator, recently had been an out-of-work inventor with nagging financial problems. None of these men would survive to see the B-24 Liberator become the most numerous aircraft ever built in America, but all were present at the creation, bonded inextricably by their mutual goal – to build a bomber better than the B-17.

Compared with the 7,366 Lancasters and 12,731 Flying Fortresses which poured from the Western industrial machine, no fewer than 19,276 Liberators were built. This figure also exceeds production totals for Dakotas, Mustangs, Thunderbolts, Mitchells and Marauders. Liberators rolled off production lines operated by Consolidated (in San Diego, California and Fort Worth, Texas), Douglas (at Tulsa, Oklahoma), Ford (at Willow Run, Michigan) and North American (at Dallas, Texas).

'Mac' Laddon's bomber

Fleet and design engineer I. M. 'Mac' Laddon approached the Army Air Corps in January 1939 with a counter-offer, after having been asked to create a new production outlet for the B-17. This triggered a period of frenetic work at Consolidated, culminating in the creation of a mock-up of the new and different bomber they wanted to build.

Early versions of Consolidated's Model 32 bomber design

This view of the Liberator prototype shows its 'Davis wing' to good effect. The five slots on the outer panels of each wing were deemed unnecessary and remained unique to the prototype. Other changes included the relocation of the pitot tubes from the wings to the forward fuselage and a 2-ft increase in tailplane span. As a result of wind tunnel testing, wing and tail fillets, redesigned engine nacelles and fully retracting main landing gear were all proposed, but time pressures meant that none of these changes was made.

Above: The XB-24's 17-minute inaugural flight took place on the afternoon of 29 December 1939, one day earlier than contracted, with Bill Wheatley at the controls. Test flights were remarkably trouble-free, one of the few 'glitches' concerning an up-latch mechanism failure in the nose landing gear. The aircraft's hydraulic system kept the gear retracted but small movements fooled its crew into thinking that the XB-24 had wing flutter problem!

looked a little like the Flash Gordon spaceships of late 1930s movie serials: a February 1939 version had a bulbous nose and fuselage with no step for the pilots' compartment and a stepped ventral area. By the following month, the proposed bomber was more portly, but still had a long way to go before resembling the real B-24. The company kept the model designation and, by May 1939, had come very close to finalising the design.

On 1 February 1939, the Army issued Type Specification C-212. This amounted to choosing Consolidated, since any other aircraft manufacturer would have only three weeks to respond. On 21 February, the Army chose the Consolidated Model 32. A month later, the Army ordered a single XB-24 prototype (allocated the serial 39-556) with the proviso that it be flown by 30 December – a requirement that was only barely met.

The XB-24 was powered by Pratt & Whitney R-1830-33 Twin Wasp radial engines rated at 1,200 hp (895 kW) on take-off and 1,000 hp (746 kW) at 14,500 ft (4420 m). For a time, it had wing slats and solid propeller spinners that did not appear on production Liberators.

The seven-place XB-24 had the first tricycle landing gear flown on a large bomber. The nose wheel permitted faster landings and take-offs, thus allowing for a heavier wing loading on the low-drag Davis airfoil. A bombardier's enclosure began the nose of the deep fuselage, which terminated behind twin rudders with a tail gunner's position. Armament included three 0.5-in (12.7-mm) guns and four 0.3-in (7.62-mm) hand-operated guns fired through openings in the fuselage sides, top, and bottom, and in a nose socket. The XB-24 could accommodate eight 1,100-lb (499-kg) bombs, twice the capacity of the B-17.

Eventually, this bomber would prove faster than its Boeing counterpart, too, but the prototype's speed was measured at only 273 mph (439 km/h) instead of the Army's specified 311 mph (500 km/h). So, on 26 July 1940 the Army ordered the prototype to be fitted with turbo-superchargers and leak-proof fuel tanks. The R-1830-41 Twin Wasp engines provided 1,200 hp (895 kW) at 25,000 ft (7620 m). Wing slots were deleted. This first ship in a long series was redesignated XB-24B, and the reworked prototype flew on 1 February 1941. To confuse everyone

forever, this airframe received a new serial number (39-680) and its original serial (39-556) was cancelled.

In March 1939, before the 'XB' ever flew, the Army ordered seven service-test YB-24 bombers with turbo-superchargers for high-altitude flight. Meanwhile, France, in dire need of warplanes, placed an order, in June 1940, for an export model, known as the LB-30. The French government collapsed before the aircraft could be delivered and the LB-30 production run was divided between Britain and the AAF (Army Air Forces), as the US Army's air arm became known in 1941.

Six of the seven YB-24s were completed as LB-30A aircraft for Britain (AM258/263) and eventually operated as transatlantic ferry transports. The seventh YB-24 had armour and self-sealing fuel tanks and was accepted by the Army in May 1941. All seven of these bombers were identical to the re-engined 'XB' including the deletion of wing slots and the addition of de-icers.

In later years, when their warplane had been eclipsed by the less capable B-17, men who flew the B-24 wondered if it was a Boeing publicity machine that did them in. More likely, it was chance – the accident that produced the evocative name 'Flying Fortress'. The popular name assigned to the B-24 was more accurate and seems an excellent choice, but it never captured the imagination.

The comparison between the two bombers has been made often. The B-17 could reach higher altitude, was easier to fly, may have been slightly more resilient when hit, and was safer to ditch. The oft-overlooked Liberator,

Above: AM929 was last of the 20 LB-30Bs delivered to the RAF, in August 1941. After trials at A&AEE it was fitted with ASV radar and cannon and issued to No. 120 Sqn, Coastal Command. Its crews claimed five U-boats destroyed before the aircraft was converted as a transport for No. 231 Sqn.

Above: RAF Liberator Mk IIs were laid down as LB-30MFs for the Armée de l'Air and sported a lengthened nose, at the request of the French. For the RAF provision was made for powered dorsal and tail turrets.

No. 1653 Heavy Conversion Unit was formed in early 1942 at RAF Polebrook to train Liberator crews for service in the Middle East. LB-30 AL581, coded 'M', was one of their aircraft

although more difficult to handle, was significantly faster (a Liberator on three engines could readily overtake a Flying Fortress using all four), carried a heavier bomb load, and went farther.

Contrary to myth, the Liberator name did not originate with the RAF. Responding to a question about the name of the aircraft only 10 months after its first flight on 28 October 1940, Reuben Fleet noted, "We chose Liberator because this airplane can carry destruction to the heart of the Hun, and thus help [the British] and us to liberate those nations temporarily finding themselves under Hitler's yoke." At the time, few were predicting a war in the Pacific, where the range, bombload and staying power of the B-24 Liberator would be essential.

As early as 1940, Washington began looking into a production pool to meet expected requirements for large numbers of Liberators. The B-24 had not exactly been opti-

mised for high-volume production, but the US government forged ahead, anyway. Consolidated agreed to establish a second plant at Fort Worth, Texas to supplement production in San Diego. Douglas would set up an assembly plant at Tulsa, Oklahoma. The intent was for these government-financed plants to assemble finished B-24s from knock-down kits provided by Ford at a new factory in Willow Run, Michigan. Not until much later was a fifth production outlet added, namely the North American Aviation plant at Dallas, Texas.

Dazzled by the promise of Boeing's stratospheric B-17, with its superb defensive armament and capacious cabin, the USAAC had initially ordered only 32 Liberators, while France and Britain (seeing the Liberator's incredible payload/range performance) each ordered over 100 examples each. The French order was overtaken by the fall of France in June 1940, but production continued, since even the USAAC had become convinced as to the type's usefulness. "If the French can't take 'em, we will!" was the response.

First in service

The first Liberators in service were six LB-30As (with the LB designation suffix standing for 'Land Bomber', and not 'Liberator, British derivative' as was once claimed) diverted from the French contract to Britain in December 1940. The aircraft's lack of self-sealing fuel tanks was judged to be a major stumbling block to front-line service, and although the aircraft were delivered with added armament (six 0.5-in Browning machine-guns, as intended for the US Navy), this was subsequently stripped and replaced with cabin heaters, passenger oxygen and de-icing equipment. The aircraft (officially known simply as Liberators, but often referred to, inaccurately, as Liberator Mk Is) then entered service with RAF Ferry Command in March 1941. Serialled AM258/ AM263, some later went to BOAC.

The first truly operational Liberators were 20 'B-24A Conversions' (LB-30Bs) built for the USAAC but diverted to the RAF as Liberator Mk Is (AM910/AM929). These aircraft had self-sealing fuel tanks and were considered to be suitable for operational use over Europe. Tests had shown that by cruising at 120 kt (138 mph/222 km/h), carrying a 2,000-lb (907-kg) load of depth charges, the aircraft would be able to remain airborne for an astonishing 16 hours. This kind of performance was just what was needed to fight the growing U-boat menace, and 10 of the aircraft were therefore delivered to Coastal Command after modification by Scottish Aviation at Prestwick. These modifications included a ventral gun pack containing four 20-mm cannon, and ASV radar, with 'Stickleback' antennas on the spine, 'towel rail' antennas on the rear fuselage sides and Yagi aerials on the nose and under the outer wings. An 11th aircraft (AM929) used more advanced ASV.Mk II radar, without requiring the massive antenna masts on the spine. In all, 15 Liberator Mk Is are known to have served with Coastal Command; the rest were used for transport and

Left and below: LB-30A AM262 cruises over New York and environs some time during 1941, prior to delivery. A number of the RAF's first Liberators were used on the newly created Trans-Atlantic Return Ferry Service between Montreal and Prestwick, Scotland. Over this 3,000-mile (4828-km) route LB-30As and LB-30Bs ferried pilots to Canada to collect American-built aircraft bound for the RAF, VIPs and other important cargoes.

training duties.

The ventral armament was a relatively late addition to the Liberator Mk I. While it had long been realised that the aircraft would benefit from the addition of forward-firing armament there were two alternative schemes that came under study. The first of these was a solid nose containing 12 fixed 0.303-in machine-guns, designed by Martin-Baker and similar to the nose of the Douglas Havoc. The second was the ventral Hispano cannon tray. An aircraft was about to be modified (by Heston Aircraft Ltd) as the 12-gun nose prototype, when the decision was suddenly made in favour of the Hispano cannon fit.

RAF Coastal Command

The Liberator Mk I entered service with Coastal Command in June 1941, joining No. 120 Squadron at Nutts Corner, in Northern Ireland for VLR patrols far out into the Atlantic, beginning operational patrols on 20 September. Coastal Command's new Liberators were used to help close the gap in ASW coverage which had previously existed in the mid-Atlantic, outside the range of aircraft based in the UK, Canada and Iceland. The new Liberators quickly demonstrated a range of 2,400 miles (3862 km) (the rival Sunderland flying boat could manage only 1,300 miles/2092 km), while their radar allowed them to find the elusive U-boats, and the 20-mm cannon provided a superb weapon for dealing with their prey if it remained on the surface. Many hoped to demonstrate the armament's effectiveness against the Fw 200 Condor, and these wishes were soon fulfilled. On 4 October 1941 AM924 tangled with an Fw 200 while escorting a convoy. The two aircraft traded fire, with the Liberator successfully driving off the big Focke-Wulf. The initial Coastal Command Liberators were destined to be extremely successful, but none more so than AM929, which forced several U-boats to abort their attacks and submerge, and which destroyed U-597, U-661, U-132, U-192 and U-540, damaging two more. Two of these kills were made by Squadron Leader Terry 'Hawkeyes' Bulloch, who went on to gain a tally of three U-Boats, plus two damaged, and a DSO.

Coastal Command were lucky to get the Liberator, since although 'Bomber' Harris had yet to take over as AOC-in-C of Bomber Command, his belief that Coastal was a 'sideshow', and that Bomber Command should have a priority when it came to the allocation of four-engined bombers was already gaining ground. Perhaps it was just as well that the Liberator was perceived as being less well suited to the short-range, heavy bombload, large-weapon, night offensive being prepared by Bomber Command, and that the Boeing B-17's poor combat record with the RAF had led it to be sceptical as to the merits of US bombers.

Bomber Command's Fortress Mk Is (B-17Cs) were notably unsuccessful, and in Coastal Command service the type (in B-17C and B-17E forms) was never viewed as anything more than a second-best supplement to the much better Liberator. The introduction of the Liberator turned the tide of the War in the Atlantic, turning an escalating loss rate into one that began to fall.

Despite the Liberator's immediate impact on the war against the U-boat while in Royal Air Force Coastal Command service, RAF Bomber Command would not be denied its opportunity to use the new aircraft. No. 108 Squadron in the Middle East re-equipped with Liberators delivered direct from the USA in late 1941, beginning combat operations on 10/11 January 1942. The unit received the first of 140 Liberator Mk IIs allocated to the RAF, though 75 were requisitioned for the USAAC as LB-30s (becoming the first USAAC 'Libs' to see action when 15 were deployed to Java with the 19th Bombardment

G-AGCD was the civil registration applied to AM259 during its service with BOAC between May 1941 and its return to the RAF in July 1944. AM258, AM262 (G-AGHG) and AM263 (G-AGDS) were the other BOAC LB-30As; apart from '258 (which crashed at Prestwick in September 1943), all were returned to the RAF in 1944.

131

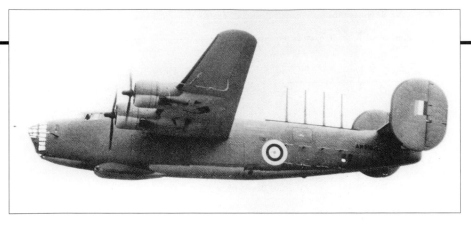

Prominent in this view of AM910 are its various ASV antennas and its four 20-mm Hispano cannon, fitted in a tray under the cockpit. The gun pack's equipment partly occupied the forward bomb bay, restricting offensive loads to four 500-lb anti-submarine bombs plus four Mk VII or six Mk V depth charges. With the latter load and 2,500 Imp gal (11365 litres) of fuel gross weight rose to 56,007 lb (25404 kg). Coastal Command was very concerned at the long take-off run required at this weight – almost 10,000 lb (4536 kg) above the XB-24's design gross weight.

Coastal beginnings

After ASV and armament clearance trials at A&AEE, the Liberator Mk I began operational patrols with Coastal Command in September 1941. At least 11 of the 20 LB-30Bs delivered were eventually converted for the maritime role, all serving with No. 120 Sqn. Liberator Mk IIs joined Coastal Command later in the year, with Mk IIIs following in mid-1942 and by the end of 1943 the Mk I had been phased out of maritime service.

Below: The lower surfaces of Coastal Command's Liberators were eventually oversprayed white for their over-water role. AM916 'OH-L' was one of No. 120 Sqn's Liberator Mk Is. An A&AEE report of June 1941 criticised the Liberator Mk I's defensive armament, which comprised just six guns – three pairs of 0.303-in Brownings mounted in the beam positions and in the tail.

AL507 was among a number of Liberator Mk IIs passed to Coastal Command to make good the inevitable attrition among Liberator Mk Is. Note the chin radome for SCR-517 centimetric ASV radar (which gave rise to the 'Dumbo' nickname applied to aircraft so-equipped, AL507 being the first) and the Boulton Paul mid-upper and tail turrets, each with four Browning 0.303-in guns. AL507 had a varied history, having been accepted in August 1941. Issued in No. 120 Sqn in March 1942, it later served with Nos 224 and 59 Sqns, before being relegated to a transport role in 1944.

Group), and only 23 of these 'hijacked' aircraft ever reached the RAF. The Liberator Mk II introduced a 3-ft (1-m) nose extension (originally requested by the French), and had Curtiss Electric propellers with distinctive long hubs. The aircraft were built with provision for a Martin A-3 mid-upper turret containing two 0.50-in machine-guns, and two more of these were mounted in the tail, behind sliding doors. In service the RAF aircraft introduced power-operated Boulton Paul tail and mid-upper turrets, each containing four 0.303-in Browning machine-guns. Single 0.303-in machine-guns were also mounted in the nose and belly, with pairs of these guns in the beam positions.

The Liberator's long-range capability (already demonstrated to Coastal Command) was shown off in January 1942, when No. 108 Squadron sent an aircraft (laden with Blenheim spares) direct from Egypt to India, and then on to Sumatra. The entry of the USA into the war halted Liberator deliveries to No. 108 Squadron, and its Liberator flight switched to long-range partisan support missions over the Balkans, with occasional long-range bombing missions like the 6 June raid on Taranto (a 13-hour mission!). Plans had initially called for the new American aircraft to re-equip the Wellington-equipped No. 1 Group,

No. 150 Squadron having received one aircraft during late 1941 and No. 142 Squadron due to re-equip in December. However, the first two complete Liberator bomber units raised in Britain (Nos 159 and 160) were instead sent overseas, after plans for a new Liberator-equipped Group (No. 8) faltered. Originally intended for service in the Middle East, the two units flew out to the Middle East from June 1942 (No. 160 after spending a month flying ASW patrols from Nutts Corner). The two units left a considerable element in the Middle East as the MEAF Liberator Squadron before flying on to India for operations against the Japanese. This became No. 178 Squadron in January 1943.

Liberator Mk III

The Liberator Mk IIs were quickly replaced by Liberator Mk IIIs, in India and in the Middle East, the survivors being relegated to transport and training duties. One of these aircraft (AL523) became embroiled in one of the great mysteries of the war when it crashed on take off from Gibraltar, killing 11 of those on board, one of the dead being the leader of the Polish Government in exile, General Vladislav Sikorski. Many still believe that the aircraft was

Nos 159 and 160 Sqns were the first RAF bomber squadrons to form on the Liberator Mk II, both units seeing service in North Africa during 1942 before transferring to India. AL579 is believed to be a No. 159 Sqn aircraft, seen during July 1942, shortly after beginning operations in the desert against targets in North Africa, Italy and Greece. Note the Boulton-Paul dorsal turret; turret shortages meant that this was omitted from some aircraft.

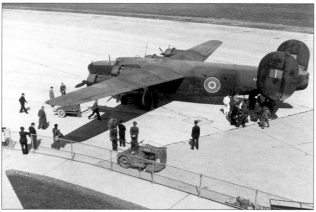

Personnel clamber from an anonymous Ferry Command Liberator Mk II at Montreal. Twenty-five Mk IIs were converted for the transport role and were used on African routes as well as trans-Atlantic services.

Perhaps the best known of Ferry Command's LB-30 Liberator Mk IIs was AL504 Commando. Issued to No. 511 Sqn, it served as Prime Minister Churchill's personal transport between October 1942 and June 1944. Luxuriously appointed, with sleeping berths, plush seats and an electric galley, the aircraft was later rebuilt by Consolidated with the lengthened forward fuselage and single fin (of the RY-3 Privateer) and stripped of its black finish. Churchill did not use the aircraft in its new guise; flying with No. 231 Sqn, it went missing in a flight between the Azores and Ottawa on 26 March 1945.

Above: AL507 (illustrated on the previous page) joined No. 511 Sqn in 1944, finally becoming G-AHYC with BOAC in August. It remained in airline use after the war, but was damaged beyond repair in a landing accident at Prestwick on 13 November 1948. In this view, though it has a BOAC 'speedbird' logo on its nose, the aircraft retains its RAF serial and markings over a polished 'natural metal' finish.

sabotaged, and that Sikorski was murdered on the instructions of Stalin, to whom he represented a particularly thorny problem. The Czech pilot was the only survivor of the crash.

The Liberator Mk III designation was applied to B-24Ds supplied to the RAF, though these aircraft tended to have the ventral ball turret deleted and faired over, and usually had British 0.303-in machine-guns and Boulton Paul tail turrets, though usually with a forward-mounted Martin dorsal turret containing two 0.50-in Brownings. Most significantly, the Liberator Mk III was the first RAF variant with turbo-supercharged engines. These had oil coolers relocated to the sides of the nacelles (from the wings), giving the nacelles a distinctive oval cross section. The new engine gave welcome extra performance, but also brought with it an unexpected disadvantage in the night bomber role, since the turbocharger exhaust glowed so brightly that the aircraft became visible over much longer ranges, and also blinded the unfortunate ball gunner when one was carried. The 0.5-in machine-guns were also a mixed blessing at night, with greater 'flash' and with their increased range being irrelevant – they fired far further than the gunner could possibly see. The Liberator Mk III saw relatively little service in the bomber role, soon giving way to the much-improved Liberator Mk VI (B-24J). It did serve in the Middle East with No. 178 Squadron, and with Nos 159, 160, 354, 355 and 357 Squadrons in India and Ceylon. The variant was extensively used in the transport role, notably with No. 148 Squadron, which specialised in supply dropping to resistance movements in the Balkans and Poland and with Nos 231 and 246 Squadrons.

Even though production of the Liberator was massive, there were never enough aircraft to meet the RAF's needs, and second-rate Stirlings, Fortresses and even Wellingtons soldiered on in roles where the Liberator would have marked a colossal improvement.

Production pool

The creation of a five-plant production pool for the B-24 Liberator symbolised the enormous capacity of the American heartland to manufacture the weaponry for modern industrial warfare. The Army Air Corps had fewer than 600 combat aircraft in 1939, but more than 60,000 by the end of 1943. The American war machine manufactured just slightly fewer than 100,000 aircraft in calendar year

1944. The B-24 Liberator was to become by far the most numerous aircraft ever manufactured in the United States (although, for reasons never clear, none was ever delivered bearing a serial number from Fiscal Year 1943): counting all orders that were later cancelled, at one time there existed contracts for 27,319 Liberators of all variants, of which 19,236 were actually built – in contrast to the Allies' 7,366 Lancasters and 12,731 Flying Fortresses.

As of 31 December 1943, projected manufacture of Liberator bombers was 20,353, including 9,104 of the popular B-24H model. By the spring of 1944, Ford's nearly miraculous Willow Run line was building a new B-24H at a median rate of more than one aircraft every 100 minutes, day and night, seven days a week. Ford even maintained the pace with no let-up while making minor configuration changes from one production block to the next. The plant actually had the capability to churn out 600 Liberators per month, but the AAF had neither enough crews nor enough bomb groups to absorb this many. For a time, bombers seemed to multiply on the ramp, sometimes stretching almost as far as the eye could see, pristine Liberators waiting for human beings to take them away.

Though 75 of the 159 LB-30s allocated to Britain were diverted to the USAAC, a number of these were later released as B-24Ds became available and, in the event, a total of 86 LB-30s (Liberator Mk IIs) was delivered to the RAF. AL613 was among the diverted aircraft and had been sent in 1941 (with AL602 and AL622) to Alaska for service with the 28th Composite Group at Kodiak, monitoring the Japanese fleet. By 1943 all had been reported lost, but in 1944 AL613 limped back to the US and was refurbished as a transport. It served Air Transport Command's Pacific Division until January 1946. Here it is shown upon completion in its new configuration, close to the standard of a C-87. As the LB-30s diverted for USAAC use had been ordered prior to the start of Lend-Lease arrangements, they had no equivalent AAF serial number and thus retained their RAF identities.

40-2376 (above) and 40-2369 (right) were among the nine B-24As delivered during June and July 1941, just as the Air Corps and GHQAF were moved under the umbrella of the newly-created US Army Air Forces.

Below: B-24A 40-2374 was one of the two Ferrying Command aircraft used to transport W. Averell Harriman (President Roosevelt's special envoy to the UK and USSR) and his staff to Moscow, to discuss Lend-Lease arrangements, in September 1941. Flown by Lt Louis T. Reicher, 40-2374 returned to the US via the Middle East, India, Singapore, Australia and Hawaii, and is seen here on 30 October 1941, arriving at Sembawang airport, Singapore. The large US 'neutrality' marking on the nose was repeated on the top surface of the wing and was a feature common to all B-24As.

B-24A model

Only one of the seven projected YB-24 service-test ships was delivered as planned, the remaining half-dozen being diverted to Britain as LB-30As (six replacement aircraft for the USAAF were completed as B-24Ds). This means that the B-24A model was essentially the first Liberator. Much of this activity took place before the bombing of Pearl Harbor in December 1941. At the time, it was clear that the US aircraft industry would increase dramatically in size and output, but it was never obvious that the United States would enter the war.

In the early, dark days – while experts hammered out the

B-24A transports

Unsuitable for use in their intended role, the nine B-24As (40-2369/2377) were handed over to Ferrying Command. Created in May 1941 for the movement of new aircraft (primarily in support of the Lend-Lease Agreement), the new command slowly took on a more general transport role, becoming Air Transport Command in June 1942. During the summer of 1941 the B-24As had inaugurated the trans-Atlantic service (with which RAF/BOAC LB-30s were to become associated) and were soon utilised on other routes worldwide. For example, three aircraft were taken off the Washington-Cairo run to fly ammunition from Darwin, Australia, to the Philippines and later assisted in the evacuation of the embattled islands in 1942. All three were eventually lost; 40-2370 was shot down, while 40-2374 was bombed by the Japanese at Broome, Western Australia. Of the other B-24As, 40-2371 and 40-2375 were prepared for the secret photographic mission to the western Pacific, thwarted by the Pearl Harbor raid in which 40-2371 was destroyed.

plan for Liberator production at five factories – Consolidated produced nine B-24A models (40-2369/2377) which were similar in appearance to the prototype but appeared in Army olive-drab. The B-24A was deemed inadequate for bombing duties and all nine examples were assigned to Ferry Command (later, Ferrying Division of Air Transport Command) as transports. One ship (40-2371) was reclaimed from the transport fleet for duty in a planned photo-reconnaissance project to snoop on a perceived Japanese build-up in the central Pacific. It never reached the central Pacific. Lieutenant Kunikiya Hira, leading No. 3 Squadron from the Japanese carrier *Shokaku*, toggled the bomb release on his Aichi D3A Type 99 'Val' dive-bomber and his bomb – the first to fall on Hickam Army Air Field, Hawaii during the 7 December 1941 attack on Pearl Harbor – hit Hangar 15 and set fire to 40-2371, the first US aircraft to be destroyed in World War II. Another B-24A (40-2370) was destroyed during the Japanese attack on Broome, Australia, on 3 March 1942.

The total number of LB-30 models accepted between 10 December 1941 and 6 January 1942 ended up being 75. Six were lost in US accidents, 23 went to Britain, and 46 served the US in various roles. The total order of Liberators for Britain ultimately included 140 Liberator Mk II bombers.

The B-24C model emerged on 19 December 1941 when the first of nine aircraft (40-2378/2386) was rolled out. The C model has been described as the 'production shakedown' variant of the Liberator. The nine B-24Cs incorporated turbo-supercharged R-1830-41 engines and the self-sealing fuel tanks earlier introduced on the XB-24B. This was also the first AAF version to be fitted with power-operated turrets. A Consolidated turret was located in the modified tail and a Martin unit was located just forward of the wing leading edge atop the fuselage.

The mature shape of the Liberator was yet to be

Nine B-24Cs preceded full-scale production of the B-24D. These incorporated the changes introduced in the XB-24B (namely turbo-supercharged R-1830-41 engines and self-sealing fuel tanks) and was the first USAAF variant equipped with power-operated turrets. Here one of this small batch of aircraft is seen during a test flight from San Diego. None saw combat, being retained in the US for training tasks under a 'restricted' RB-24C designation.

finalised. The evolution of the Liberator design over time can be seen in a glance at the changing length and shape of the bomber's fuselage – or in a comparison of the fuselage length of various models. Virtually every 'Lib' boasted a Davis wing with a span of 110 ft 0 in (33.52 m) and a wing area of 1,048 sq ft (97 m²), but only the XB-24 and XB-24B had a fuselage length of just 63 ft 9 in (19.43 m). As the Liberator grew, the B-24C through XB-24F models were lengthened to 66 ft 4 in (20.21 m). Subsequently, the B-24G and later models were stretched further to a length of 67 ft 2 in (20.47 m).

First B-24D rolled out

The first B-24D was rolled out at San Diego on 22 January 1942. It was, of course, the B-24D variant which finally introduced a combat-ready bomber. The first six aircraft in the series were 'tradebacks' for airframes originally conceived as YB-24s, so they received early serials (40-696/701). This first half-dozen D-models were taken on charge by the AAF beginning 23 January 1942. All were powered by the R-1830-43 engine.

All subsequent B-24D models were powered by R-1830-43s, except for those from Block 135 onward manufactured in San Diego, which employed the R-1830-65. The engines were in distinctive, oval cowls with air scoops on each side of the engine (providing cooling for the turbo-supercharger and intercooler). The B-24D used a three-bladed Hamilton Standard propeller 11 ft 7 in (3.53 m) in diameter. A few aircraft in the series had short ring cowls and wide-bladed or 'paddle' propellers, but most had long-ring cowlings and narrow props. The most distinctive feature of the D model, of course, apart from its increased fuselage length, was the latticed glass nose compartment, which lacked a power turret (although many B-24Ds would be retrofitted with power turrets, especially among units in the Pacific).

The B-24D introduced a power turret for the tail gunner. The first version was the A-6 model, employed on all B-24Ds through aircraft 42-40157. It was replaced by the A-6A, configured to improve gunner visibility with transparent panels on its sides. Not readily apparent to the eye is the fact that the guns on the A-6A were mounted asymmetrically, so that the left gun barrel protruded about 6 in (15 cm) farther than the right. Later Liberators, however, employed the A-6B tail turret in which the guns were symmetrical. The Motor Products Corporation manufactured A-6B turrets which became standard for later B-24H and B-24J production bombers, with the exception of some built by Douglas with the A-6A model. Some 122 B-24Js were built with the lightweight M-6A tail stinger, a hand-operated but hydraulically-assisted unit.

The 500-lb (227-kg) Briggs/Sperry A-13 44-in ball turret (with its two 0.5-in/12.7-mm Browning M2 machine-guns) made its debut with San Diego-built B-24D-140s and was a familiar feature on most remaining production Liberators, although some commanders elected to dispense with the ball turret in a combat environment, especially in the Pacific. The ball turret position gave rise to many myths. One myth holds that ball gunners were chosen for their compact size, but no evidence of this survives. Another is that the ball gunner was enclosed in his turret from take-off to landing; in fact, the gunner could enter and leave his

The B-24C, and early production B-24Ds, lacked the navigator's astrodome on the top surface of its lengthened nose. The highly polished propeller blades and gunless turrets (Martin dorsal and Consolidated A-6 tail) suggest that this aircraft had not long left the factory.

41-23754 was among the first B-24Ds flown across the Atlantic for use by the Eighth Air Force. Photographed at Alconbury in December 1942, it was on charge with the 409th BS, 93rd BG. Note the nose armament, typical of an early production B-24D – a single 0.5-in gun mounted low and a pair of ball-and-socket 'cheek' guns.

Before undertaking Liberator assembly in its own right, Consolidated's Fort Worth plant was used as a modification centre for aircraft produced at San Diego and, before 'KD' (or 'knock-down') aircraft came on stream from Ford's Willow Run facility, began to assemble aircraft using components supplied by San Diego. This explains why aircraft with B-24D-CO serials are seen in photographs of the Fort Worth production line. In this view workers put the finishing touches to a batch of aircraft destined for ASW units. The nearest aircraft are 42-40823, 42-40835 and 42-40686, all B-24D-COs.

Below: The first B-24Es off Douglas' Tulsa production line shared assembly facilities with a batch of A-24Bs for the USAAF.

Production

The B-24 was built in greater numbers than any other American type of the period. In order to produce a total of 18,482 LB-30s, B-24s, AT-22s and C-87s, plus 774 PB4Y-2s and RY-3s, between 1940 and 1945, one of the most complex production organisations was established, involving five factories run by four different companies – Consolidated, Douglas, Ford and North American.

Below: The role played by the Ford factory at Willow Grove, Michigan, in the B-24 production pool was a crucial one. Willow Run turned out large numbers of 'KD' aircraft which were shipped, as large components, by road for assembly at Fort Worth and Tulsa (by the Douglas Aircraft Company). 'KDs' were also produced at San Diego; these cockpit sections for late-model Liberators await departure from the California factory.

turret at will. As a safety measure, during take-off everyone except the two pilots and the engineer were required to be strapped-in in the rear of the fuselage. The A-13 also was rarely a problem on landing because it was retractable, unlike the A-2 model used on most B-17s.

The story of Liberator armament continues at this point, even though it gets ahead of much of this narrative. The A-6A and A-6B were the first nose power turrets fitted on Liberators and remained standard until well into the B-24J era. Emerson, located in St Louis, Missouri, manufactured the A-15 power turret which replaced the A-6 as the nose turret of choice for late B-24J and other late-model Liberators. Installation of turret types was not consistent from one factory to another, since decisions were made in part based on availability. To further confuse things, many Liberators (especially B-24Ds in the Pacific) had power turrets installed long after being manufactured, while being readied en route to the combat zone at one AAF depot or another.

A power turret in the nose was a mixed blessing, at best, for every Liberator that had one, from the modified B-24Ds onward through the eventual B-24M. The Liberator always suffered from handling and visibility problems, and these were very much the fault of the nose turret, which was weighty and induced excessive drag. The AAF ran numerous experiments, both in the field and at test facilities in the US, aimed at reshaping the Liberator's nose. In the end, none was adopted operationally. Several experimental Liberators took to the air in the US, including one to which a B-17 Flying Fortress nose had been grafted. Eighth Air Force equipped two B-24s with Bell power-boost nose turrets and sent one back to the United States for evaluation, where it was promptly rejected. As a sidenote, it should be pointed out that visibility from the Liberator's nose, like that from the cockpit, was never good: to see anything, the bombardier had to get down on all fours; the navigator, behind him, could not help with external visual cues because he had insufficient window space.

B-24D production

Production of the widely-employed B-24D model included 2,381 from San Diego, 305 from Fort Worth, and 10 from Tulsa, for a total of 2,696. A single B-24D (41-11678) was converted to the sole XB-24F with a thermal anti-icing system, evaluated by NACA (National Advisory Committee for Aeronautics) at NAS Moffett Field, California. The system included a fairing about 3 ft (0.91 m) in height and 5 ft (1.52 m) in length which protruded above the centre fuselage. Ford modified a B-24D to incorporate a single vertical tail and 1,350-hp (1007-kW) R-1830-75 engines under the designation XB-24K.

A close relative of the B-24D was the AT-22 Liberator advanced trainer. Consolidated turned out five of these at Fort Worth, the first being delivered to the AAF on 11 June 1943. AT-22s were configured with special compartments where flight engineers could be trained. They were used alongside B-24D models which lacked any special designation when used in the training role, and in 1944 were themselves redesignated TB-24Ds.

The designation B-24E applies to the aircraft nearly identical to the B-24D, built by Ford and Consolidated-Fort Worth and final-assembled by Douglas in Tulsa. The Ford B-24E differed from the B-24D in having R-1830-65 engines, a different propeller, and a tunnel gun in the ventral position in lieu of a ball turret.

This version of the Liberator was virtually a training exercise for the hastily-assembled Ford work force: in fact, it took so long to standardise production methods that the entire production run of E models was officially labelled unfit for combat. Almost all B-24Es served in a limited training capacity in the US. Notwithstanding the mercurial nature of 79-year-old Henry Ford – who, among other things, refused to hire women – the automobile company quickly adapted its 'moving line' manufacturing technique. Ford was one of the great success stories of American

These views of B-24Es on the Douglas assembly line show stockpiled sub-assemblies supplied by Ford. Douglas was the only manufacturer to produce B-17s and B-24s.

industry, tapping the methods used to mass-produce the 'Tin Lizzie' Model T automobile and applying them to the Liberator. This bold approach to building many aircraft in a short time did have its drawbacks: Ford was never able to adjust easily to the incremental changes in design of B-24 Liberator models. By adjusting slowly to model changes, Ford continued to turn out B-24E models long after they had been found to be relatively useless.

In all, there were 801 B-24E models: 490 from Willow Run, 167 from Tulsa, and 144 from Fort Worth. One B-24E was converted from 8 September 1943 to become the prototype XC-109 gasoline hauler, the first of more than 200 such conversions of Liberators of various models.

B-24G, H and J

The B-24G was unique to the North American plant at Dallas. It was similar to the B-24Ds produced at other factories, albeit with a nose turret fitted from No. 26 onward. Late B-24Gs also introduced a ball turret.

A latecomer to the production pool of five factories, the North American Dallas facility was a complete and independent B-24 fabrication and assembly plant. The factory ultimately delivered 750 B-24G Liberators and a subsequent 430 B-24Gs, followed by 536 B-24Js. Most of the B-24Gs, which were delivered in natural metal after no. 271, showed up in the Mediterranean area with the Fifteenth Air Force.

The G model initially came off the production line with R-1830-43 engines but, from aircraft No. 111, the R-1830-65 (which was fitted with a different brand of carburettor) was installed.

The first B-24H, which was also the first production Liberator with a factory-installed nose turret, was rolled out

at Willow Run on 30 June 1943. The B-24H model was probably the best-performing and best-liked Liberator. It had most of the features of the later B-24J, L and M variants, but was lighter and, thus, capable of being coaxed 1,000 ft (305 m) higher. This became the most popular Liberator in the hands of crews of the Eighth Air Force in England and the Fifteenth Air Force in Italy; while no figures exist to prove the point, AAF bomber crews felt the H model gave them the best chance for survival. The B-24Hs emerging from Ford at Willow Run used the Emerson nose turret and the Sperry A-13 retractable ball turret. The H model also had improved waist gun positions.

Production of the B-24H model included 1,780 aircraft from Willow Run, 738 from Fort Worth, and 582 from Tulsa, for a total of 3,100.

The B-24J was the definitive Liberator. It was the most used, the most photographed, and the most recaptured, later, in films and models; it also was built in greater numbers than any other. It was the only variant to be

Some accounts have suggested that the 26th B-24G – the first with a nose turret – made its maiden flight before the first B-24H, but this appears not to have been the case and that it was the B-24H that introduced the nose turret. 42-78349 was the 305th of 430 Gs built by North American's Dallas plant, many of which were delivered to units in the Mediterranean. Like the B-24H, the B-24G was equipped with an Emerson A-15 nose turret.

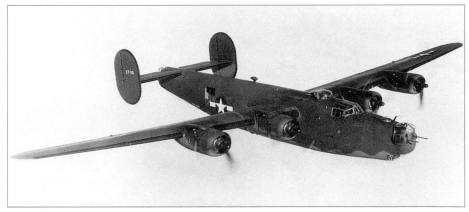

Left: Often described as the most popular of the Liberators among those that flew it, the B-24H was generally lighter than its contemporary, the B-24J, and was therefore able to reach higher cruising altitudes. 42-7718, from the first batch, was a Ford product; others were assembled at Fort Worth and Dallas.

View from the cockpit

A well-known axiom says that the source of complaints from crews concerning the lack of aircraft performance could often by traced to the fact that they were being operated in conditions far in excess of those envisaged by the aircraft's designer. This was as true of the Liberator as for any other type – B-24Js were regularly flown at gross weights of over 36 tons and flying characteristics suffered. The same axiom concludes that even when changes were made to improve performance, squadrons in the field invariably took the opportunity this presented to further increase maximum loads!

As built the Liberator had always had a reputation for possessing heavy controls and increasing gross weights simply made the problem worse. The B-24J was difficult to fly in close formation (a feature not helped by the limited visibility from the cockpit) and so tiring that many pilots found it impossible after a few hours. The long-span Davis wing was responsible for an inherently slow roll rate, made worse by the addition of fuel tanks in the outer wings, and climb rates were notoriously slow with combat loads; an altitude of 1,000 ft (305 m) typically took some six minutes to reach. The visibility problem was a product of the cockpit's inward-sloping side windows, which forced the pilot and co-pilot to sit low in order to have enough headroom. Thus, it was difficult to see over the instrument panel, a situation made worse by the presence of the compass atop the instrument panel and the astrodome and nose turret.

ered by one version or other of the Twin Wasp 14-cylinder, two-row radial engine
-47, Consolidated PBY and Grumman F4F. Twin Wasp production totalled some
rers Buick and Chevrolet undertaking production alongside P&W. Consolidated
use of its smaller frontal area, which incurred a lesser drag penalty.
rbo-supercharged and typically rated at 1,200 hp (895 kW) for take-off. Nine variants
ily, differences being largely confined to cooling, carburation and turbocharging
ne with the PB4Y- 2 Privateer, which utilised four R-1830-94s. These were rated at
tted with turbochargers, as the Privateer was expected to spend much of its time
changes meant that the -94 engine was housed in new cowlings which differed in
r cooling and intercooler intakes either side of the B-24's engines gave their
rbocharger (and, therefore, no intercooler) the R-1830-94 used a more circular
the RY-3 (Liberator C.Mk IX) and Convair Model 39 (XR2Y-1) transports.
nat, compared with the Wright R-1820 Cyclone (which powered the B-17), the
e Cyclone had a reputation as an 'oil slinger'), while its various systems were more
ce. Engine changes were also more straightforward in a B-24 – a complete 'power
norm and simply required the removal of four bolts holding the pack to the airframe.

Consolidated-Vultee B-24H-1-FO Liberator

726th Bombardment Squadron (Heavy)
451st Bombardment Group (Heavy)
49th Bombardment Wing
15th Air Force, US Army Air Forces
Castelluccio, Italy
Summer 1944

The 451st Bombardment Group (comprising the 724th, 725th, 726th and 727th Bomb Squadrons) began operations with the 47th Bombardment Wing, 15th Air Force from Gioia del Colle, Italy in January 1944. Moving to San Pancrazio in March, the Group transferred, in April, to the 49th BW – the fourth and last 15th Air Force B-24 wing to enter combat. Concurrent with this change came a move to Castelluccio, which remained the group's base for the duration of hostilities. Campaigns in which the group fought were: Europe, Naples-Foggia, Anzio, Rome-Arno, Normandy, Northern France, Southern France, North Apennines, Rhineland, Central Europe and Po Valley. By VE-Day the four squadron had completed 216 missions, 5,568 effective sorties, dropped 11,961 tons (12153 tonnes) of bombs and lost 135 aircraft. Though these losses were comparatively high, the 451st claimed 98 enemy aircraft destroyed, with an additional 41 probables and 13 damaged. For missions against Regensburg (25 February 1944), Ploesti (5 April 1944) and Vienna (23 August 1944), the Group received three Distinguished Unit Citations.

Revised nose armament
Though the B-24G-1-NT is often credited with introducing the nose turret to production Liberators, the first 25 examples of this North American-built variant were completed with B-24G-NTs, with a 'glass' nose as per the B-24D. The first 'nose-turreted G' was not delivered to the USAAF until November 1943, some five months after the first B-24H (42-7465) had been handed over. The first B-24Hs were also the first complete (as opposed to knock-down) Liberators built by Ford, the Willow Run factory eventually completing 1,780 in all. Total B-24H production reached 3,100 (of 9,104 ordered), Consolidated's Fort Worth plant completing 738 aircraft, and Douglas at Tulsa another 582.

While the bulk of the first Liberators equipped with nose turrets were equipped with CAC Model 5800-3 or CAC (Motor Products Corporation) Model 5800-5 turrets (Army designations A-6A or A-6B, respectively) as previously used in the tail position, the B-24H was an early user of the Emerson Model 127 (Army designation A-15) turret, which was later universally adopted for production B-24s. The Emerson design had also been derived from a design for a tail turret, though the Model 111 was never installed in the rear of a B-24 as planned. Instead, with the need for a nose turret-equipped Liberator for issue to front-line squadrons gaining urgency, a crash modification programme was ordered and tests carried out in early 1943, resulting in the A-15 nose turret as installed in B-24H-FOs. Though 190 lb (86.2 kg) heavier than the A-6 and with a slightly smaller field of fire, the Emerson turret was judged the superior as it was more responsive, being electrically powered rather than hydraulically powered via an electric pump, as was the CAC/Motor Products A-6.

Other defensive armament in the B-24H-FO comprised a Briggs A-13 retractable ball turret (as first introduced in the B-24D-140-CO), a Martin A-3C electric top turret (replaced by the A-3D 'high hat' at Block 25) and a CAC/Motor Products A-6A tail turret (replaced by the A-6B from B-24H-5-FO onwards). Each waist window carried a flexible 0.5-in machine-gun and was completely open to the air, though a small wind deflector was fitted just forward of the opening. Blamed for the high incidence of frostbite among waist gunners, the waist window was the subject of modification by Ford and Northwest Airlines' St Paul Modification Center. Ford's K-6 mount, consisting of a single Perspex sheet covering the opening and incorporating a universal swivel gun mount, was adopted for production aircraft and was introduced at Block 20 at all three factories building the B-24H.

Pratt & Whitney R-1830 Twin Wasp
All members of the Liberator family were pow that also powered such types as the Douglas 173,618 examples, with automobile manufact designer 'Mac' Laddon chose the R-1830 beca

Engines fitted to B-24s and PB4Y-1s were t were employed across the whole Liberator fan details. The only major powerplant variation ca 1,350 hp (1007 kW) for take-off but were not f on patrol at comparatively low altitudes. Thes shape (viewed from the front). The turbocharg cowlings their distinctive oval shape. With no cowling. This Twin Wasp variant also powered

Consensus among ground crews suggests R-1830 was a cleaner engine to work on (as th modern and easier to access during maintenar pack' (engine plus accessories) swap was the

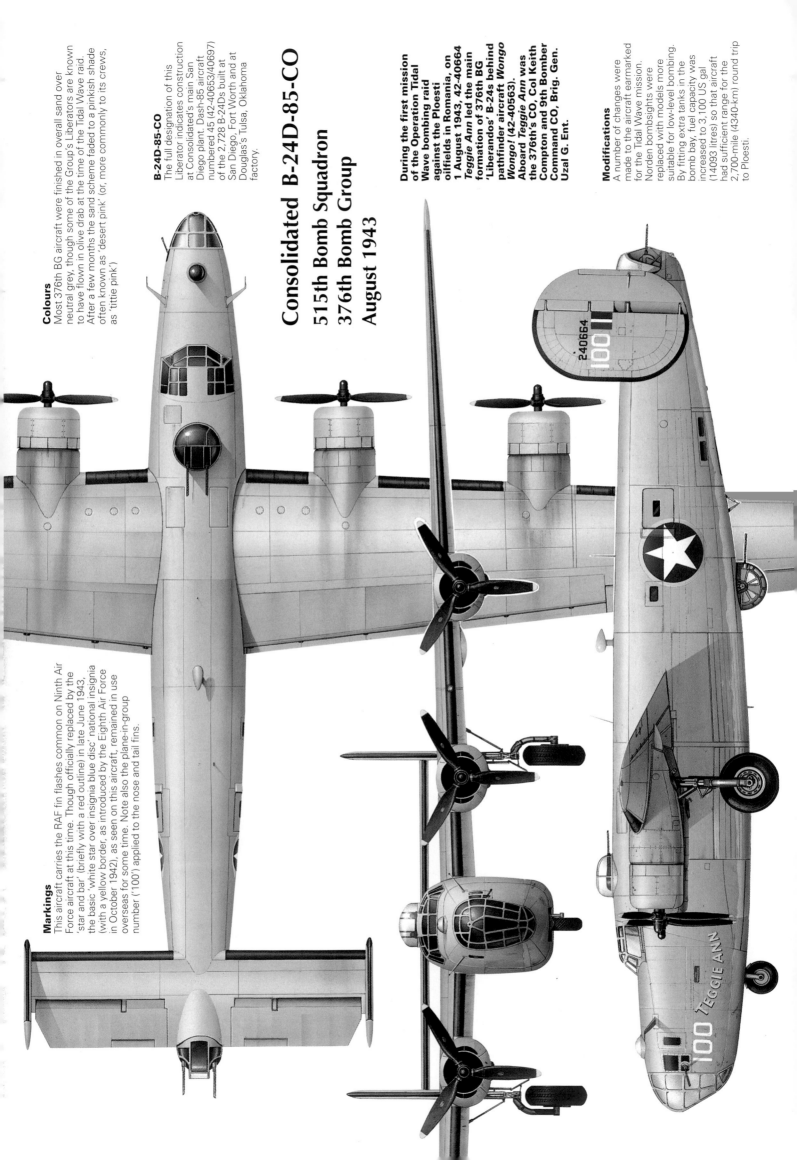

Colours

Most 376th BG aircraft were finished in overall sand over neutral grey, though some of the Group's Liberators are known to have flown in olive drab at the time of the Tidal Wave raid. After a few months the sand scheme faded to a pinkish shade often known as 'desert pink' (or, more commonly to its crews, as 'tittie pink')

B-24D-85-CO

The full designation of this Liberator indicates construction at Consolidated's main San Diego plant. Dash-85 aircraft numbered 45 (42-40653/40697) of the 2,728 B-24Ds built at San Diego, Fort Worth and at Douglas's Tulsa, Oklahoma factory.

Consolidated B-24D-85-CO
515th Bomb Squadron
376th Bomb Group
August 1943

During the first mission of the Operation Tidal Wave bombing raid against the Ploesti oilfields in Romania, on 1 August 1943, 42-40664 *Teggie Ann* led the main formation of 376th BG 'Liberandos' B-24s behind pathfinder aircraft *Wongo Wongo!* (42-40563). Aboard *Teggie Ann* was the 376th's CO, Col Keith Compton and 9th Bomber Command CO, Brig. Gen. Uzal G. Ent.

Modifications

A number of changes were made to the aircraft earmarked for the Tidal Wave mission. Norden bombsights were replaced with models more suitable for low-level bombing. By fitting extra tanks in the bomb bay, fuel capacity was increased to 3,100 US gal (14093 litres) so that aircraft had sufficient range for the 2,700-mile (4340-km) round trip to Ploesti.

Markings

This aircraft carries the RAF fin flashes common on Ninth Air Force aircraft at this time. Though officially replaced by the 'star and bar' (briefly with a red outline) in late June 1943, the basic 'white star over insignia blue disc' national insignia (with a yellow border, as introduced by the Eighth Air Force in October 1942), as seen on this aircraft, remained in use overseas for some time. Note also the plane-in-group number ('100') applied to the nose and tail fins.

Left: This USAF diagram shows the positions occupied by the 10 crew in a B-24D, from the front: bombardier, navigator, two pilots, radio operator and dorsal gunner, ball gunner, two waist gunners and a rear gunner.

Right: Liberator nose turrets brought complaints from bombardiers bemoaning the now limited visibility from the nose of their aircraft; with little head room the only way they could see forward was by getting down on hands and knees. With this in mind B-24Ds fitted with nose turrets often had extra windows fitted either side of the nose, aft of the turret. Above the bombardier's position on this 30th Bomb Group B-24J-CO-176 is a Consolidated/Motor Products ('Consair') A-6A turret; some late B-24Js used Emerson A-15 turrets.

Consolidated B-24J Liberator

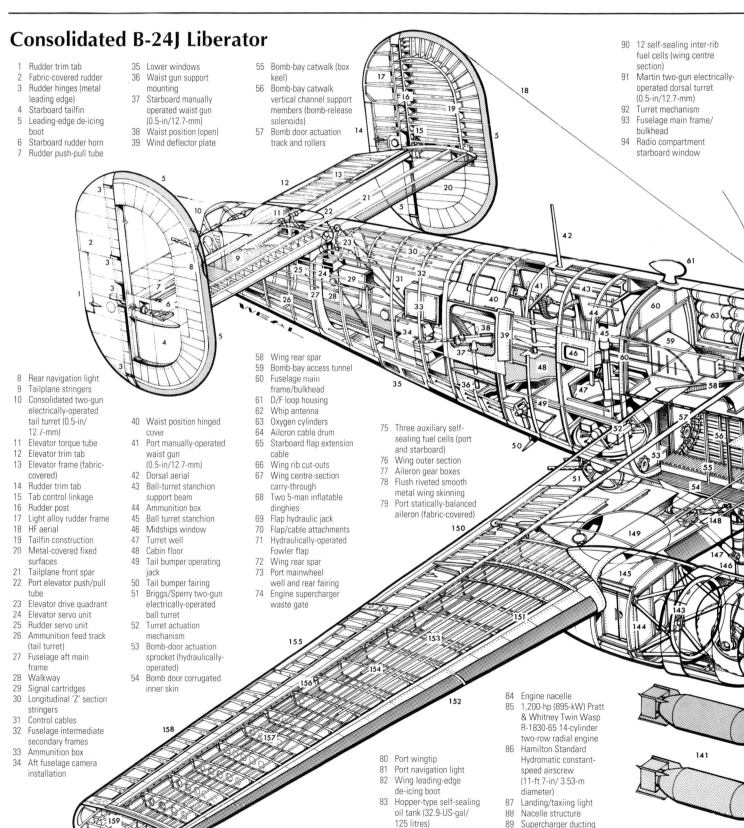

1 Rudder trim tab
2 Fabric-covered rudder
3 Rudder hinges (metal leading edge)
4 Starboard tailfin
5 Leading-edge de-icing boot
6 Starboard rudder horn
7 Rudder push-pull tube
8 Rear navigation light
9 Tailplane stringers
10 Consolidated two-gun electrically-operated tail turret (0.5-in/ 12./-mm)
11 Elevator torque tube
12 Elevator trim tab
13 Elevator frame (fabric-covered)
14 Rudder trim tab
15 Tab control linkage
16 Rudder post
17 Light alloy rudder frame
18 HF aerial
19 Tailfin construction
20 Metal-covered fixed surfaces
21 Tailplane front spar
22 Port elevator push/pull tube
23 Elevator drive quadrant
24 Elevator servo unit
25 Rudder servo unit
26 Ammunition feed track (tail turret)
27 Fuselage aft main frame
28 Walkway
29 Signal cartridges
30 Longitudinal 'Z' section stringers
31 Control cables
32 Fuselage intermediate secondary frames
33 Ammunition box
34 Aft fuselage camera installation

35 Lower windows
36 Waist gun support mounting
37 Starboard manually operated waist gun (0.5-in/12.7-mm)
38 Waist position (open)
39 Wind deflector plate
40 Waist position hinged cover
41 Port manually-operated waist gun (0.5-in/12.7-mm)
42 Dorsal aerial
43 Ball-turret stanchion support beam
44 Ammunition box
45 Ball turret stanchion
46 Midships window
47 Turret well
48 Cabin floor
49 Tail bumper operating jack
50 Tail bumper fairing
51 Briggs/Sperry two-gun electrically-operated ball turret
52 Turret actuation mechanism
53 Bomb-door actuation sprocket (hydraulically-operated)
54 Bomb door corrugated inner skin

55 Bomb-bay catwalk (box keel)
56 Bomb-bay catwalk vertical channel support members (bomb-release solenoids)
57 Bomb door actuation track and rollers
58 Wing rear spar
59 Bomb-bay access tunnel
60 Fuselage main frame/bulkhead
61 D/F loop housing
62 Whip antenna
63 Oxygen cylinders
64 Aileron cable drum
65 Starboard flap extension cable
66 Wing rib cut-outs
67 Wing centre-section carry-through
68 Two 5-man inflatable dinghies
69 Flap hydraulic jack
70 Flap/cable attachments
71 Hydraulically-operated Fowler flap
72 Wing rear spar
73 Port mainwheel well and rear fairing
74 Engine supercharger waste gate
75 Three auxiliary self-sealing fuel cells (port and starboard)
76 Wing outer section
77 Aileron gear boxes
78 Flush riveted smooth metal wing skinning
79 Port statically-balanced aileron (fabric-covered)

80 Port wingtip
81 Port navigation light
82 Wing leading-edge de-icing boot
83 Hopper-type self-sealing oil tank (32.9-US-gal/ 125 litres)

90 12 self-sealing inter-rib fuel cells (wing centre section)
91 Martin two-gun electrically-operated dorsal turret (0.5-in/12.7-mm)
92 Turret mechanism
93 Fuselage main frame/ bulkhead
94 Radio compartment starboard window

84 Engine nacelle
85 1,200-hp (895-kW) Pratt & Whitney Twin Wasp R-1830-65 14-cylinder two-row radial engine
86 Hamilton Standard Hydromatic constant-speed airscrew (11-ft 7-in/ 3.53-m diameter)
87 Landing/taxiing light
88 Nacelle structure
89 Supercharger ducting

Iain Wyllie

Ford's Liberator's – production at Willow Run

B-24H-1-FO Liberator 42-7697 was among the first batch of 305 B-24Hs built by the Ford Motor Company, as part of an initial contract for 795 aircraft (including 490 B-24Es) placed with the auto manufacturer in September 1941. Ford was initially contracted to supply complete B-24s in knock-down (or KD) condition at the rate of 100 per month. Each KD included all the major airframe components – fuselage, wings, tail and landing gear – which were shipped for final assembly to one of three companies at one of four locations, namely Consolidated at San Diego or Fort Worth, Douglas at Tulsa, Oklahoma or North American at Dallas, Texas. The San Diego and Dallas plants would also fabricate as well as assemble B-24s. From the beginning Ford had expressed an interest in producing finished aircraft and had built the giant Willow Run, Michigan plant with this in mind. The company's enthusiasm was finally rewarded with the September 1941 contract and Willow Run eventually produced 6,792 B-24s on top of 1,893 in KD form, though the factory never ran at full capacity. (In March 1944 it was producing better than one aircraft every 100 minutes, 24-hours-a-day, seven-days-a-week. It could have built 600 aircraft per month, but the USAAF had difficulty assimilating completed aircraft at the same rate.)

From the outset Ford was keen to apply its production line methods to the manufacture of the Liberator but soon experienced difficulties. Thirty thousand drawings needed to be copied for interpretation by a relatively unskilled labour force – a time-consuming exercise made worse by the fact that Consolidated used fractional measurements, while Ford used a decimal system. Working with aluminium, rather than steel demanded new skills and the constant design changes inherent in the production of combat aircraft, especially in wartime, were very much at odds with a car industry more used to year-long production runs without significant changes to a product. For example, Ford built 21,000 production jigs and related fixtures for Liberator production, but only used 11,000, the remainder having been rendered useless by changes to the aircraft's design.

In time Ford got to grips with these problems, but delays meant that as the first Ford-built KD B-24E left the Douglas assembly plant in early 1943, the entire production run of this variant was only months away from being declared obsolete. In the event, the last 'E' was delivered in September; very few served in action overseas (examples were noted in use by the 8th Air Force as late as 1944) most having been reduced to training roles.

ctric' aircraft, the B-24 employed hydraulic landing gear, wheel
s. The main hydraulic system was powered by a pump driven off
liary electric pump. If both pumps failed a hand pump was provided
s seat.
em powered the engines' primers and starters, fuel boost pumps,
feathering pumps, cowling flaps, intercooler shutter motors, oil
ntation, lighting and communications. Fed by generators on each
two storage batteries and an auxiliary petrol-driven generator,
ht. This novel device, equivalent to the modern auxiliary power unit
nofficially as the 'putt-putt') allowed a B-24 to be started by one
tarter cart. The APU also allowed pre-flight checks, turret running
out having to start the aircraft's engines.

B-24H in service

Ford production at Willow Run transitioned from the B-24E to the H in early 1943, referring to the process as the '801 change', as it occurred as the company produced its 801st airframe (including knock-down aircraft). Keen to provide its squadrons with turret-equipped aircraft as soon as possible, the USAAF requested the fitting of a nose turret and 55 other so-called 'master changes' in mid-March 1943 with a demand that 50 of the new aircraft be completed by 30 June. However, detail design hold-ups and supply problems conspired against Ford and by the end of June only six B-24Hs had been delivered. Forty-two followed in July and during that month most were delivered to the 392nd Bomb Group, 8th Air Force in England.

The H was possibly the most popular Liberator bomber variant from the crews' point-of-view. Used extensively in Europe, it was equipped with the all important nose turret, was lighter than the contemporary B-24J (and therefore able to attain an extra 1,000-ft of altitude in the hands of an experienced pilot) and was fitted with a tab on its left aileron which made it possible to trim either wing more effectively. The 392nd Bomb Group actively attempted to (unofficially) standardise on the B-24H and became well-known for their willingness to swap a brand-new B-24J for an H in good condition. This did mean, however, that pilots were forced to sacrifice the preferred C-1 autopilot of the J-model for the B-24H's A-5 equipment.

Not all the B-24H's features met with wholesale approval in the ETO; Ford's armoured pilot's seat did not allow the occupant to wear a parachute and hindered a fast exit from the aircraft. Many were removed and replaced with a 'flak curtain' fitted to the rear of the flight deck.

Offensive payload

Any discussion regarding the relative merits of the B-24 over the B-17 invariably came around to the size of the Liberator's bomb bays. Measuring 17 ft 10 in (5.44 m) in overall length, they was almost twice the size of the B-17's bomb bay. Its roller-shutter doors were also unique, designed and patented by Consolidated employee Emric Berger and licensed to the US Government without royalty payment. These retracted upwards along the outside of the fuselage, rather than dropping down into the airstream, as with conventional bomb doors. With normal tyre deflection, ground clearance under Liberator's fuselage was less than 20 in (50.8 cm). This, plus a level attitude afforded by tricycle undercarriage, made loading much easier.

Provision was made for the following bomb loads: 20 x 100-lb, 12 x 300-lb, 500-lb or 600-lb, eight 1,000-lb or 1,100-lb or four 2,000-lb. From B-24D-30-CO, stations were provided to carry eight 1,600-lb armour-piercing bombs, though with this load the bomb bay doors could not be fully closed, limiting speed to 275 mph (443 km/h). A demountable underwing rack located between the fuselage and inner engine on each wing could be employed to carry a 4,000-lb bomb, though with this load flight characteristics were adversely effected and the capability was rarely used. The Liberator's bomb bay was always loaded from the front in deference to centre-of-gravity considerations.

Hydraulic systems

While the B-17 was essentially an 'ele
brakes, wing flaps and bomb bay doo
No. 3 engine and backed up by an au
on the floor to the right of the co-pilot

The B-24's 24-volt DC electrical sys
the auxiliary hydraulic pump, propelle
dilution solenoids, as well as instrume
engine, the system was backed up b
located in the nose wheel compartme
(and known officially as the APU, but
man without the need for an external
and radio checks to be carried out wit

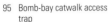

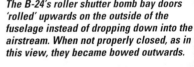

Right: This unidentified USAAF B-24 appears to be about to receive a new (or newly overhauled) No. 2 engine. The P&W R-1830 Twin Wasp had an enviable reliability record and was cleaner to work on than the B-17's Wright Cyclone, which had a reputation as an 'oil thrower'.

95 Bomb-bay catwalk access trap
96 Radio-operator's position
97 Sound-insulation wall padding
98 Emergency escape hatch
99 Pilot's seat
100 Co-pilot's seat
101 Co-pilot's rudder pedals
102 Instrument panel
103 Windscreen panels
104 Compass housing
105 Control wheel
106 Control wheel mounting
107 Control linkage chain
108 Fuselage forward main frame bulkhead

114 Consolidated (or Emerson) two-gun electrically-operated nose turret (0.5-in/12.7-mm)
115 Turret seating
116 Optically-flat bomb-aiming panel
117 Nose side-glazing
118 Bombardier's prone couch

146 Mainwheel oleo (Bendix pneudraulic strut)
147 Side brace (jointed)
148 Undercarriage actuating cylinder

B-24J Specifications

Powerplant: four Pratt & Whitney R-1830-65 Twin Wasp 14-cylinder radial piston engines, each rated at 1,200 hp (895 kW)

Performance
maximum speed: 290 mph (467 km/h) at 25,000 ft (7620 m)
cruising speed: 215 mph (346 km/h)
range: 2,100 miles (3379 km) with 5,000-lb (2270-kg) bombload
maximum range: 3,300 miles (5310 km)
ceiling: 28,000 ft (8534 m)

Weights
empty: 36,500 lb (16556 kg)
maximum gross: 65,000 lb (29484 kg)

Dimensions
span: 110 ft (33.53 m)
length: 67 ft 2 in (20.47 m)
height: 18 ft (5.48 m)
wing area: 1,048 sq ft (97.36 m²)

Armament
defensive (as built): 10 0.5-in machine-guns positioned in pairs in nose, dorsal, tail and ball turrets and in waist positions (one each)
bombload: various types to a total weight of up to 8,800 lb (3990 kg)

The B-24's roller shutter bomb bay doors 'rolled' upwards on the outside of the fuselage instead of dropping down into the airstream. When not properly closed, as in this view, they became bowed outwards.

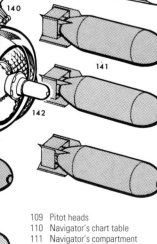

119 Ammunition boxes
120 Navigator's swivel seat
121 Navigator's compartment entry hatch (via nosewheel well)
122 Nosewheel well
123 Nosewheel door
124 Forward-retracting free-swivelling nosewheel (self-aligning)
125 Mudguard
126 Torque links
127 Nosewheel oleo strut
128 Angled bulkhead
129 Cockpit floor support structure
130 Nosewheel retraction jack
131 Smooth-stressed Alclad fuselage skinning
132 Underfloor electrics bay
133 Roll top desk-type bomb-bay doors (four)

109 Pitot heads
110 Navigator's chart table
111 Navigator's compartment starboard window
112 Chart table lighting
113 Astro-dome

134 Supercharger nacelle cheek intakes
135 Ventral aerial (beneath bomb-bay catwalk)
136 Nacelle/wing attachment cut-out
137 Wing front spar nacelle support
138 Undercarriage front-pivoting shaft
139 Drag strut
140 Bendix scissors
141 Internal bomb load (max 8,000 lb/3629 kg)
142 Starboard mainwheel
143 Engine-mounting ring
144 Firewall
145 Monocoque oil tank

149 Starboard mainwheel well and rear fairing
150 Fowler flap structure
151 Wing front spar
152 Wing leading-edge de-icing boot
153 All-metal wing structure
154 Spanwise wing stringers
155 Aileron trim tab (starboard only)
156 Wing rear spar
157 Wing ribs (pressed and built-up former)
158 Statically-balanced aileron (metal frame)
159 Starboard navigation light
160 Wingtip structure

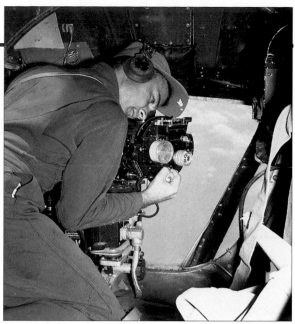

Above: B-24 bombardiers typically used a Mk 14 bomb sight. This view illustrates the bombardier's cramped conditions in aircraft equipped with a nose turret. Note the pad above his head, positioned to afford some protection in turbulence.

Above: Early Liberators were equipped with 'D-shaped' control yokes, but these were replaced by 'U-shaped' controls (as shown here) in later production B-24s. Note the magnetic compass above the control panel, impeding the pilot's already limited visibility, C-1 autopilot controls in the cut-out in the centre of the instrument panel and a Honeywell turbo-supercharger selector box, mounted to the left of the throttles.

Above: Hit by either flak or fire from an enemy fighter, the Plexiglas on the A-6 tail turret of this B-24H has been blasted away, leaving the armoured central glass panel shattered but intact.

Below: The waist gunners' staggered positions were open to the elements on the B-24D; on later variants they were enclosed. A single 0.5-in machine-gun in each position was standard, though twin mounts were sometimes fitted in the field.

Above: In a bid to reduce weight the tail turret in some B-24Js and Ls was deleted and an M-6A 'stinger' mount fitted in its place. This consisted of a pair of hand-held guns in a hydraulically-assisted mount. However, in the field it was not unusual for this to be replaced with an A-6 turret or an open twin-gun mount, as illustrated here. Note that each pair of guns is mounted closer together than in a standard turret.

The tunnel gun (above) fitted to B-24Cs and early Ds was awkward to aim and fire and largely ineffective. It was replaced by the Bendix belly turret that proved equally troublesome and which, in turn, gave way to the ball turret fitted to most later B-24s. The latter was the Briggs/Sperry A-13/A-13A ball turret (below) introduced in late-production B-24Ds and fitted to most subsequent models.

Built in greater numbers than any other Liberator variant, the B-24J was the only model assembled by all five factories in the production pool. The first examples appeared from the San Diego plant in August 1943, at Fort Worth in September 1943, Willow Run in April 1944 and Dallas and Tulsa in May 1944. Production totalled some 6,678 aircraft. 42-73250 (above) is a B-24J-25-CO with Consolidated A-6A nose and tail turrets, A-3C dorsal turret and A-13 ball turret; B-24J-150-CO 44-40188 (right) differs in having later A-6B turrets in the nose and tail positions.

Navy Liberators

Consolidated also developed the naval patrol version, the PB4Y-1. This happened after Consolidated elected not to proceed with a twin-engined Navy variant, known in company parlance as the LB-29. There was need for a rugged, long-range craft that could have some impact on the damage being inflicted by German U-boats in the North Atlantic, although, as it would turn out, the PB4Y-1 Liberator became especially effective in the Pacific. The Navy took charge of the first example in September 1942.

The advent of the naval Liberator highlighted an Army-Navy dispute in which key players were the AAF's General Henry H. 'Hap' Arnold and Navy boss Admiral Ernest J. King. Arnold and the Army wanted to keep a lock on land-based bombers, even if that meant taking on the unpopular (for soldiers) long-range anti-submarine mission. The Navy had a chip with which to bargain: it agreed to give up its plans to manufacture the Boeing PBB-1 Sea Ranger patrol flying-boat at Boeing's Renton, Washington plant (a factory sorely needed by the Army for B-29 Superfortress production). As part of the deal, the Navy received B-24Ds, B-25 Mitchells and B-34 Venturas straight off the production line, designated PB4Y, PBJ and PV.

The first Navy Liberators were little more than B-24Ds reserialled with a Bureau Number. More than 160 of these PB4Y-1 models were wearing Navy colours in August 1943 when the AAF gave up its long-standing anti-submarine patrol mission and transferred another 45 Liberators to the Navy. Most of these were B-24Ds, but all twin-tailed Liberators flown by the Navy received the PB4Y-1 designation regardless of which AAF model they represented.

The Navy modified many of its B-24D/PB4Y-1 aircraft with an Erco (Engineering and Research Co.) powered ball turret in the nose. It continued to receive PB4Y-1 Liberators until January 1943, when the sea service acquired its last ship, a B-24M. A single North American B-24G (42-78271) became a PB4Y-1 with BuNo. 38782, the only aircraft among the 977 PB4Y-1s not manufactured at San Diego.

As part of several programmes aimed at attacking German submarine pens and V-1 'buzz bomb' sites, the Navy converted several PB4Y-1s into flying bombs. They carried huge explosive charges and were to be taken almost to the target, where the pilot would bail out. Then, this early version of a cruise missile would continue, unmanned, to its target and wreak destruction. The programme proceeded in parallel with the AAF's Project Aphrodite, which used B-17 Flying Fortresses for the same job. No real success was achieved.

manufactured by all five facilities in the production pool. Consolidated in San Diego began B-24J production (directly following the B-24D model) on 31 August 1943. The B-24J was distinguished by the C-1 autopilot and M-9 bomb sight. It incorporated Consolidated-designed nose and tail turrets in lieu of the Emerson turrets on earlier Liberators. B-24J production totalled 6,678 bombers, including 2,792 from San Diego, 1,587 from Willow Run, 1,558 from Fort Worth, 536 from Dallas, and 205 from Tulsa.

By the time the Allies landed in Normandy on 6 June 1944, it was obvious that Consolidated's San Diego plant and Ford's Willow Run outlet were more than sufficient to provide all the Liberators that would be needed by the United States and its allies. In fact, American industry manufactured just a few hundred less than 100,000 aircraft in 1944, including dozens of experimental prototypes, many of which contributed neither to the enhancement of aeronautical knowledge nor to the war effort. Belatedly, it was determined to end Liberator production by Douglas in Tulsa, North American in Dallas, and Consolidated in Fort Worth, although the Fort Worth facility continued to turn out 'Libs' until the end of the year.

Modifications

To understand how Liberators were finished, and in many cases modified, en route to the combat zone, it must be remembered that nine field centres were in operation during the war. At these locations, technicians worked on the typical B-24 to transform it into a finished product, ready for action. The best known was the Consolidated

Later production B-24Js are exemplified by B-24J-65-CF 44-10560, which features an Emerson A-15 nose turret though, like all Fort Worth-built Js, it retains redundant tunnel gun scanning windows in the lower rear fuselage.

San Diego's mechanised assembly line moved at a rate of 8½ in (21.6 cm) per minute. B-24J 44-40114 nears completion during 1944.

facility in Tucson, Arizona, which was involved in the American war effort from the start – LB-30 crews stopped there to pick up aircraft within days of the Pearl Harbor attack, thinking they were en route to the Philippines, though they actually ended up in Java – and which processed 3,888 Liberators between their factories and their operational units.

The other facilities were operated by American Airlines in New York (99 Liberators modified), Bechtel in Birmingham, Alabama (3,857), Consolidated in Fort Worth (854), Consolidated in Louisville, Kentucky (1,746), Douglas in Tulsa, Oklahoma (642), Lockheed in Dallas, Texas (22), Martin in Omaha, Nebraska (18), Northwest Airlines in St Paul, Minnesota (3,114), and United Airlines in Cheyenne, Wyoming (14,380).

Some of the modifications were simple and obvious, such as installing appropriate radio equipment depending on the war theatre to which a B-24 was bound. Others were more complex, such as installing an H$_2$X radar unit on a Liberator that became a radar pathfinder, or 'Mickey' aircraft; or converting a B-24 to an F 7 photo-reconnaissance variant; or (in the case of four bombers that went through St Paul) adding both radar and photo capability.

In addition to the centres, AAF bases sometimes modified Liberators en route to combat units. For example, the AAF's Hawaiian air depot installed many dozens of power nose turrets on early B-24D models that were manufactured without them. By one estimate, by the time B-24 production reached its height in 1944, there were 50 versions of the Liberator, and they were not interchangeable. In some cases, the cockpits were so different that a pilot could not simply move from one to the other.

Then there were the one-off modifications. Dozens of early B-24D models were modified for specific projects, and each became unique. One Liberator (42-40200) became a test ship for Project Yehudi, the programme aimed at rendering an aircraft invisible through external illumination. Another (41-11707) went to Wright Field to evaluate wide-bladed, or paddle, propellers. Yet another B-24D Liberator (40-2352) became one of the first American air-refuelling tankers, using a hose, reel, and line system to refuel a B-17E Flying Fortress in tests at Eglin Army Air Field, Florida. A D-model (42-41003) was retained at the San Diego factory to evaluate thermal de-icing features.

One of the more ambitious modifications reflected the AAF interest in finding a way to escort bombers to their targets. Air Proving Ground Command at Eglin Army Airfield converted a B-24D Liberator (41-11822) into the sole XB-41 bomber escort, a programme that was undertaken after development of the more successful YB-40

B-24J-195-CO 44-41064 V GRAND was San Diego's 5,000th B-24. 'Autographed' by the plant's employees, it served with the 783rd BS, 465th BG of the 15th Air Force.

Flying Fortress intended for the same job.

The XB-41 was equipped with an added Bendix chin turret, a different tail gun installation, a second Martin turret aft of the wing, a revised Sperry ball turret, and dual, power-operated N5 gun units in the waist positions. The result was something like a porcupine that stuck out in all directions – but carried a mind-boggling 12,830 rounds of ammunition.

In February 1943, the XB-41 was put through its paces at Eglin. Initial experiments, including firing of all weapons, seemed promising, but a full evaluation was never conducted to see whether the heavier XB-41 could keep up with a Liberator formation. By then, a dozen YB-40s were operating in combat with the 92nd BG in England, but they were of little use in fending off Luftwaffe attacks. The idea of a heavy bomber being fitted with guns to escort other bombers was pushed out of the way when it became apparent that the P-51 Mustang was fully capable of going all the way with the bombers.

South Pacific fighting

When the United States entered World War II, the first B-24D had not yet emerged from the factory, but American leaders and doctrine were firmly fixed on the idea that giant waves of four-engined bombers would be the backbone of US air power.

The features of the B-24, principally its 'Davis' wing, that bestowed upon it such impressive range made it ideal for naval use and far more suitable than the B-17 (PB-1) for patrol duties. Here an anonymous PB4Y-1 is seen in its element, on a doubtless long, and probably uneventful, over-water patrol.

Right: This PB4Y-1 (ex-B-24D) carries its designation, Bureau Number and AAF serial on its tail fin. Like their USAAF counterparts in the Pacific, a number of these early PB4Y-1s had a nose turret fitted – more often than not of the Erco ball variety.

Lower right: At the other end of the PB4Y-1 spectrum in terms of its origin was Brown Bagger's Retreat, former B-24L 44-41800, BuNo. 65385. Note the Erco 250SH nose ball turret, which was also fitted to some B-24Ds and was to be a feature of the PB4Y-2 Privateer.

PB4Y-1 – the Navy's Liberators

When, in the autumn of 1943, the USAAF relinquished its ASW role it handed over 45 B-24Ds to the US Navy. These joined 167 Liberators already in Navy service (for ASW and other duties), the first of which had been acquired in September 1942 and issued to VB-101 at NAS Barber's Point, Hawaii. All were designated PB4Y-1, regardless of their AAF designation and, by January 1945, totalled some 977 aircraft. With the exception of a single B-24G acquired as a test ship, all PB4Y-1s were San Diego-built machines.

The loss of a B-24A at Pearl Harbor (described previously) resulted in two crew members killed and three wounded, the first American casualties of the war. Staff Sergeant Burton R. Grinyer, the crew's photographer, was walking toward the Liberator with two other members of the crew at the time of the attack. One of the men ran, and was cut down by another Japanese aircraft on a strafing run. A second went in a different direction and was wounded. Grinyer figured that lightning couldn't strike twice and dived into the crater made by the bomb. He survived.

LB-30 and B-24A in combat

The LB-30 and B-24A were the first American Liberators to be blooded in combat, during the early 1942 fight for Java and the raid on Broome, Australia. The 7th Bombardment Group (Heavy) fought on Java in January and February 1942 with one squadron each of B-17E Flying Fortresses and LB-30 Liberators, before being overwhelmed by numerically superior Japanese forces.

The LB-30s that eventually went to Java with the 7th BG were flown only by the group's 11th BS. Crews picked them up at the Consolidated modification plant at Tucson, Arizona. After a hasty crew checkout, they departed for Wright Field, Ohio, for further modification, and from there to the port of debarkation, MacDill Army Air Field at Tampa, Florida, where they were processed for overseas movement and sent forth.

The AAF ordered the 11th BS to the Philippines, but it was too late and the Liberators never got there. On 8 December 1941, an unprepared General Douglas MacArthur caused much of the United States' air power in the region to be caught and destroyed on the ground. By January 1942, the Philippines was being overrun. The 7th BG ended up instead on Java, where the Americans flew the first combat mission by the 7th BG on the night of 16-17 January 1942.

That night, Major Austin Straubel led three LB-30s piloted by himself (AL609), First Lieutenant Jack Dougherty (AL535) and First Lieutenant William E. Bayse (AL576). These Liberators, along with two B-17Es, took off from Malang and flew northeast, where they stopped overnight at Kendari, refuelled, and loaded bombs. Then, Straubel's bombers proceeded to two targets on Celebes. The first target was an airfield at Langoan for the LB-30s, while the two B-17Es struck at shipping in the harbour of Menado.

"When we took off," remembered Straubel's bombardier, Captain Raymond O. Carr, "we had to use our landing lights, as no runway lights were on. This field [Kendari] was the objective of the Japanese who had landed, so we fully expected they were already around the field and would start firing at us when they saw our lights. I had the .50-calibre [12.7-mm] gun in the nose ready to shoot the minute I saw any sign of them. Nothing happened on the outskirts of the field, so we breathed a sigh of relief and began concentrating on the work before us."

At Langoan, Straubel's flight bombed the airfield from 19,500 ft (5945 m), arriving over target at 05.36. The men saw no aircraft on the ground. They noted no anti-aircraft fire. "We didn't come in contact with any pursuits," Carr recalled. "We took a straight course for home and slowly lost our altitude, as we thought we didn't have anything else to worry about. "Suddenly, somebody yelled, 'Pursuits!'" Five Mitsubishi A6M Zeros attacked the three LB-30s. One of the gunners was credited with shooting down one Japanese fighter. Aboard Bayse's aircraft, First Lieutenant Victor Poncik saw Dougherty suddenly peel off to the right into a huge cumulus cloud with an engine smoking – "...and we saw no more of him" – although he later reappeared.

Bayse's bomber took hits in two engines; he feathered one and shut it down. He struggled to keep formation with Straubel but fell back, even at full power on the remaining engines. Finally, with two badly wounded gunners onboard, with damaged flight controls and hundreds of

Nose turret conversions

The lack of suitable nose armament in the B-24D was a source of considerable criticism from crews and led to several in-the-field modifications and, eventually, to the fitting of nose turrets on the production line. Before the latter were available, AAF air depots in Oklahoma and Hawaii fitted surplus A-6 tail turrets in the noses of B-24Ds bound for the Pacific theatre, though with a nose turret installed crews remained unhappy as forward visibility became the poor relation. The Eighth Air Force, in particular, designed new nose configurations to address the problems of accommodation and visibility in the nose of the aircraft but, for various reasons, none reached front-line use and this shortcoming in the B-24 was never satisfactorily resolved.

miles yet to go to Java, Bayse elected to try to land at a Dutch airfield at Macassar in the southwestern Celebes. He made an emergency crash landing. The wounded men were removed and taken to a Dutch hospital. The remainder of the crew was unharmed but the aircraft was, as Poncik put it, "only food for salvage". Eventually, with a Japanese landing at Macassar imminent, two Navy PBY Catalinas picked up some of the downed airmen, who took a launch out to the waiting PBYs while explosions and fires lit up the Macassar waterfront. Two weeks were to pass before a Navy PBY picked up the crew of Dougherty's LB-30 (AL535), which had ditched near the shore at Maselembo Island in the Java Sea.

Java, as it turned out, was becoming a rout for the Americans. The 7th BG, with its 9th BS at Madioen (B-17E) and the 11th BS at Jogjakarta (LB-30), struggled to put bombers into the air to strike Japanese airfields on the

XB-41 bomber escort

With a view to adopting the type as a bomber escort, B-24D 41-11822 was modified by Convair San Diego with extra armament. However, as experience with the similar XB-40 (B-17) in the ETO was to show, the extra weight of its additional guns meant that the XB-41 was unable to maintain speed. The aircraft sported twin 0.5-in guns in a chin turret (below), an extra dorsal turret behind the wing and twin 0.5-in guns in each waist position (below right). The existing dorsal turret behind the cockpit was raised to maintain its rearward field of fire. Delivered for testing at Eglin Field in January 1943, it was judged unsuitable and the project was abandoned.

Malay peninsula and convoys in the waters nearby. The air commander, Major General Lewis H. Brereton – who, two months earlier, had been rebuffed upon urging MacArthur to disperse his air power in the Philippines – arrived in Java on 28 January 1942 aboard a Ferry Command B-24A (40-2373). On the trip, he was buffeted by rough weather, a reminder that this was the worst time of the year to be flying. Brereton concluded that Java could not be held. By then, the Japanese had occupied the airfield at Kendari, and a strike by two LB-30s on the night of 28 January produced no apparent result.

Liberators continued to fight the Japanese onslaught. With Borneo, Celebes and Timor also under assault, the island of Java (which belonged to the Dutch) was the final Allied possession not yet in enemy hands.

On 17 February 1942, Brereton and Major General George Brett decided to divide the surviving forces in Java: Brett took US aircraft and crews to Australia, Brereton moved to India to build a force to strike at Japan through China. Two days later, a trio of LB-30s bombed Japanese ships off Bali. Zero fighters attacked the formation at 3,000 ft (914 m). The bomber crews claimed two Japanese fighters shot down, although Japanese records indicate only damage. A day later, an LB-30 bombed a Japanese destroyer off Java.

On 20 February 1942, three LB-30s of the 11th BS left Jogjakarta to bomb Japanese ships off Bali once more. The flight was led by Captain Wade. He pressed home the attack from 13,500 ft (4115 m) and each aircraft dropped eight of the British 300-kg (661-lb) bombs the Americans had been using. Wade's bombardier, Sergeant Charles Schierholz, was certain they sank a Japanese cruiser. Bomber crews reported three direct hits and eight waterline hits on a cruiser which was left burning, but Japanese records do not indicate damage to *Nagara*, the only light cruiser in Bali operations. Wade's LB-30 force encountered no Japanese fighters, only heavy but inaccurate AA fire.

Above: As nose turrets were fitted by at least two modification centres, there were detail design differences between aircraft treated at different facilities. Those B-24Ds modified at the AAF's Oklahoma City air depot had a pronounced 'droop' and extra side windows for the bombardier – an attempt to address the latter's cramped conditions and poor visibility. This is a 308th BG aircraft.

Left: The nose turret installation designed by the Hawaiian Air Depot followed the configuration of the B-24J more closely but, while being neater in appearance than the Oklahoma City equivalent, offered even worse visibility and cramped conditions than the B-24J. SEXY SUE, MOTHER OF TEN was B-24D 42-40078 of the 98th BS, 11th BG.

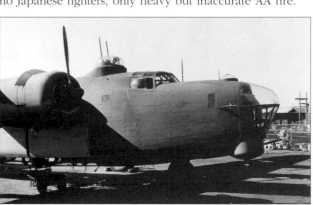

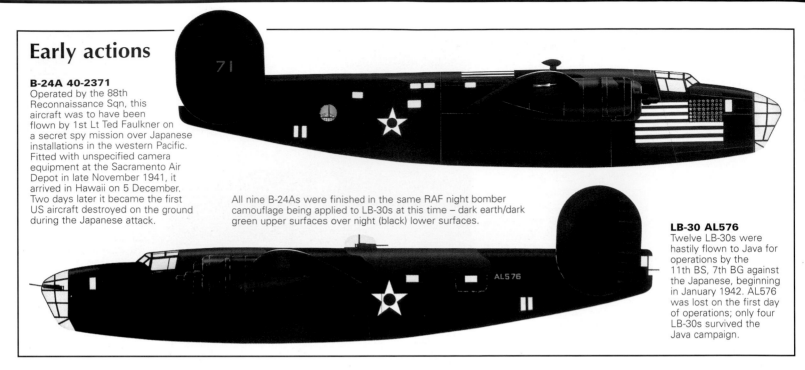

Early actions

B-24A 40-2371
Operated by the 88th Reconnaissance Sqn, this aircraft was to have been flown by 1st Lt Ted Faulkner on a secret spy mission over Japanese installations in the western Pacific. Fitted with unspecified camera equipment at the Sacramento Air Depot in late November 1941, it arrived in Hawaii on 5 December. Two days later it became the first US aircraft destroyed on the ground during the Japanese attack.

All nine B-24As were finished in the same RAF night bomber camouflage being applied to LB-30s at this time – dark earth/dark green upper surfaces over night (black) lower surfaces.

LB-30 AL576
Twelve LB-30s were hastily flown to Java for operations by the 11th BS, 7th BG against the Japanese, beginning in January 1942. AL576 was lost on the first day of operations; only four LB-30s survived the Java campaign.

Losing battle

Elsewhere on Java, B-17s and a handful of newly-uncrated A-24 Dauntlesses were busy, but it was a struggle to keep the 11th BS's LB-30s in action. On 21 February 1942, two LB-30s again went after enemy ships in the vicinity of Bali, with newly-seized Denpasar as a secondary target. One aircraft dropped its bombs on a tent camp, the other on what was apparently the airport, with no visible results. Records show that this was the 18th mission launched by LB-30s during the Java campaign.

A strafing attack on Jogjakarta on 22 February 1942 destroyed an LB-30 (AL567) on the ground and hinted that the Japanese would soon be able to overrun the embattled 7th and 19th BGs. That day, US Army chief of staff General George Marshall ordered Brereton to withdraw from Java.

Pilot Crowder flew General Brereton and staff members from Bandung to Colombo, Ceylon, in his LB-30 (AL533) on 24 February 1942. Crowder returned to Java – and thereafter took his LB-30 to Australia – while Brereton went on to New Delhi to set up a new Tenth Air Force headquarters. In the weeks ahead, the 7th BG would relocate to India, but with B-17Es only, while many of its members would be absorbed by the B-17C/D-equipped 19th BG in Australia. Only much later, after a Middle East episode in Flying Fortresses, would the 7th BG return to the China-Burma-India theatre and operate four squadrons of B-24s.

Captain Wade and his crew flew *Minnie From Trinidad* (AL608) from Jogjakarta to Ceylon on the evening of 25 February 1942, carrying Brereton's new boss, British General Sir Archibald Wavell and other officials. At Colombo, Wade picked up General Brereton and carried him on to Asansol, India, where the 7th BG relocated with B-17Es.

Back in Java, the end was near. On the night of 27 February, First Lieutenant 'Butch' Helton took off in an LB-30 to bomb Denpasar airport on Bali. He bombed by moonlight at 00.30 from 7,000 ft (2133 m) and reportedly scored five hits with his 300-kg bombs. One sign of the chaos that overcame the 7th BG's 11th BS at Jogjakarta is that a carefully-kept record listing aircraft serials and crews on particular missions ended the previous day.

On the night of 28 February 1942, one B-17E and one LB-30 launching from Jogjakarta combined with five B-17Es from Madioen to bomb Japanese ships just off the north coast of Java. As was frequently the case in these early, difficult missions, each bomber proceeded to the target individually. This marked the 21st LB-30 combat mission. A day later, an LB-30 pilot had to bring back his bombs when he was unable to find the Japanese ships.

On 1 March 1942, it was decided that three LB-30s at Jogjakarta (AL533, AL572 and AL609) would have to be written off. Dutch soldiers began blowing up everything at the base and the evacuation got underway: four transport B-24As, a B-17 and an LB-30 hauled men and equipment out of Jogjakarta en route to Australia. Loaded to capacity,

Aleutian Liberators

The contribution of the B-24 Liberator to the Aleutian Islands campaign has often been overlooked. The 28th BG, originally a Composite Group, initially included the 36th BS, with B-17s and a single LB-30. The 21st and 404th Bomb Sqns joined the 28th during 1942, bringing with them B-24Ds. It received little attention at the time, but 28th BG Liberators bombed Japan's home islands as early as 18 July 1943. The Japanese naval base at Paramushiro became one of those perennial targets, often damaged, never neutralised. In addition to its bombing duties, the 28th BG specialised in photo and radar reconnaissance as well as sea search and rescue. The Group operated from Ladd Field, Alaska, beginning on 12 July 1942 and from Nome Municipal Airport beginning later the same month. In all the 28th BG flew 2,578 sorties during the war, suffered 33 aircraft losses and received credit for 29 Japanese aircraft shot down. Twenty-two more Liberators were lost in non-combat mishaps.

From about August 1942 AAF Liberators began raids on occupied Kiska from Alaska; Attu was not within range until airstrips on Adak and Amchitka were established. These bases also allowed raids on Paramushiro in the Kurils. Equipped with Canadian-built ASV Mk II radar (note the Yagi antennas under the wings and on the extreme nose, plus 'washing line' arrays on the upper fuselage), a B-24D taxies across muddy PSP on an Alaskan airfield after another long-range sortie. The last of the bomb squadrons to join the 28th, the 404th BS, was known as the 'Pink Elephants' as it was equipped with sand-coloured aircraft that had been destined for service in North Africa!

It is 9 December 1942 and after only three weeks of active duty in the region, 90th Bomb Group B-24Ds are seen at a dusty Jackson Airstrip, near Port Moresby. Note that the nose armament on the nearest aircraft comprises four guns, the additional upper '50-caliber' having been added in the field.

these aircraft made harrowing flights from Java to the evacuation base at Broome on the Australian northwest coast. The last two LB-30s got out on the night of 2-3 March 1942, piloted by Crowder and Helton. Many ground personnel left on a Dutch ship. On 3 March 1942, the Japanese mounted a surprise air attack on Broome that destroyed two B-17s and three B-24As. A B-24A (40-2370) flown by Lieutenant Kester of Air Transport Command was shot down and all but one person killed.

In the battle for Java, B-17s withdrawn from the Philippines had been joined by 39 more Flying Fortresses and by exactly a dozen LB-30 Liberators. Of the 12 LB-30s that had reached Java and fought on that embattled island in early 1942, only four survived. Seven aircraft (AL533, AL535, AL567, AL572, AL576, AL609 and AL612) were written off and lost in action, and one other (AL521) was destroyed during a Japanese air raid on Darwin. Three LB-30 bombers (AL508, AL515 and AL570) went to Australia to join Brett's forces. A single aircraft (AL608, *Minnie From Trinidad*) went to India.

Fifth Air Force

General Douglas MacArthur ran the southwest Pacific war, and oversaw Fifth Air Force, from his headquarters in Australia. Activated in February 1942, Fifth Air Force was responsible for New Guinea and operated cheek-by-jowl with Thirteenth Air Force, which came under the Navy's command and was responsible for Guadalcanal. Fifth Air Force leapfrogged forward against the Japanese in New

Fifth Air Force

Lower left: Hell's Belle of the 400th BS, 90th BG was photographed after scoring a direct hit on a 5,000-ton Japanese cargo vessel near Kairiru Island off the north coast of New Guinea, probably in late 1942 or early 1943. Though the censor has removed the aircraft's serial number, it is known to be 41-24290, a B-24D that was later transferred to the RAAF.

Guinea, the Bismarck Archipelago, Netherlands East Indies and Philippines; by 1945, Fifth was moving north and preparing for the invasion of Japan. General George C. Kenney, Fifth Air Force commander, also commanded the Allied Air Forces in the southwest Pacific, including those of the Royal Australian Air Force (RAAF), which became a Liberator operator late in the war, and the Netherlands East Indies Air Force.

Kenney regarded the B-24 as ideal for his mission, seeing it as a key tool to drive the Japanese out of New Guinea

Bomb Groups of the Fifth Air Force

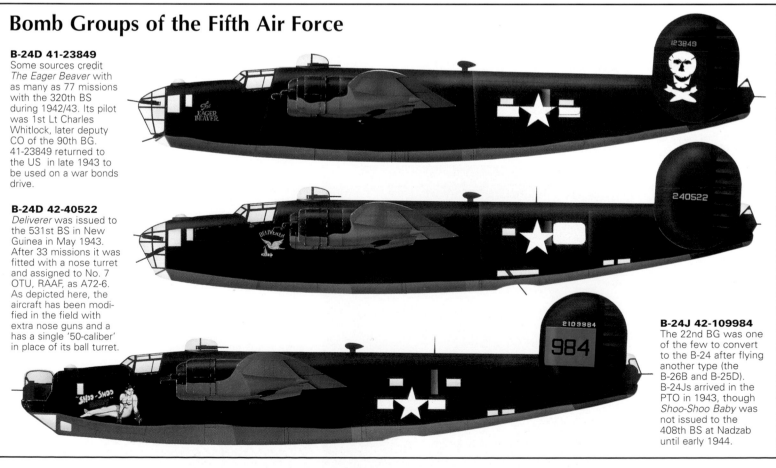

B-24D 41-23849
Some sources credit *The Eager Beaver* with as many as 77 missions with the 320th BS during 1942/43. Its pilot was 1st Lt Charles Whitlock, later deputy CO of the 90th BG. 41-23849 returned to the US in late 1943 to be used on a war bonds drive.

B-24D 42-40522
Deliverer was issued to the 531st BS in New Guinea in May 1943. After 33 missions it was fitted with a nose turret and assigned to No. 7 OTU, RAAF, as A72-6. As depicted here, the aircraft has been modified in the field with extra nose guns and a has a single '50-caliber' in place of its ball turret.

B-24J 42-109984
The 22nd BG was one of the few to convert to the B-24 after flying another type (the B-26B and B-25D). B-24Js arrived in the PTO in 1943, though *Shoo-Shoo Baby* was not issued to the 408th BS at Nadzab until early 1944.

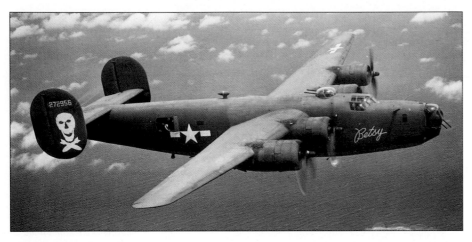

B-24Ds destined for the PTO were modified with nose turrets during 1943. 42-72956 *Betsy* of the 321st BS was photographed in February 1944, on a mission to Wawak, New Guinea. The 'skull and crossed bombs' logo of the 90th BG was applied over a green background on the tails of 321st BS aircraft.

and to retake the Philippines. Instead of insisting that his heavy bombers operate at high altitudes, he used them in low-level attacks on Japanese shipping; strategic missions were flown at lower altitudes than in Europe, thus allowing the crews to achieve greater bombing accuracy.

Early on, much of the burden in the southwest Pacific had fallen on the 90th BG. The group's commander, Colonel Arthur H. Rogers – who gave his name to the group's 'Jolly Rogers' appellation – described his outfit as "the first heavy bomber unit to reinforce a losing army in the jungles of New Guinea". En route to war, the group

Right: The 380th BG arrived in Darwin in May 1943 and quickly moved to New Guinea. This 528th BS, 380th BG B-24D (equipped with a nose turret) makes a bombing run over a target at Amahal on Ceram Island, in the Moluccas, Netherlands East Indies.

Below: A Japanese fighter (circled) attacks Liberators of the 530th BS, 380th BG in the vicinity of Kolono during a raid on the Celebes Islands, Netherlands East Indies on 8 January 1944.

paused in September 1942 at Hickam Field, Hawaii. There, Rogers and Colonel Marion D. Unruh planned the installation of Consair A-6 tail turrets to the nose of the B-24D, which had been manufactured with glass noses and no nose turret.

A near mutiny by a handful of officers who preferred the B-17 Flying Fortress was nipped in the bud in October 1942. AAF boss General Henry H. 'Hap' Arnold wrote letters to Kenney and Hawaii-based Major General Willis Hale of Seventh Air Force acknowledging "a real and acute problem in psychology and leadership" and confirming no change in his plan "to replace B-17 type aircraft with B-24 type aircraft in all of the combat theatres throughout the world, except the United Kingdom". This was Arnold's response to smouldering in the ranks, some of which occurred in the 90th BG while Rogers was a member of the group but not yet its commander.

The B-17, Arnold said in the letter, was "a fine heavy bomber which has been lavishly built up by the press". As the 90th BG headed to war, Arnold wanted to debunk any myth that the 90th was replacing the B-17C/D-equipped 19th BG "with an airplane [the B-24] that is an inferior weapon". Arnold noted that "a number of pilots thoroughly experienced in B-17s" were then commanding B-24 units "and showing real pleasure and pride [in] their new B-24 aircraft" – but not in the 90th BG, where a mutinous attitude prevailed. While the 90th was still on his turf in Hawaii, Hale sacked group commander Colonel Eugene Mussett and a squadron commander. Responding to a complaint from troops of 'congestion' in the Liberator's nose – the crowded condition remained a problem throughout the war – Hale had the navigator's position moved from the nose to a new location behind the pilot. And he ramrodded the installation of nose turrets in the B-24D. Later-model Liberators received MPC or Emerson nose turrets at the factory.

Ready for war

From Hawaii, the 90th BG moved in November 1942 to Iron Range, Australia, a place so isolated that no roads connected it to other towns. Troops finished the job of equipping the B-24D with a nose turret. Rogers wrote in his diary, "I performed this test and was pleased to find out the airplane had actually picked up from 8-12 mph [13-19 km/h] at the different settings. This had been due to the shifting forward of the centre of gravity which made the airplane fly on an even keel. Before this, the plane had been tail heavy and had a tendency to fly nose up, which increased the resistance and slowed down the speed. After completing these tests I was anxious to get back up North in the combat zone to actually find out the results of the nose turret against the Japs' head-on attack."

Japanese forces had landed at Buna and Goma on the north coast of New Guinea the previous July, and at Milne Bay, the eastern tip of New Guinea, in August. The 90th BG began attacking Japanese airfields, troop concentrations, ground installations and shipping. By year's end, they had pushed the Japanese back over the Owen Stanley range and recaptured Buna and Goma, shifting attention to Guadalcanal and the main Japanese base at Rabaul, New Britain.

Although the group had spent considerable time and effort honing its men and aircraft, when the 90th BG moved to Port Moresby on 10 February 1943, Kenney worried that the group had not had sufficient training. He wrote, "This job here calls for night take-offs with maximum loads and often with cross-winds, climbing through overcast to 15 and sometimes 20,000 ft [6096 m] to get on top in order to navigate. It is then normally necessary to come on down to see the target under a ceiling which may range from 2,000 to 10,000 ft [609 to 3048 m], pulling back up after the attack to navigate home and on arrival back at Moresby in breaking through again to land.'

New Guinea was a stinking hell for B-24 crews. And they rarely had company. They flew small-scale missions typi-

cally consisting of two or three bombers, rarely even accompanied by the other diverse AAF warplanes in the area – P-39 and P-400 Airacobras, P-40 Warhawks, B-25 Mitchells. At Iron Ridge and again at Port Moresby, 90th BG crews began painting nicknames and caricatures on their olive-drab, turret-equipped B-24Ds. In later decades, this would be known as nose art, although the term did not exist during the war.

General Kenney received a second Liberator group in May 1943 when the 380th Bomb Group 'Flying Circus' reached Darwin and moved to New Guinea. Until early 1945, the 380th BG flew almost alone against airfields and shipping in the Netherlands East Indies and the Timor Sea, and often called on Balikpapan, the well-defended Japanese oil refinery complex on Borneo. The 'Circus' began its war with a 23 June 1943 mission to Makassar on Celebes, travelling 2,000 miles (3218 km) to strike shipping facilities and docks.

The 43rd BG 'Ken's Men' (named in honour of Kenney) finished its conversion from the B-17E/F Flying Fortress to the B-24 in September 1943 and became the third Liberator outfit in Fifth Air Force. A typical mission of the period occurred on 12 October 1943 when Kenney sent 63 Liberators, 107 Mitchells and 106 Lightnings to Rabaul. The Japanese bastion there remained a difficult target, for many months to come. In January 1944, the 22nd BG 'Red Raiders' began flying B-24s at Nadzab. Ultimately, the 'Red Raiders' of the 22nd BG flew and fought from airfields at Nadzab, Port Moresby, Biak, Clark Field and Okinawa – but that gets ahead of the story.

The 90th BG moved again to Nadzab on 23 February 1944 (joining the 22nd BG and 20th CMS). Moving westward along the northern coast of New Guinea, Allied forces were to bypass Wewak and go after Hollandia. The first softening-up attack on Hollandia was delivered by 22 Liberators on 4 March 1944. Two weeks later, as the Japanese withdrew from Wewak and concentrated at Hollandia, 40 Liberators bombed shipping from medium altitude and sank a freighter. Some 75 Liberators hit Hollandia on 30 March 1944, and 71 the following day. These were huge raids for the conditions of the theatre.

Finally, on 22 April 1944, American troops seized Hollandia. The war moved on. That week, 47 Liberators of the 43rd and 90th Bomb Groups bombed Japanese-held Biak, claiming credit for three Japanese fighters downed in the air and 13 destroyed on the ground.

Thirteenth Air Force

The 307th BG 'Long Rangers' was the first Liberator combat group in the remaining numbered air force in the Pacific region. Thirteenth Air Force was constituted on 14 December 1942, commanded by Major General Nathan F. Twining. The 307th BG established its base at Carney Field on Guadalcanal, then moved forward to Henderson Field to begin harrying Japanese shipping near Guadalcanal and Bougainville. Soon after came the 'Bomber Barons' of the 5th BG.

B-24 pilot 1st Lt John Lance remembered the contrast between the two numbered air forces:

"The speciality of Thirteenth had developed into flying

Seventh Air Force

Above: Smoke rises from the bombed-out phosphate plant on Nauru on 22 April 1943, after the attentions of 307th BG Liberators. The island was some 900 miles (1448 km) from Funafuti and well defended by Japanese fighters. Though stiff resistance was met the raids on Nauru, which spanned five days, were judged a success.

Liberators returned to Funafuti in November 1943, when 11th BG aircraft used the island as a base for attacks on Tarawa in support of Operation Galvanic – the invasion and occupation of the Gilbert Islands. The island was not immune from retaliatory attacks; B-24D 42-40688 Wicked Witch of the 42nd BS was burnt out after one such raid.

Above: An early production B-24J of the 431st BS, 11th BG cruises over Kwajalein in the Marshall Islands during June 1944. The 11th BG arrived on the atoll in April and remained until October, when it moved to Guam.

Right: 7th Air Force aircraft (possibly from the 98th BS, 11th BG) make for Truk on a mission from the Marshalls in 1944. Note the shadow of Liberator on top of another in the lower right corner of the picture.

special cohort of night-fighting Liberators attacked dozens of Japanese ships and claimed to have achieved direct hits no less than 24 per cent of the time. 'The Snoopers', the 868th Bombardment Squadron (Heavy), operated autonomously. Their radar equipment and techniques came from the confusingly-named Colonel Wright's Project located at Langley Field, Virginia, and their radar-equipped Liberators were usually known as SB-24s. The SB-24s were equipped with SRC-717-B search and navigation radar, AN/APQ-5 LABS bombing radar, SCR-729 IFF and an AN/ARN-1 radio altimeter. The SC-717-B radar antenna was placed in the ball turret location of the SB-24.

Thirteenth Air Force was always in the thick of the fighting. Throughout the early and middle months of 1943, Liberators flew long-distance missions with small numbers of aircraft. On 18 June 1943, the 307th BG flew a 1,700-mile (2736-km) round-trip to bomb Nauru. The bombers arrived as the sun came up and their bombs touched off large fires. US forces began the process of taking New Georgia in late June. While fighting near the airfield at Munda persisted in July, the 'Long Rangers' bombed airfields at Kahili and Buka. These targets, together with Buin, were hit repeatedly, not only by Liberators but by SBD Dauntlesses and TBF Avengers that took on shipping just offshore. On 25 July 1943, Liberators joined with SBDs and TBFs to fly a mission supporting troops on the ground. After a bitter fight, US troops seized Munda – future home of many Liberators – on 6 August.

Marines established a beachhead on Bougainville on 1 November 1943. Some of the Japanese bases in the region became ineffectual as the advance continued, but Rabaul was always a problem – the source of Japanese warships and aircraft that confounded the Allies at every turn – and Liberators went to visit frequently. During the early fighting at Bougainville, a 'Snooper' SB-24 apparently disabled the Japanese heavy cruiser *Haguro*, which was the flagship of the Emperor's formidable forces.

Thirteenth Air Force bombers mounted a major effort against Rabaul on 11 November 1943. The initial wave consisted of two dozen Fifth Air Force Liberators sent to bomb Lakunai airfield. Warplanes from no fewer than five aircraft-carriers also struck in and around Rabaul, before a giant force – by Pacific standards – of 42 Liberators from the 5th BG and 307th BG arrived. The 'Bomber Barons' and 'Long Rangers' staged through Munda and most of them bombed through the clouds from heights as great as 23,000 ft (7010 m). Four fighters were claimed by Liberator gunners. The siege of Rabaul continued well into 1944, one of the longest ongoing air campaigns.

In a major reorganisation on 15 June 1944, Seventh and Thirteenth Air Forces were merged into a single command, designated Far East Air Forces (FEAF). On that date, Lieutenant General Ennis C. Whitehead took over Thirteenth and the FEAF command went to Kenney, who was perhaps the most beloved combat commander of the war.

B-24s in the CBI

If the story of the Liberator in American hands began with the LB-30 on Java, it soon moved to the CBI (China-Burma-India) theatre. The first B-24D model to arrive in an operational theatre was 40-698, flagship of Force Aquila, the fore-

long-range unescorted formation missions, and we held all the records for that sort of thing. [We] had these specially designed planes with belly turrets, which gave more protective power for a group that was flying unescorted."

It should be noted that the Pacific was the theatre which produced the low-level, blind bombing SB-24, or 'Snooper'. Ten SB-24Ds began operations on Guadalcanal on 22 August 1943, each equipped with a special radar scope for the bombardier and a computer that translated and transmitted the information on the scope to the bomb release controls. Assigned first to the 5th BG, the SB-24Ds flew over 100 missions in two months, averaging over 11 hours each. With flame dampeners shrouding the glare of their engine exhausts, the black-painted SB-24s were able to bomb targets from low altitudes in darkness or fog. This

---

San Diego's last B-24D – 11th BG

B-24D 42-72963 was the last B-24D built at Convair's San Diego plant, completed in August 1943. It was fitted with a Consair A-6 nose turret at Hawaiian Air Depot en route to the PTO.

B-24D 42-72963
Delivered to the 42nd BS at Funafuti, this aircraft flew its first mission on 21 November 1943, but failed to return from its second on 2 December.

272963

The legend on the tailfin of this aircraft reads 'The Long Rangers', an appropriate name for the 307th BG and later represented by a simple 'LR' symbol on its aircraft. Frenisi, a nose turret-equipped B-24D 42-40323, is seen here re-enacting its return from its 100th mission in July 1944. Note the repeating of the aircraft's 'last three' on the aircraft's nose – a not uncommon feature of B-24s in the PTO.

Lower left: This black-bellied aircraft of the 370th BS, 307th BG is believed to have been among the SB-24D 'Snooper' aircraft later operated by the 868th BS. Fitted with radars, IFF, a radio altimeter and a bombing computer, the 'Snoopers' were initially assigned (from August 1943) to the 5th BG, but gained independence as the 868th BS and reported directly to the 13th AF from January 1944. Highly successful in their low-level, 'blind' bombing role, the unit remained active until VJ-Day. Munda Belle was SB-24D 42-40144.

Thirteenth Air Force

runner of Tenth Air Force in India. This aircraft reached India on 7 April 1942, before there was a Liberator in England or North Africa. By then, the 7th BG was in India with a single LB-30 that had escaped from Java, and a handful of B-17Es. Much later in the war, the 7th BG operated four Liberator squadrons from bases in India, bombing Japanese forces in autumn, winter and spring, and ferrying gasoline over the Himalayas into China during the summer. There were transports flying the 'Hump,' as well as C-109 tankers, the version of the Liberator modified for fuel-hauling, but much of the fuel for Chiang Kai-shek's forces on the Chinese mainland was carried by straight Liberator bombers. It was a seasonal tasking and it became as familiar to Liberator crews as dropping bombs.

The CBI saw many innovations by Liberator pilots and crews. Here was where the AAF introduced the Azon bomb, a 1,000-lb M65 bomb equipped with a set of radio-steerable fins and a flare installed in the tail of the bomb. Using a control box with a control stick, the Azon was handled initially by 10 specially-trained 7th BG Liberator crews.

The theatre was also the introductory stage for ECM (electronic countermeasures) Liberators, which carried an extra crew member and were equipped with twin SCR-587 receivers, AN/APA-6 pulse analysers, two AN/AAR-5 receivers, and two AN/APR-2 auto-search receivers, and direction-finding antennas. A handful of these ferret Liberators were operated by the 7th BG on missions to Burma from bases in India.

The CBI was fertile ground for the 24th Combat Mapping Squadron, which operated F-7A and F-7B photo-reconnaissance Liberators. The 24th CMS photographed a 6,000-sq mile (15538-km²) area in northeastern India and

Burma, helping aircrews to understand terrain and facilities that previously had been little-known to the West.

Fourteenth Air Force was the outgrowth of Major General Clair Chennault's famous 'Flying Tigers' that took over combat operations inside China (while Tenth Air Force remained in India). The 308th BG soon began operating under the Fourteenth in China, from early 1943. Some of the group's B-24s had been modified to carry four 0.5-in (12.7-mm) machine-guns in the forward bomb bay. The pilot controlled these guns and used them for strafing.

Eighth Air Force

Liberators were performing anti-submarine duty in American and British hands, and fighting in the Pacific and Mediterranean theatres long before the Consolidated bomber became an important part of the Eighth Air Force in England – but the Eighth ultimately became the largest aerial armada ever assembled, and inevitably its achievements spawned more headlines and used more press ink than any of the other B-24 operators. The 'Mighty Eighth' had very humble beginnings, however.

The US Army Air Forces were well ahead of most of the rest of the world in planning long-range bombing missions by huge, four-engined bombers, and even before the war AAF officers envisioned putting 1,000 bombers in the sky at

China-Burma-India

Only one Liberator group was based in China, the 308th Bomb Group arriving at Kunming in early 1943. Ranging over enemy territory from Saigon to Shanghai, the 308th supported Chinese ground forces, attacked airfields and fuel supplies in French Indo-China, bombed ports in Burma and hit Japanese shipping in the East and South China Seas, Formosa Strait and Gulf of Tonkin. Largely self-sufficient, it flew its own supplies of fuel, ammunition and spare parts 'over the Hump' from India.

Stripped of the OD paint in which it would have operated upon arrival in the CBI, this B-24D of the 308th Bomb Group, 14th Air Force is seen departing Kunming in China. C-46s and C-47s on the ground were undoubtedly involved in the airlift 'over the Hump' from India.

155

Above: Shipdham was not long completed when these 44th BG(H) 'The Flying Eightballs' B-24Ds taxied out on a practice mission in late 1942. The 44th flew its first operational mission on 7 November.

Right: 42-63959 is one of the comparatively few (295) B-24Ds built at Fort Worth. Its 'C' tail marking is that of the 389th BG, the third B-24 group to form up in England, and dates the photograph to late 1943/early 1944. Though the last San Diego-built B-24Ds had ventral ball turrets, all -CF aircraft were equipped with the inadequate 'tunnel' gun.

B-24D-1-CO 41-23737 Eager Beaver of the 328th BS, seen here being loaded with 1,000-lb bombs, was one of the first 17 Liberators assigned to the 93rd BG, formed at Alconbury in September 1942. Note the early nose armament, comprising three 0.50-calibre machine-guns. The lower weapon proved largely ineffective as it was not able to be elevated. In some aircraft it was moved further up the nose, while in others a second gun was added.

Eighth Air Force genesis

Below: Relieved to be home again, this Liberator crew was pictured upon its return from Wilhelmshaven on 27 January 1943. VIII Bomber Command's first raid on a German target, the mission included B-17s from four groups, plus the Liberators of the 44th and 93rd BGs. Bad weather over the target and poor navigation meant that none of the B-24s was able to find the target; one B-17 and a pair of B-24s were lost. Jenny was B-24D 41-23778 of the 66th BS, 44th BG.

a time. But when General Ira Eaker went to England in 1942 to create Eighth Air Force, he had five airmen with him. It was not Eaker, of course, but Major General Carl 'Tooey' Spaatz, who first commanded Eighth Air Force (from 5 May 1942), but Eaker, who headed up Bomber Command under Spaatz before relieving him in the top job (on 1 December 1942), was the bomber boss in England from the start. It is unclear how Eaker viewed the Flying Fortress vs. Liberator controversy, but he clearly believed – wrongly, as it turned out – that a big bomber like the B-24 Liberator could fight its way to the target without help from any escort.

The founders of Eighth Air Force came to England prepared to use B-17s, and the B-24 Liberators that soon followed, to bomb German-occupied Europe by daylight. At first, they intended to do this with minimal or even no fighter escort. The Royal Air Force, whose Lancaster and Halifax crews already possessed considerable experience pounding the continent during the nocturnal hours, scoffed at daylight bombing. But nothing other than daylight bombing was seriously contemplated at Eighth Air Force headquarters at Bushey Park (codename Widewing), where Spaatz set up shop. Here, Spaatz established the 1st Bomb Wing (B-17s), then the 2nd Bomb Wing, which became the US parent unit

for B-24 Liberator groups, beginning in September 1942. The first of these was the 93rd BG, 'Ted's Traveling Circus', named for commander Colonel Edward J. ('Ted') Timberlake, Jr and flying from Alconbury.

Baptism of fire

The debut of the Liberator on the continent came on 9 October 1942 when the 93rd BG contributed 24 B-24Ds to the 108 bombers (the others being B-17s) in a five-group assault against the French city of Lille. Timberlake led the Liberator force at the controls of his *Teggie Ann* (B-24D-5-CO Liberator 41-23754). Sergeant Arthur Crandall, a gunner of the 93rd BG, shot down an Fw 190 near Lille that day, in a B-24D named *Ball of Fire* (apparently B-24D-1-CO 41-23667) piloted by Captain Joseph Tate. It was the first aerial victory for an Eighth Air Force Liberator.

The next Liberator group to reach England was the 44th BG 'Flying Eight Balls', arriving at Shipdham on 7 November 1942. The group's first mission consisted merely of seven Liberators creating a diversion for an attack elsewhere by Flying Fortresses. The 44th was destined to participate in 343 missions between that date and 25 April 1945, a larger number of missions and more tons of bombs (18,980) than any other Liberator group apart from the 93rd. Over the course of the war, this group lost an extraordinary 192 Liberators and claimed 330 Luftwaffe fighters destroyed.

The Liberator was getting off to a slow start, with two squadrons from the 93rd BG briefly on loan to Royal Air Force Coastal Command for anti-submarine patrols in the Bay of Biscay. On 21 October 1942, 'Ted's Traveling Circus' launched 24 Liberators on an intended low-level raid against U-boat pens at Lorient, France, but because of 100 per cent cloud cover they were unable to bomb the target. A mission to Brest by 12 Liberators on 7 November 1942 also produced little result, although it marked the first combat by the 44th BG, which launched seven aircraft in a diversionary effort.

On 13 December 1942, the 93rd BG was uprooted from England and sent off to North Africa. On this occasion, it was shifted to the Twelfth Air Force, although a later sojourn to the same region would place it under the purview of the Ninth Air Force, to add its Liberators to the aerial bombardment of Axis supply ports. In North Africa, the men found primitive facilities, furious winds, rain, and mud. At one point, it was impossible to taxi a B-24D Liberator because the mud created such an obstacle. The 93rd BG eventually flew 22 missions over 81 days before returning to England. Another trip to North Africa lay in the future of 'Ted's Traveling Circus': soon after, the 93rd and 44th BGs were uprooted to North Africa for the low-level Ploesti raid.

The MTO (Mediterranean Theatre of Operations) evokes

Founders of the 2nd Air Division

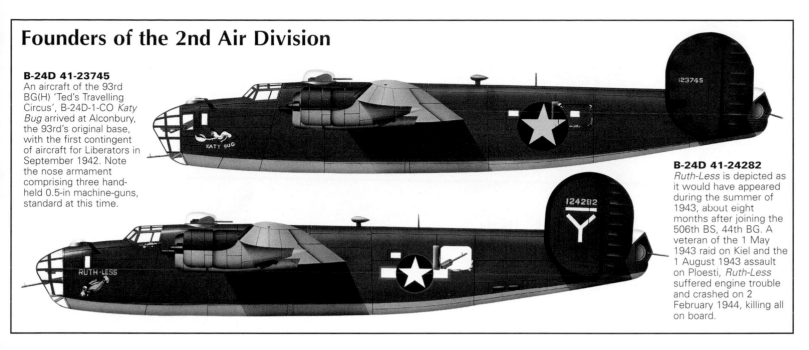

B-24D 41-23745
An aircraft of the 93rd BG(H) 'Ted's Travelling Circus', B-24D-1-CO *Katy Bug* arrived at Alconbury, the 93rd's original base, with the first contingent of aircraft for Liberators in September 1942. Note the nose armament comprising three hand-held 0.5-in machine-guns, standard at this time.

B-24D 41-24282
Ruth-Less is depicted as it would have appeared during the summer of 1943, about eight months after joining the 506th BS, 44th BG. A veteran of the 1 May 1943 raid on Kiel and the 1 August 1943 assault on Ploesti, *Ruth-Less* suffered engine trouble and crashed on 2 February 1944, killing all on board.

many images, from the early Allied landings in North Africa to heavy bomber raids mounted from Italian air bases and aimed at the heartland of the Third Reich. To many, the term MTO means Fifteenth Air Fore, created on 1 November 1943 to oversee the theatre's air operations. In fact, the B-24 and the American airmen who maintained and flew it were fighting in the Mediterranean much earlier, sometimes in squadron-sized units, sometimes in concert with other air forces, and for a period were part of Ninth and Twelfth Air Forces.

Ploesti – Operation Tidal Wave

The most important prelude to the subsequent formation of Fifteenth Air Force was the 1 August 1943 low-level mission against Ploesti. Carried out by five American B-24 Liberator groups, the raid was launched from five bases around Benghazi, Libya.

Taking part were three B-24 outfits detached from the Eighth Air Force in England: the 93rd Bomb Group, 'Ted's Traveling Circus', commanded by Lieutenant Colonel Addison Baker, plus the 44th 'Eight Balls' and the 389th 'Sky

Halverson Provisional detachment (Halpro)

In June 1942, a specially-trained Liberator outfit equipped with 23 early B-24Ds reached the Middle East, commanded by Colonel Harry A. Halverson. The Halpro unit, or Halverson Provisional detachment, was made up of members of the 98th Bombardment Group (Heavy) – the term will be abbreviated BG throughout – and was heading east with the intention of bombing Japan. After these B-24Ds reached Fayid, Egypt, it became apparent that they were needed to combat German forces in the immediate vicinity, so they were kept in-theatre.

To Halpro fell the honour of making the first USAAF bombing mission against German forces in Europe. It happened on the night of 11-12 June 1942, when 13 Liberators struck the oil refineries at Ploesti, Romania. It was a symbolic achievement but it is unclear whether the Liberators' bombs did much damage. Four of the bombers became the first US Liberator casualties in the theatre when they landed near Ankara, Turkey. Turkish technicians actually repaired and flew one of the bombers (B-24D-CF, 41-11596, *Brooklyn Rambler*) and it was eventually repatriated. As for Ploesti, the name of the place was to become synonymous with the Liberator – but only later.

Halpro flew several bombing missions before 28 June 1942, when Major General Louis H. Brereton arrived from the Far East with B-17 Flying Fortress-equipped elements of the 7th BG (which had already fought on Java with B-17Es and LB-30s). Brereton formed the USMEAF (US Army Middle East Air Force), which absorbed Halpro. Now dubbed the Hal Bomb Squadron, Halpro bombed the harbour at Benghazi, Libya, on 20 July 1942. By late 1942, the 343rd and 344th Bombardment Squadrons (Heavy) were operating as flying components of the 98th BG. A second Liberator group, the 376th BG 'Liberandos', was constituted at Lydda, Palestine on 19 October 1942, also equipped with B-24Ds and later relocating to Abu Sueir, Egypt.

Scorpions'. These groups had come from England (leaving the British Isles bereft of Liberators) and prepared for Ploesti with about 10 missions to other MTO targets in what was dubbed Operation Husky.

For the main event at Ploesti, the two Eighth Air Force groups joined 'KK' Compton's 'Liberandos' and the 98th BG 'Pyramiders' under Colonel John 'Killer' Kane, which was now also subordinate to Ninth Air Force. Kane was a spirited and volatile leader, much-admired, but a contrast to the quiet, experienced and much-loved Colonel Leon Johnson of the 44th group.

The Allies believed that sending heavy bombers against German oil production would strike a major blow and alter the course of the war. The AAF considered the B-24

41-11603 Malicious, the 99th B-24D off the San Diego production line, was one of the 23 Halpro detachment aircraft, though it is pictured here after the unit's absorption by the 376th BG, when the aircraft joined the 512th BS.

Left: American bombers struck a German target for the first time on 11 and 12 June 1942, when 13 aircraft of the Halpro Detachment bombed the oil refinery at Ploesti, Romania. B-24D Edna Elizabeth was among these aircraft and landed afterwards in Turkey. Just visible in the lower nose of 41-11622 is a pair of fixed forward-firing guns, fitted specifically for use in these low-level raids and fired by the pilot.

123744

B-24D-1-CO 41-23744 Geronimo *of the 328th BS is pictured at Tafaroui, Algeria having suffered a collapsed nose wheel on landing, a not uncommon occurance among Liberators. The date was 14 December 1942, during the 93rd BG's first detachment to North Africa.*

Liberator to be the only heavy bomber for this mission. The aircraft could carry a greater bomb load than the B-17 Flying Fortress, was faster and had greater range. The B-24 was championed by Brereton, who was viewed as the brain – or, in the view of some, the culprit – behind the raid.

Some 179 Liberators made up the Tidal Wave – the formal name of the operation – which set forth from dusty desert air bases around Benghazi. One aircraft crashed on take-off, and one flew into the sea. The lead aircraft was attacked by a Messerschmitt Bf 109, jettisoned its bombs early, and crashed. No fewer than 10 Liberators had to abort and return to base, their engines fouled by the troublesome North African sand.

The strike force was to head north to Corfu, then swing to the northeast. At Corfu, the lead bomber, *Wongo Wongo!* (B-24D-120-CO 42-40563 of the 512th BS/376th BG), piloted by First Lieutenant Brian Flavelle, inexplicably began pitching violently. This Liberator stood abruptly on its tail in mid-air, shuddered, then dived suddenly straight into the sea. Radio silence was being observed and no one knew what had happened. *Wongo Wongo!*'s wingman went down to investigate, found no sign of survivors, and ultimately had to abort and return to North Africa, carrying the deputy mission navigator. Contrary to myth, *Wongo-Wongo!* was not carrying the lead navigator for the mission. For decades to come it would be reported, erroneously, that the loss of this Liberator led to the navigational problems which followed.

On approach to the target, the 376th BG mistook the IP (initial point) at Floresti and turned south too soon. The 93rd BG followed but, thanks to second-thinking and prompt action by pilot Lieutenant Colonel Addison Baker and co-pilot Major John L. Jerstad in *Hell's Wench* (B-24D-120-CO 42-40994 of the 328th BS/93rd BG), the 93rd made another turn which took the group back in the direction of Ploesti. Among the trailing groups, the 389th flew northeast toward its target at Campina, 17 miles (27 km) north of the main refinery. Some confusion persisted as the 44th BG and 98th BG pressed on, but both reached the correct IP at Floresti and proceeded to their assigned targets.

Right: Twelve days of training preceded the 1 August 1943 Ploesti raid, culminating in a full dress rehearsal over the Libyan desert involving the entire force of 175 aircraft.

Below: B-24D-85-CO Teggie Ann *(42-40664) of the 515th BS, 376th BG carried Brig. Gen. Uzal G. Ent and Col Keith K. Compton to Ploesti and back on 1 August 1943. A fortnight later it crashed at Melfi, Italy returning from a raid on Foggia.*

Navigational mix-ups

Over the target, navigational mix-ups caused some of the refineries to be attacked by too many Liberators, others by too few. Dodging fighters and flak, many of the bombers flew into cables raised as barriers in their path by balloons.

Despite potent defences and heavy losses, the Ploesti bombing inflicted heavy damage on the Romanian oil fields. A final tally showed that 179 Liberators took off, 14 aborted and 165 attacked. Of the B-24 losses, 33 were to flak and 10 to fighters. Some 56 Liberators were damaged. Eight aircraft recovered in Turkey. Of those B-24s which returned to North Africa, 99 recovered at their bases and 15 landed elsewhere. Some 532 American flyers died.

It will forever be debated how these bombings affected Germany's ability to fuel its combat forces, but no one doubted that the Third Reich was dealt a heavy blow. Baker, Jerstad, Johnson, Kane and pilot Second Lieutenant Lloyd H. Hughes were awarded the Medal of Honor, all but Johnson

Right: Flying at low level, in gusting turbulence, confronting enemy fighters and gunfire – no one could doubt the courage of the Liberator crews who went against Ploesti. The Ploesti attackers were going into nothing less than a fiery furnace. This famous view shows one of the Tidal Wave aircraft at chimney-top height on 1 August 1943.

With the arrival, in North Africa, of the 98th and 376th BGs in the second half of 1942, the Liberator force could concentrate on attacking supply lines feeding Rommel's Afrika Korps and did so for the rest of the North African campaign. These groups were joined by a detachment of the 93rd BG in support of Operation Torch in November. Pictured is B-24D (41-11761/'R') of the 98th BG sometime during 1942/43.

October 1943. The Mk Is were in many ways a hard act to follow, being true VLR aircraft with a range in excess of 2,400 miles (3862 km). The Mk II, by contrast, had a range of only 1,800 miles (2897 km), while the Mk III would reach only 1,680 miles (2704 km). The effort to close the Atlantic gap took another turn in September 1942, when No. 120 Squadron sent a detachment to Reykjavik in Iceland. The rest of the squadron moved north in February 1943, and remained in Iceland until replaced by No. 86 Squadron, and later No. 53. 1942 also saw the introduction of new, shallower-bursting Torpex-filled depth charges fitted with Mk.XIIIQ pistol fuses.

Coastal Command was extremely pleased with the Liberator, and would have re-equipped all GR squadrons with the type had sufficient aircraft been available. But far from being able to re-equip its GR force with Liberators, Coastal Command faced a dramatic shortfall in aircraft numbers. Early in 1943 the Admiralty had assessed that at

Far left: Targets in Italy were among those struck by Liberators based in North Africa. This was the scene after USAAF B-24s, B-25s, B-26s and P-38s had undertaken daylight raids on marshalling yards, factories and an oil dump in Naples, sometime during August 1943. RAF and RCAF Wellingtons had visited the Italian city the night before.

and Kane posthumously. It was the only raid in history for which five Medals of Honor were awarded.

When the Liberators returned to Europe and the Eighth Air Force, a rehearsal for an eventual Allied invasion of the continent – Operation Starkey, launched in September 1943 – brought the 44th, 93rd and 389th BGs back to the business of dropping bombs in northern Europe, together with the newly arrived 392nd BG. The last-named group participated in a 6 September 1943 effort that sent 69 Liberators from the four groups against four targets in France. The main event came the next day when a maximum effort directed 330 heavy bombers over France, accompanied by 144 B-26 Marauder medium bombers directed against a dozen scattered targets. Eighth Air Force planners had hoped to see how the Luftwaffe would react to such an armada, but the Germans refused to co-operate and did not engage in significant numbers.

That same month, the heavy bomber component of Eighth Air Force, VIII Bomber Command, underwent a shake-up on 13 September 1943. The higher headquarters that had been named the 1st, 2nd and 4th Bomb Wings were redesignated the 1st, 2nd and 3rd Air Divisions. The 1st Air Division remained all-B-17, the 2nd Air Division all-B-24, and the 3rd Air Division operated both heavy bombers before becoming all-B-17. The Eighth Air Force was a long way from the strength it would eventually reach. In Europe, the B-24 Liberator's war was only beginning.

RAF anti-submarine operations

When the US entered the war in December 1941, German U-boats were the scourge of the Atlantic. Within a year, the Allies were losing 700,000 tons (688800 tonnes) of shipping each month. Inevitable attrition among the Coastal Command Liberator Mk Is had been made good in November 1941 by the delivery of some Liberator Mk IIs. In April 1942, the Command had benefited from the transfer of six ex-Bomber Command Liberator Mk IIs, while in May, the Command received its first Liberator Mk IIIs. The acquisition of the Liberator Mk II allowed the re-equipment of No. 224 Squadron at Tiree in July 1942, completing a process that had begun in February, but halted as a result of Pearl Harbor and the resulting disruption in Liberator deliveries. Tiree was a poor base for the Liberators, with a poor weather record and thin runways laid straight onto loose sandy soil. These soon developed troughs at the touchdown points, and had to be filled and re-filled with wood chips.

No. 120 Squadron began to re-equip with Mk IIs in June, though the last of the Liberator Mk Is soldiered on until

Liberator Mk III and IIIA – the RAF's B-24Ds

The RAF's first B-24Ds were 11 early production aircraft transferred from the USAAF in March/April 1942 after Prime Minister Churchill appealed for more aircraft with which to tackle the U-boat menace. Though they were considered non-standard by the RAF, most were equipped with ASV radar and rushed into service with Nos 59, 86 and 120 Sqns. When 'true' Lend-Lease B-24Ds (Liberator Mk IIIs) began to arrive shortly afterwards, the 11 so-called 'Battle of the Atlantic' aircraft were redesignated Mk IIIA, though confusion among British officials in Washington appears to have resulted in the later Lend-Lease aircraft also being described (erroneously) as Mk IIIAs. In the event all the early B-24Ds were to be known as Mk IIIs, though the Mk IIIA designation was later revived (briefly) to describe Mk III aircraft fitted with ASV radar. Later B-24D deliveries were known as Liberator Mk Vs.

With a serial number in the LV3xx series, this aircraft is one of the 11 Lend-Lease B-24Ds hurriedly delivered in 1942 for use by RAF Coastal Command. Fitted with American armament and other equipment, these machines were treated as non-standard by the RAF and though most saw service with Coastal squadrons, another 89 examples (for which RAF serials were allotted) were not delivered.

Left: Liberator FL910, the first Mk III to enter service and shown here shortly after delivery (without radar or unit markings), was an early U-boat victor, claiming U-216 near Ireland on 20 October 1942, while serving with No. 224 Sqn. Unfortunately, the ensuing explosion damaged the aircraft's controls and it crashed spectacularly at RAF Predannack upon its return.

No. 120 Squadron, based at Aldergrove, Northern Ireland, was one of the most successful of Coastal Command's Liberator squadrons, claiming 19 U-boats sunk between October 1942 and VE-Day. Pictured in April 1943, FK228/'M' is a GR.Mk III equipped with early ASV Mk II radar, with its various nose-, wing- and fuselage-mounted antennas. In the background is an aircraft equipped with the later centimetric ASV Mk III in a 'dumbo' nose radome.

Liberator GR.Mk V FL927/G sports a number of the modifications made to Coastal Command aircraft. As well as ASV Mk III radar in the nose radome, the aircraft appears to retain the antennas associated with the earlier ASV Mk II installation. Also evident are rocket rails and a Leigh Light under the starboard wing. The 'G' in the aircraft's serial number (indicating that it was to be provided with an armed guard while on the ground) was applied while the aircraft was at the A&AEE, Boscombe Down.

least 260 long-range aircraft were needed to combat the U-boat menace, yet only 70 such aircraft (using the term long range fairly loosely) were available. Coastal Command requested 190 Lancasters from Bomber Command (which was of course unwilling to supply any aircraft, let alone its precious Lancasters). It was impossible to increase Liberator orders (which were already sufficient to eventually re-equip all VLR squadrons), but Air Vice Marshal Slessor himself personally visited the USA in an effort to expedite deliveries, helping to unleash what would become a flood of aircraft – mostly new Mk IIIs.

The Liberator Mk III (and the closely related Mk V) was destined to form the backbone of Coastal Command's VLR squadrons until late 1944, and was the Coastal Command

Below: The 'dumbo' ASV installation later gave way to a retractable radome fitted in the space normally occupied by the redundant ball turret. Here the installation is seen in an unidentified Mk III/V.

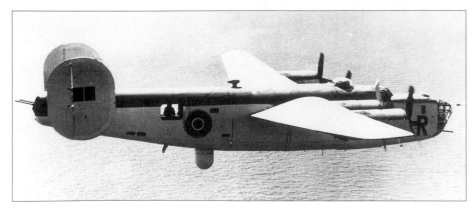

equivalent to the USAAC B-24D. There is some confusion over some RAF Liberator designations, with contradictory information in different official documents and in post-war histories of the aircraft. The Liberator Mk IIIA designation, for example, was initially applied to to 11 Lend-Lease B-24Ds delivered to Coastal Command with standard US armament. It was subsequently applied (erroneously) to all Lend-Lease B-24Ds, which should have been known simply as Mk IIIs. Later, after the Mk IIIA designation was dropped, it was revived and applied (possibly unofficially) to ASV-equipped Mk IIIs. Some sources suggest that the designation may have also applied to a number of aircraft modified to carry 600 Imp gal (2728 litres) of extra fuel. Liberator Mk IIIs destined for Coastal Command were modified by Scottish Aviation at Prestwick.

Some official documents describe the Liberator Mk V as being the radar-equipped Coastal Command equivalent to the Mk III, although some Mk IIIs were also radar-equipped. The difference may have been that the Mk V used ASV Mk III or converted H_2S centimetric radar in an undernose 'Dumbo' radome (or later in a retractable dustbin radome aft of the bomb bay, in place of the ball turret), rather than the ASV.Mk II of the GR.Mk II and GR.Mk III. The change of radar was prompted by Germany's development of the Metox 600 radar warning receiver. This was developed following the capture of an ASV.Mk II set from a downed Coastal Command aircraft. Those U-boats equipped with Metox were able to detect ASV transmissions from 30 miles, (48 km) giving them ample time to dive and disappear. Centimetric radar was needed in order for Coastal Command's 'hunters' to remain undetected by their prey.

To confuse matters, there is some evidence to suggest that some Mk IIIs and even Mk IIs were retrofitted with the later radar, without any change in designation.

Liberator GR.Mk V

Others have suggested that the GR.Mk V designation applied to a dedicated Very Long-Range (VLR) version of the Mk III, with reduced armament and armour, and increased fuel tankage, able to lift well over 2,000 Imp gal (9092 litres) of fuel and eight 250-lb depth charges. The fuel tankage of the GR.Mk V was split between the main tanks (1,950 Imp gal/8865 litres), wing auxiliary tanks (370 Imp gal/ 1687 litres), two bomb-bay tanks (330 Imp gal/1500 litres) and a Catalina-type overload tank (75 Imp gal/341 litres) – a maximum total of 2,725 Imp gal (12388 litres). Some official documents give the type a maximum tankage of 2,935 Imp gal (13343 litres), though this was a theoretical figure, not allowing the aircraft to carry any offensive armament. With 2,560 Imp gal (11638 litres) of fuel and six depth charges, the Liberator Mk V had a range of 3,440 miles

Left: Closing the Atlantic gap from 'the other side', 15 Liberator Mk Vs, diverted from the RAF's allocation, were delivered to the RCAF from April 1943. These equipped No. 10 (Bomber Reconnaissance) Sqn at Gander, replacing the obsolete Digby (B-18 Bolo). Known by the designation GR.Mk V(Can), the aircraft had American armament and a different autopilot fitted.

(5536 km) (or 3,890 miles/6260 km, according to other sources), and a maximum endurance of 15 hours and 20 minutes when cruising at 150 kts and carrying a 1,500-lb (680-kg) bombload, though the addition of a Leigh Light reduced the maximum fuel load to 2,224 Imp gal (10110 litres) and dramatically increased drag, with a significant effect on range. But at the other end of the spectrum, it was possible to fly even longer sorties with a reduced bombload. One No. 59 Squadron Liberator GR.Mk V, for example, flew an 18-hour patrol from Iceland, landing at Goose Bay in Canada.

Although they were delivered with mid-upper and tail turrets, many of the GR.Mk IIIs and GR.Mk Vs were stripped of these in an effort to reduce weight and drag, although other modifications soon restored both! One such modification applied to radar-equipped Mk IIIs and Mk Vs in service was the Leigh Light, a podded 19 million candlepower searchlight in an 870-lb (395-kg) nacelle, shining a narrow (3°-12°), focussed beam about one mile ahead of the aircraft. The tactic used was to switch on the Leigh Light as the target 'blip' disappeared into the clutter of sea returns, which happened at about three-quarters of a mile. To equip four UK-based squadrons the RAF ordered 60 units, but fitment was a slow process, and by D-Day, only Nos 53, 120 and 224 Squadrons had received any Leigh Light-equipped aircraft. The equipment did prove extremely successful, however, and helped to drive the U-boats under water all of the time, depriving them of the cover of darkness.

The Liberator Mk III and V together accounted for 38 of the 65 U-boats sunk by Liberators of all marks, and equipped seven squadrons, making it the most important home-based Coastal Command Liberator variant. The first Liberator GR.Mk III went to No. 224 Squadron in May 1942, though the unit would retain Mk IIs until these were replaced by GR.Mk Vs. The GR.Mk III did re-equip No. 120 Squadron from June 1942, and allowed No. 86 Squadron to replace its obsolete Beauforts from October 1942, converting at Thorney Island before moving to Aldergrove in March 1943.

No. 59 Squadron began to replace its Hudson Mk IIIs with Liberator Mk IIIs in August 1942, but conversion was halted,

USAAF anti-submarine Liberators

The USAAF established its Antisubmarine Command on 13 October 1942 and called on the B-24D Liberator, among other types, for patrol work from US bases and in Europe, North Africa and the Middle East. The AAF ultimately operated two anti-submarine wings, an anti-submarine group, and a search attack group, all with land-based bombers being flown on exhausting, offshore patrols in search of the elusive U-boat. Among those who pulled this duty were the unsung heroes of Sixth Air Force in the Panama Canal Zone, which was sovereign US territory. A Liberator crew sank U-654 in Atlantic waters, flying from Panama on 22 August 1942. Panama-based Liberator crews flew more than 3,000 sorties over Pacific waters without ever coming face-to-face with the foe. Perhaps because of their vigilance, Japan never challenged the Panama Canal.

Elsewhere in the world of anti-submarine operations, US Liberator crews were credited with all or part of the sinkings of nine additional U-boats between October 1942 and August 1943, while RAF Coastal Command Liberators killed another 19 U-boats during the same period. More important than this figure is the extent to which they made the U-boat skipper's job more difficult: in many respects, the B-24 crews did their jobs merely by being there, and thus posing one extra obstacle to enemy submarine operations.

and the aircraft were replaced by Fortresses during December 1942 and January 1943, before Liberator conversion recommenced, first to the Mk III, and then, from March, to the centimetric radar-equipped Mk V. GR.Mk Vs were also delivered to Nos 53 and 311 Squadron at Beaulieu (after conversion at Thorney Island) in May and June 1943, No. 120 Squadron in September, and to No. 547 in November.

While Liberators would continue to fly ASW sweeps and convoy escorts in the Atlantic until the end of the war, the type soon began to be used more widely. The second area to come under the Liberator's scrutiny was the Bay of Biscay, across which U-boats passed en route from their French bases to their patrol areas. Liberators of the USAAF's 93rd Bomb Group had started ASW operations from St Eval (the 409th Bomb Squadron) and Holmsley South (the 330th Bomb Squadron) in October 1942, on temporary loan from the Eighth Air Force. During this attachment, one USAAF B-24 was attacked by five Ju 88s, but the pilot, Major Ramsey D. Potts managed to evade the worst of their fire while his gunners coolly despatched two of the attackers and

No. 10 (BR) Sqn's Liberators were stripped of their paint in early 1944; the aircraft's original finish had proved very difficult to maintain in acceptable condition. All but two of the RCAF's Liberator GR.Mk V(Can)s were delivered with 'dumbo' ASV radar installations; the remaining pair had retractable ventral radomes.

Below: White Savage of the 479th Anti-Submarine Group, based in the UK, was one of those B-24Ds fitted with a nose turret by the Oklahoma City Air Depot. It also has an ASV radome in the former ball turret position.

Left: The USAAF's first anti-submarine B-24s were standard B-24Ds hastily sprayed white on their under surfaces for low-level operations. B-24D 42-40328 Lulu's Ole Lady was photographed at Gander, Newfoundland.

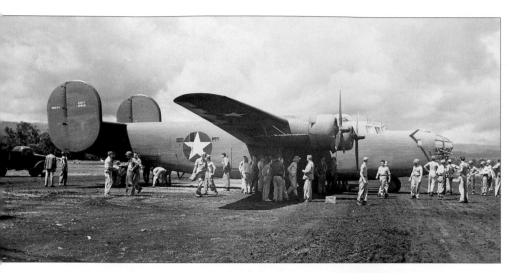

Above: Pictured at Henderson Field, Guadalcanal, this early PB4Y-1 is almost certainly one of the VB-101 aircraft sent to the Solomon Islands in early 1943 to harry Japanese shipping resupplying Japanese forces in the island group.

Right: Thunder Mug was PB4Y-1 BuNo. 32108 of VB-109. Flown by the squadron's CO, Cdr Norman 'Bus' Miller, Thunder Mug made the first raid by a land-based aircraft on Iwo Jima, taking-off from Saipan on 14 July 1944. One of the most decorated naval units of the Pacific war, its crews receiving a total of 301 awards for their exploits.

Lower right: A PB4Y-1 of VPB-104 waits for a burning Koshu to sink after two low-level bombing and strafing runs on the 3000-ton vessel, December 1944. Targets such as this were typical of those sought by Navy Liberators patrolling vast areas of the Pacific.

US Navy PB4Y-1s

Hurriedly ordered to Guadalcanal in early 1943, VB-101 – the Navy's first PB4Y-1 unit – spent the next seven and a half months undertaking patrol and bombing missions against Japanese targets in the region. PB4Y-1 numbers in the southwest Pacific quickly built up, the type equipping (reconnaissance squadrons) VD-1, VD-3 and VB-104 during the first half of 1943 and another two VD- and five VB- squadrons before year's end. Three additional VB- units (redesignated VPB- units in 1944) were equipped with PB4Y-1s during 1944/45.

In the Atlantic VB-103 joined the campaign against German U-boats in May 1943, based at Argentia, Newfoundland. In August the squadron moved to England as the USAAF relinquished its anti-submarine role and by early 1944 there were 10 Navy Liberator units patrolling the Atlantic, from bases on both sides of the Atlantic and more southerly bases such Parnatmarin (Natal, Brazil) and Wideawake Field, Ascension Island.

and the Azores. The Navy had felt that ASW should be its job, and the USAAF's Antisubmarine Command had been disbanded on 24 August 1943. Unfortunately, there were criticisms that Admiral Ernest King took an altogether too parochial view of the Battle of the Atlantic, and husbanded most of his aircraft to protect the Eastern seaboard of the USA (where short-range and training aircraft were quite capable of meeting the threat), otherwise concentrating only where US convoys (to their forces in North Africa, for example) were threatened. These criticisms were essentially based in fact, though they were probably over-stated.

The focus on the Channel and Bay of Biscay intensified as preparations for the liberation of occupied Europe gathered pace. By 5 June, St Eval housed a wing of four Liberator squadrons (Nos 53, 206, 224 and 547), while No. 311 was based at Predannack. On D-Day itself, of course, Coastal Command would be heavily stretched in keeping U-boats and enemy surface combatants (from E-boats to destroyers) out of the Channel, and in contributing to the various deception operations intended to mislead the Germans as to the likely target of the invasion. Some 49 U-boats were assigned to anti-invasion duties, including 24 operating from Brest and 19 from St Nazaire. A rapid advance through France followed the success of D-Day, and the U-boats began to lose the sanctuary of their French bases, which were captured or became untenable. The focus of operations shifted back to the Atlantic and to the U-boat bases and coastal shipping in Norway's fjords. No. 206 Squadron led the exodus from the south of England, moving to Leuchars in July 1944. No. 311 Squadron moved to Tain in August 1944, and in September No. 224 moved to Milltown, No. 53 to Reykjavik, and No. 547 to Leuchars.

New weapons

The Liberator Mk III and V introduced new weapons for use against the U-boats. As U-boats tended more and more to try to fight it out on the surface, with weapons having longer and longer reach, the Liberators became more vulnerable to return fire, and several were shot down. A single 0.50-in machine-gun in the nose glazing was barely adequate, and instead some aircraft were fitted with a pair of braced stub wings on the sides of the nose, from each of which were suspended four 5-in HVARs. Another alternative rocket fit mounted four rockets on lighter, flimsier rails, which could retract, into the rear part of the bomb bay.

damaged a third. RAF Liberators (from Nos 59 and 160 Squadrons) were flying increasing numbers of sorties over the Bay and in the Channel Approaches by the end of 1942, and began recording an increasing number of contacts and attacks.

The need for increased activity in the bay of Biscay intensified in the run up to the Allied invasion of French North Africa (Operation Torch) and its aftermath, with a need to protect the growing number of southbound convoys. Sleepy St Eval and Chivenor were upgraded to accommodate much larger numbers of aircraft, and patrols were stepped up. From early 1943, the RAF Liberators were augmented by US Navy Liberators operating from bases in Devon, Gibraltar

Guadalcanal PB4Y-1

PB4Y-1 BuNo. 32081
USAAF B-24D-90-CO 42-40726 was delivered to the US Navy in 1943. After being fitted with a nose-mounted Erco ball turret it was issued to VB-104 'Buccaneers' on Guadalcanal and named *Whitsshits*.

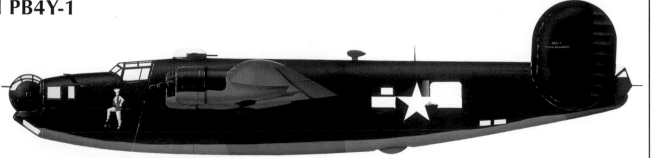

When U-boats did crash dive, they were tending to do so more quickly, benefiting from better training and procedures and crews who kept a much better lookout. The U-boats often managed to escape and there was an urgent need for a weapon that could follow the U-boat below the surface. This materialised in the form of the American Mk 24 Acoustic Homing Torpedo, 10 of which could be carried by the Liberator. These weapons were extremely secret, and even to talk about them was a criminal offence. Popularly known as FIDO or 'Wandering Annie', the torpedo was usually referred to as the Mk 24 Mine in official documents, and entered service in May 1943. On 12 May, Flight Lieutenant Wright scored a hit on the *U-456* with a torpedo, causing damage which kept it on the surface and vulnerable to further attack. It was sunk the next day by a prowling Sunderland. On 14 May, Pilot Officer Gaston used another torpedo to sink *U-266*, the first kill by a homing torpedo. Other new innovations introduced during 1943 included sonobuoys and even small anti-tank bombs, which proved grimly effective against surfaced U-boats.

The first Liberator GR.Mk VIs had started arriving in Britain in December 1943, but the type did not enter service until March 1944. Only one unit had wholly equipped with the type by D-Day, and the Mk III and Mk V would remain in front-line Coastal Command use until early 1945.

Africa and India

The importance of the Liberator Mk III/V to home-based Coastal Command units has sometimes tended to obscure the type's use in the anti-submarine role in other theatres. When the first two bomber units were sent to India at the end of 1942, one of them (No. 160 Squadron) actually reverted to the long-range ASW role, operating from Ceylon. Two more Liberator ASW squadrons operated in India, No. 354 forming at Drigh Road (near Karachi) in May 1943 and receiving Liberator GR.Mk Vs and IIIAs from August 1943. No. 200 Squadron moved to St Thomas Mount (near Madras) with its GR.Mk Vs in February 1944. Both units began converting to the GR.Mk VI in the spring of 1944, with No. 160 following in June. No. 160 began phasing out some Mk IIIs and Vs until June 1946.

The long range of the GR.Mk V made it particularly useful in the Far East, where there were huge areas of ocean to cover, and where distances were huge. This necessitated flying with all defensive armament deleted (apart from two

guns in the tail) and No. 160 Squadron began to fly very long range mine-laying ('Nutting') sorties from Kankesanturai and later Minneriya. Mission times of 17 and 18 hours became common as the squadron mined Penang and Pankalang Brandon, and reached 21 hours (a round trip of 3,500 miles) from March 1945, when the unit first mined Singapore harbour. The absolute duration record for an RAF landplane fell to a No. 160 Squadron Liberator on 31 July/1 August 1945, when Flt Lt Muir captained BZ862 on a 24-hour and 10-minute mission to drop supplies to clandestine forces behind enemy lines in Malaya. This mission illustrated the multi-role nature of the maritime units in the Far East, which was unusual. The maritime Liberator squadrons in the Far East flew ASW patrols, anti-shipping strikes, SAR and meteorological missions and Special Duties sorties.

No. 200 Squadron had previously served in West Africa, trading its Hudsons for Liberator GR.Mk Vs in July 1943. The unit was based at Yundum in the Gambia, and patrolled the coast of West Africa, protecting Cape-bound convoys from a growing U-boat nuisance. On 11 August, the first day of Liberator operations, one aircraft (BZ832) failed to return. Evidence from U-boat survivors told the story of how the aircraft, captained by a young New Zealander, Flying Officer Lloyd Trigg, had pressed home an attack on the *U-468* after being mortally hit by the submarine's 20-mm cannon armament. The attack sank the U-boat, but the burning Liberator crashed into the sea and exploded. Trigg was subsequently gazetted with a Victoria Cross.

Robert F. Dorr and Jon Lake

Above: This view of VB-103 PB4Y-1 BuNo. 32032/'C' over the Bay of Biscay was one of a series of images of this aircraft made by the USN during 1943. It was lost on 12 November after being hit by defensive fire from U-508. Not visible in this view is the retractable ASV radar set installed in the customary position in the lower fuselage.

Left: Another VB-103 aircraft, this PB4Y-1 has been field-modified to accept an Erco ball turret in the bow position.

Part 2 of this report will be featured in a forthcoming volume, and will include development of the B-24K, L, M, N, freighter variants and the Privateer, operations during 1944/45, post-war US and foreign service and a list of operators

320 Squadron, MLD

'Animo Libero Dirigimur'
(Guided by the spirit of liberty)

Above: No. 320 Squadron's RAF badge is officially described thus: "In front of a fountain an orange tree fracted and eradicated". The squadron was the first formed within the RAF with foreign personnel.

No. 320 Sqn's earliest and most important wartime aircraft were the Fokker T-VIIIWs (left) and Mitchell Mk II/IIIs (below), respectively. Its current mount is the Lockheed P-3C Orion (bottom).

Famous as the wartime RAF squadron equipped with Fokker T-VIIIW floatplanes that had escaped the Netherlands ahead of the German invasion, No. 320 Squadron went on to give yeoman service, first as a Coastal Command unit and, later, as a tactical bomber unit in 2nd TAF, supporting the Allied armies invading northern Europe in 1944. Today the squadron is, once more, part of the MLD, flying P-3C Orions.

Following the German invasion of Holland in May 1940 and the subsequent defeat of the Dutch military, many Dutch pilots and maintenance crews fled to the UK. No. 320 (Netherlands) Squadron was the first of a number of foreign squadrons established in England during the early days of World War II,

paving the way for more squadrons manned by personnel from not only the Netherlands, but other occupied continental countries.

On 10 May 1940, the day of the German invasion, the MLD (*Marineluchtvaartdienst*, Naval Air Service) had a total of 43 operational aircraft. Although its air force counterparts were

Fokker T-VIIIW

Intended to replace the Fokker T-IV in MLD service, the T-VIIIW torpedo-bomber/reconnaissance floatplane was built in three variants: the **T-VIIIW/g** of mixed wood/metal construction, the all-metal **T-VIIIW/m** and the larger **T-VIIIW/c**, again of mixed construction.

In all 36 T-VIIIWs were ordered, including 19 T-VIIIW/gs, 12 T-VIIIW/ms (for service in the Dutch East Indies) and five T-VIIIW/cs, the latter an enlarged variant with 890-hp (664-kW) Bristol Mercury XI engines for Finland. Of these, 11 T-VIIIW/gs were completed along with the five T-VIIIW/cs, though the latter were captured by advancing German forces before delivery and put to work as SAR/reconnaissance aircraft in the Mediterranean.

The T-VIIIW/g was powered by a pair of 450-hp (336-kW) Wright R-975 Whirlwind engines and was able to carry up to 750 kg (1653 lb) of bombs or torpedoes. A forward-firing 7.62-mm machine-gun in the forward fuselage and a hand-held 12.7-mm weapon in the rear cockpit were provided for self-defence. Generally operated with a crew of three or four, the type had a range of 1,700 miles (2735 km).

Pictured (below) is the first of the T-VIIIWs – R-1 – the first of the type taken on charge by the MLD on 25 April 1939. As one of the aircraft that escaped to the UK, R-1 was serialled AV958 in RAF service and remained in use until the Fokkers were phased out in November 1940.

Above: Upon its formation No. 320 (Netherlands) Squadron, RAF spent four months at RAF Pembroke Dock, in south Wales. From here the squadron flew convoy protection sorties until the Fokker T-VIIIWs were withdrawn and replaced by Lockheed Hudsons.

heavily involved in the air battle above Holland, the aircraft of the MLD did not contribute significantly. Their wartime taskings consisted of reconnaissance of Dutch territorial waters and the destruction of enemy boats and submarines, and therefore played little part in the defence of Holland. Faced with imminent defeat by the invading Germans, all remaining flight and maintenance crews received the order to leave Holland for France.

Refuge in France

On 14 May 1940 all remaining MLD aircraft plus 76 crew and personnel left their base at Kudelstaart for Cherbourg. A total of 24 aircraft – one Fokker C-XIW, 11 Fokker C-XIVW reconnaissance aircraft, five Fokker C-VIIIW reconnaissance aircraft and seven Fokker T-VIIIW torpedo-bombers – survived the German invasion. All other MLD aircraft in Holland were lost in combat or to German bombing, or were destroyed by their own crews. The sole exception was Fokker T-VIIIW serial R-3, which had left Holland for Brighton, England, on the first day of the war, carrying the Dutch Minister of

K6175 was one of the Anson Mk Is with which No. 320 Sqn re-equipped after the withdrawal of the Fokker T-VIIIWs. Note the pre-war Dutch uniform worn by the groundcrewman facing the camera.

Foreign Affairs and the minister for the colonies.

A number of reconnaissance flights were made by Fokker T-VIIIWs operating from Cherbourg, but the other aircraft headed for Brest. Finally, all were ordered to divert to the United Kingdom on 19 May and five days later arrived at the British seaplane base at Calshot. There, only eight Fokker T-VIIIW torpedo-bombers were retained in service, together with a single Fokker C-XIVW.

Dutch naval staff urged the British authorities to form two squadrons manned by Dutch crews, but the Royal Air Force had an urgent need for pilots to equip its own squadrons and requested that Dutch personnel fly in regular RAF units. However, Dutch insistence that a true Dutch squadron be formed prevailed and, on 1 June, 1 and 2 Eskadrilles (Squadrons) were formed. 1 Eskadrille was the first foreign squadron in the RAF to fly operational sorties during World War II.

1 Eskadrille was initially commanded by Lieutenant 1st Class J.M. van Olm and based at RAF Pembroke Dock in Wales, operating the eight Fokker T-VIIIWs and single Fokker C-XIVW, the latter as a target tug. The T-VIIIWs were painted in RAF colours during July and serials AV958 to AV965 were applied; a small orange triangle marking on the forward fuselage indicated their Dutch origins. Five other MLD aircraft that had survived the invasion were deployed to Felixstowe to act as decoys, and were withdrawn from use during 1940. None of these decoys was ever attacked and all were soon scrapped. All remaining aircraft were shipped to the MLD's Naval Air Station Morokrembangan in the Netherlands East Indies for use as training aircraft. The second eskadrille was equipped with obsolete RAF Avro Ansons and based at RAF Carew Cheriton, Pembroke.

Soon after their arrival in England, the Fokker T-VIIIWs of 1 Eskadrille flew their first

No. 320 Sqn operated five Hudson variants between 1940 and 1943. The first examples were a batch of 14 Mk Is (left) which entered service in October 1940, some examples remaining in use until September 1942. Hudson Mk IIs were employed from March 1941 for just six months, while Mk IIIs (above, acquired in July 1941) served alongside the Mk Is and were phased out at the same time as the earlier machines. From January 1942 Hudson Mk Vs were employed, though all had left the squadron by May. The Mk V differed from the earlier aircraft in being powered by Pratt & Whitney R-1830 Twin Wasp radials (as opposed to Wright GR-1820 Cyclones). Twin Wasp-powered Mk VIs (some with ASV radar, right) were the squadron's last Hudsons, replacing the surviving Mk I/III aircraft in August 1942.

operational sortie, undertaking convoy protection. During June, 13 sorties were flown.

In August the Dutch squadrons were renumbered as RAF units, 1 and 2 Eskadrilles becoming Nos 320 and 321 (Netherlands) Squadrons, respectively.

No. 320 Squadron suffered its first casualties on 26 July, when Fokker T-VIIIW AV964 (formerly R-10) and its crew of four were lost during a routine sortie. A second Fokker (AV963, formerly R-9) was lost during a convoy protection flight on 26 September, again with all four crew members. In late October 1940, after 133 operational sorties, the remaining six Fokker aircraft were finally withdrawn from use; all were scrapped in June 1941.

However, before their final withdrawal, a number of Fokker aircraft and crew made special one-off sorties. In one of the last of these, Operation Windmill took a Dutch crew in their T-VIIIW to Frisian Tjeukemeer, a lake in the Netherlands, to pick up an agent and three other passengers. When the Fokker touched down on the lake on the night of 16 October 1940, it became immediately apparent that the operation had been betrayed to the Germans. The aircraft was hit by enemy fire. To make matters worse, on the return flight it was

shelled by a British coastal AA battery. The Fokker's crew managed to land safely in Felixstowe, but its aircraft was riddled with over 40 shell holes.

The aircraft's commander, Flying Officer 2nd Class H. Schaper, was later awarded the Distinguished Flying Cross (DFC), the first Dutch serviceman to receive the decoration.

Replacements for Fokkers

Although the Dutch Naval Commander had hoped to secure Consolidated PBY Catalinas with which to re-equip the squadron, 14 Lockheed Hudson Mk Is were ordered, these being the only suitable aircraft available at the time. Additional examples were donated by the Prince Bernhard Fund, which raised money for the fight in Europe among Dutch settlers in the East Indies.

Pending the arrival of the Hudsons from the United States, a flight of 10 Anson Mk Is formed the operational component of No. 320 Squadron, and four aircraft were employed in an observer/air gunner training flight. On 1 October, the squadron moved to RAF Leuchars in Scotland, where aircrew began training on the Hudson. Two weeks later, the first of the squadron's own Hudsons arrived,

allowing the squadron to train on its own aircraft.

Meanwhile, No. 321 Squadron was temporarily disbanded on 18 January 1941 because of lack of personnel and its remaining staff and aircraft were transferred to No. 320 to man 'B' (training) Flight. A number of RAF personnel and members of the former Dutch army air force also joined the squadron at this time. The Ansons remained on the squadron until October 1941.

With the introduction of the Hudson, No. 320 Squadron's role in Coastal Command was the provision of convoy protection and reconnaissance of the Irish Sea, the Bristol Channel and St George's Channel, the unit becoming operational on the new type from February 1941. Most of its sorties were flown at night, typically at an average altitude of 4,000 ft (1219 m).

The first Hudson was lost on 25 February 1941 when it crashed during take-off, destroying a hangar in the process. Four crew were killed and an Anson in the hangar was destroyed by the ensuing fire. On 8 March, another Hudson crashed in the Irish Sea just prior to landing, this time only wounding three crew members.

As a prelude to more offensive operations, which began in earnest in July, two No. 320 Squadron Hudsons, as part of a combined force with six RAF Ansons, flew a successful bombing mission to Mandal airfield in Norway in the early hours of 10 May 1941.

On 23 November a notable mission involved a Hudson, flown by Sgt C. van Otterloo, on a patrol near the Norwegian coast. Discovering a small German convoy, he attacked, two bombs exploding on a 6,000-Imp ton (6096-tonne)

No. 320 Sqn was the third RAF unit to receive the Mitchell, a hastily acquired replacement for No. 2 Group's outmoded Bristol Blenheims. Though No. 2 Group judged the Douglas Boston a better aircraft foir its needs, the Mitchell had a superior bomb load capacity, namely up to four 1,000-pounders.

Above: While at Dunsfold the squadron's servicing personnel were reorganised as No. 6320 Servicing Squadron. Here a Mitchell undergoes maintenance at the Surrey station.

Right: Squadron personnel assemble in front of one of the unit's Mitchells early in 1944. Seated, front and centre, is No. 320's then CO, Captain J.N. Mulder.

freighter, which immediately caught fire and capsized. Anti-aircraft fire from one of the vessels hit the aircraft, fatally injuring van Otterloo. The aircraft went into a dive but the observer, Flying Officer van Rossum – despite having never flown a Hudson before – was able to level the aircraft and return it to the UK. Van Otterloo and 2nd Observer van der Meer were later awarded the Flyerscross, and van Rossum received the DFC.

On 21 April 1942 the squadron moved to RAF Bircham Newton in Norfolk, although most sorties were flown from RAF Docking, a satellite airfield of Bircham Newton, as the latter was not suitable for night operations. Over the next 12 months the squadron's tasks included participation in the 'thousand-bomber raids' over Germany and in the first of these, on 24 June 1942, six of the squadron's Hudsons flew to Bremen, one being lost with all crew.

On 30 January 1943 the squadron received orders for a sortie in search of the German battleship *Tirpitz*. Given the expected duration of the mission and the need to carry as much fuel as possible, the aircraft involved were only to carry a single 250-lb (113-kg) bomb. The lengthy flight near the Danish and Norwegian border failed to find *Tirpitz*.

By 1943 No. 320 Squadron had bombed German air bases, carried out reconnaissance and patrol flights along the Norwegian coast and undertaken SAR sorties. The last operational Hudson sortie was performed on the night of 9/10 March 1943. The squadron had operated a total of 61 Hudsons of five variants, though most had been Mk Is, IIIs and VIs. The Dutch Hudsons had performed a total of 1,234 operational sorties, including 289 convoy protection sorties, 94 reconnaissance flights, 94 SAR sorties and 401 attack sorties on ships and airfields. During their missions, the Hudsons destroyed almost 40 German ships. Losses totalled 27 aircraft to all causes.

Mitchells and a new role

After lengthy discussions during 1942 the decision was made to replace the Hudson with a new type. Douglas DB-7Cs were earmarked for the squadron, but were transferred to Russia instead. By the end of November it had been decided that No. 320 Squadron's Hudsons would be transferred to the Mediterranean, making the search for a replacement type all the more urgent. A replacement was eventually found in the North American B-25 Mitchell, a type already serving with the Royal Dutch Indian Army in Australia. Beginning in March 1943, a total of 67 Mitchell Mk IIs was assigned to the unit, with the RAF serials FR141/207.

With this change in equipment, No. 320 Squadron joined the RAF's other Mitchell tactical bomber units in No. 2 (Medium Bomber) Group, Bomber Command, and moved to RAF Methwold, Norfolk. On 1 June 1943, with the removal of tactical units from Bomber Command, No. 2 Group was transferred to 2nd Tactical Air Force, Allied Expeditionary Air Force.

In the meantime, a lengthy training process had begun that took more than five months to complete. During this period, No. 320 Squadron moved again, this time to RAF Attlebridge at 30 March 1943. The last of the squadron's Mitchell Mk IIs to be ferried across the Atlantic, FR142, arrived shortly afterwards,

Above: During May 1944, in the run-up to the D-Day landings during which No. 320 Sqn would provide air support, General Dwight D. Eisenhower, Supreme Commander Allied Forces Europe, addresses squadron personnel at RAF Dunsfold. No. 320 Sqn shared Dunsfold with Nos 98 and 180 Sqns, the first two RAF Mitchell squadrons. These units, with No. 226 Sqn at Hartfordbridge, were airborne on the night of 5/6 June 1944, with the intention of hitting roads and rail lines in northern France, though poor weather forced No. 320 Sqn's aircraft to return with their bombs.

Right: Adorned with AEAF 'invasion' stripes, No. 320 Sqn Mitchell Mk IIs depart Dunsfold on a post-invasion bombing sortie.

Left: In October 1944 the squadron moved across the English Channel to landing ground B.58 (Melsbroek, Belgium). The somewhat weather-beaten aircraft nearest the camera carries a sizeable collection of mission symbols – a testament to the intensive operations demanded of the 2nd TAF after D-Day.

Below: Pictured shortly after 6 June, No. 320 Sqn aircraft 'NO-K' flies a typically hazardous daylight medium-altitude carpet bombing sortie, over Caen's steelworks complex.

flown by HRH Prince Bernhard of the Netherlands.

The squadron's first operational sortie during the training phase was a search and rescue flight to locate aircraft and crew that had ditched on 12 June 1943. One of its first offensive sorties was flown on 17 August 1943, when six Mitchells flew to Calais, but this sortie was not a great success. A second attempt with their new aircraft saw nine Mitchells fly to Vlissingen in Holland on 20 August 1943, where they attacked the De Schelde Dornier aircraft factories from a height of 12,000 ft (3657 m) with 500-lb (227-kg) bombs. This caused enough damage to ensure that part of the factory never returned to production. One Mitchell ditched during the return flight to the UK, but fortunately all were rescued by an RAF Walrus.

On 18 October, the commander of No. 320 Squadron, Overste E. Bakker, was killed with his crew when his Mitchell, FR178, was hit by anti-aircraft fire during the bombing run as part of a 12-ship Mitchell assault on the occupied airfield of Lanvéoc near Morlaix. Other sorties included the 8 November 1943 'Noball' attack on a V-1 construction site in Mimoucques using 2,000-lb (907-kg) bombs. During the following months over 400 sorties were made, mainly against V-1 launch sites. Most of No. 320 Squadron's targets were located in France and consisted of fuel depots, bridges, V-1 launch sites, troop concentrations, railway emplacements, harbours and factories.

After the training period, No. 320 Squadron moved to RAF Lasham on 30 August to become fully operational. About 20 Mitchells were

assigned to No. 320 Squadron, which meant that some 12 aircraft were normally serviceable each day. However, during big sorties most aircraft were airworthy and able to participate. During daylight, No. 320 Squadron normally operated in 'boxes' of six aircraft, flying in two 'V' formations of three aircraft. To cover a greater target area, larger formations were flown with as many as 24 aircraft in a diamond formation. Typically, 4,000 lb (1815 kg) of bombs were carried.

Another advantage of flying in a box was increased safety from enemy action. A formation normally consisted of six to eight boxes, often protected by as many as five fighter squadrons that operated in a wide circle around the attack formation, flying close cover, high cover and distant cover.

The greatest enemy of the Mitchell, and the weapon that caused the most casualties to No. 320, was flak. During a bombing run, aircraft never maintained the same course for more than 20 seconds at a time, but losses were still high and it became necessary to draft in personnel of other nationalities, including Australians, Belgians, British and New Zealanders.

New personnel; D-Day approaching

On 18 February 1944, with the Allied invasion of France less than four months away, No. 320 Squadron moved again, this time to RAF Dunsfold in Surrey.

Unfortunately not all the squadron's losses were the result of enemy action. On 26 April 1944 Mitchell FR124 'F' was flying a so-called Batseye Dawn sortie, monitoring activity in a number of occupied harbours. Flying without lights, the Mitchell was mistaken for a German Do 217 by the pilot of Mosquito FB.Mk VI NS903 (Flt Lt McLurg), who fired on the aircraft. The pilot ordered the crew to bail out of the heavily damaged Mitchell; one crew member did so, but the second did not hear the command and the third had failed to bring his parachute. The pilot (Jopie Mulder, later a commander of the Dutch air force) was forced to make a crash landing near Maidstone in Kent. The crew survived, but the Mitchell was a write-off.

Above: In 1949, a number of the remaining Mitchells with the squadron (of 21 acquired from the RAF) were withdrawn from use and the nine remaining aircraft refurbished to see out their last months in service. Mitchell 2-6, one of No. 320 Sqn's post-war Mitchells, was photographed at NAS Valkenburg.

Below: On 12 July 1951 34 MLD officers and airmen, headed by Rear Admiral H. Schaper, arrived at NAS Norfolk, Virginia, to train on PV-2s. A total of 18 was purchased (serials 19-1/ 9-18), all through the Mutual Defense Assistance Program.

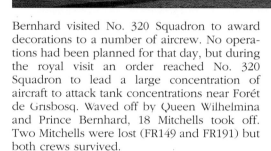

Left: P2V-5 19-21, the first of the MLD's Neptunes takes to the air in the US prior to delivery. All 12 examples were refurbished and passed to the Portuguese air force in the early 1960s.

Below: Later in their career the P2V-5s adopted this 'anti-flash' grey/white finish in deference to their wartime nuclear role, though larger areas of grey were later added to decrease the aircraft's conspicuity in the maritime patrol role.

In the last days before D-Day, the squadron's aircraft were painted with Allied Expeditionary Air Force (AEAF) 'invasion' stripes and the crews were made aware of the upcoming action. In the early hours of 6 June 1944, 12 Mitchells of No. 320 Squadron were tasked with destroying bridges over the Orne and Dives river, thereby becoming one of the first AEAF units over Normandy. However, the attack was aborted as bad weather made identifying the targets impossible; all aircraft returned safely to Dunsfold. Over following days, the beachhead well established, No. 320 Squadron attacked fuel and ammunition supplies.

Tragedy struck early on 8 June when a group of 14 Mitchells set course for France and two Mitchells (FR150 and FR182) collided above Surrey. The aircraft caught fire, their weapons (four 1,000-lb bombs each) exploding with the loss of both crews. On the same day a third Mitchell (FR179) crashed near Constances.

Two days later a significant success was achieved, when a force of about 70 Mitchell aircraft, headed by the commander of No. 320 Squadron, H.V.B. Burgerhout, and 40 Typhoons escorted by Spitfires, destroyed the headquarters of the Panzer Gruppe based in La Caine. The commanding German Chief of Staff (General Major Ritter von Dawans) and most of his staff were killed in the operation.

On 12 June, Queen Wilhelmina and Prince

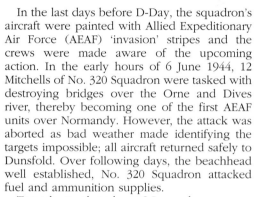

The MLD's first 15 P2V-7s were purchased directly for use by No. 321 Sqn at Hato, Curaçao. Among those aircraft passed to No. 320 Sqn in 1962, aircraft 204/'V' was to be the last MLD Neptune in service, finally retiring in 1982 as the first P-3C Orions were delivered. It was later flown to England and has since been displayed at the RAF Museum, Cosford.

Bernhard visited No. 320 Squadron to award decorations to a number of aircrew. No operations had been planned for that day, but during the royal visit an order reached No. 320 Squadron to lead a large concentration of aircraft to attack tank concentrations near Forét de Grisbosq. Waved off by Queen Wilhelmina and Prince Bernhard, 18 Mitchells took off. Two Mitchells were lost (FR149 and FR191) but both crews survived.

Performing sorties almost every day took a considerable toll on No. 320 Squadron: eight Mitchells and 25 crew were lost on the days immediately after D-Day. Over 340 sorties were performed and 650 Imp tons (660 tonnes) of bombs and other ordnance were delivered.

Falaise and Market Garden

Important sorties during the ensuing months included action around Falaise and the support of the Allied advance into Belgium, Holland and Germany. During Operation Market Garden, the airborne operation to retake a number of Dutch bridges near the town of Arnhem in southern Holland, No. 320 Squadron attacked Wehrmacht barracks in Ede. Unfortunately, the Market Garden operation failed, and during the retreat No. 320 Squadron attacked targets around Arnhem.

No. 320 Squadron moved to Melsbroek AB in

Belgium in October 1944. Over the following months, the Dutch Mitchells regularly flew sorties over occupied Holland, Belgium and Germany, against targets that included bridges, railway complexes, airfields and V-1/V-2 installations. In November 1944 the first Mitchell Mk IIIs reached the squadron, replacing the earlier Mk II version. In December 1944, the Germans began their Ardennes offensive, the last major effort by the Wehrmacht to halt the advancing Allies. Due to bad weather, neither side could fly, but from 23 December, with the weather improving, No. 320 Squadron was able to attack German strongholds.

On 1 January 1945, the Luftwaffe launched Operation Bodenplatte – a major attack on Allied airfields, including Melsbroek. Fortunately, No. 320 Squadron was on a sortie at the time, their aircraft suffering little damage. During February 1945 more than 260 sorties were flown.

On 3 March 1945 No. 320 Squadron was tasked with attacking V-2 launch installations near The Hague. Due to errors in the target's co-ordinates and a much stronger wind than anticipated, the bombs fell on a densely populated area of The Hague, tragically killing many civilians.

The final wartime move took No. 320 Squadron to Achmer, Germany on 30 April

In 1974 320 Squadron renewed its acquaintance with the Grumman Tracker, forming a flight of four target-towing US-2Ns (transferred from 5 Squadron), which remained in use until 30 September 1975. On this day, the last two aircraft, 159 and 160 (pictured, with underwing target-towing equipment), were flown into storage at NAS De Kooij. The S-2 saw no further service in the MLD; 159 and 160 have both since been preserved.

1945. Only two days later, on 2 May 1945, No. 320 Squadron performed its last wartime operation, attacking Itzehoe in Germany. Germany capitulated six days later.

Some 3,252 sorties comprising 8,750 flight hours were flown by Mitchells between March 1943 and May 1945, for the loss of 33 aircraft and 88 aircrew members. Over 5,500 Imp tons (5588 tonnes) of bombs were dropped. During the whole war, 157 aircrew and 61 aircraft were lost: three Fokker T-VIIIWs, four Ansons, 21 Hudsons and 33 Mitchells.

On 2 August, shortly after the squadron had moved back to the UK, control of No. 320 Sqaudron was passed to the Dutch navy.

Post-war revival

No. 320 Squadron (known simply as 320 Squadron in the post-war MLD) began training on the de Havilland Mosquito, but in 1946, during the second phase of training at Twenthe air base in Holland, the decision was taken to disband the unit, effective from 1 May 1946. No Mosquitoes ever flew in Dutch service. Squadron personnel were divided between 860 and 861 Squadrons, then operating the Fairey Firefly.

As the MLD was rebuilt in the immediate post-war years, Valkenburg airfield was loaned to the MLD and its first new equipment consisted of Fairey Fireflies and North American Harvards. Three years later, on 22 March 1949, 320 Squadron was reformed with Auster Mk IIIs, B-25B/C Mitchells, Airspeed Oxfords and Supermarine Sea Otters.

During 1947, 21 Mitchells stored at RAF Kirkbride were inspected with a view to returning them to service with the MLD. Three were assigned to search and rescue, two were sent to the Technical School (CLO) at Deelen, three were cannibalised, and the remaining 13 used for anti-submarine warfare and SAR tasks. However, the Mitchell was far from ideal for the maritime role and, by 1950, it had been decided to replace it with the Lockheed PV-2 Harpoon.

The Harpoon, which entered service in 1951, proved to be a vast improvement, designed specifically for the maritime bomber role. As well as operations over home waters, the Harpoons made deployments to Agadir, Morocco, the last in 1954.

As early as October 1953 the first Dutch aircrews were training at NAS Whidbey Island in the US on the Harpoon's replacement, the P2V-5 Neptune. On 30 December 1953 the first four of 12 P2V-5 Neptunes were delivered. By the end of 1954, 13 Harpoons had been returned to the US, followed by the remainder in the first months of 1955. All 18 ex-MLD Harpoons were refurbished and delivered to the Portuguese air force.

The P2V-5s remained in Dutch service until 1960, when they returned to the US and, after modernisation, were also transferred to Portugal.

During 1960 the first of 45 Grumman S2F-1 and CS2F-1 Tracker ASW aircraft were purchased for service aboard the Dutch aircraft-carrier, HrMs Karel Doorman. These equipped No. 4 Squadron, assinged to the carrier, and shore-based No. 320 Squadron, which operated the aircraft for two years.

More Neptunes

Meanwhile, P2V-7B Neptunes were acquired in 1960 for use by 321 Squadron, for operations in Dutch New Guinea. Fifteen Neptunes entered service as replacements for the unfortunate Martin PBM-5A Mariner, six of which had crashed in just over two years. With the independence of Indonesia in 1962 the P2V-7Bs were flown back to the Netherlands and were issued to 320 Squadron. Although their roles were long-range patrol and ASW, the aircraft were not equipped with ASW devices. This was rectified the following year, when the first Neptune was fitted with ASW equipment. At this point nose armament was deleted and the aircraft's designation was changed to SP-2H.

To augment the Neptune fleet, four ex-Aéronavale aircraft were purchased in 1965, the year in which the squadron celebrated its 25th birthday with special ceremonies at NAS Valkenburg.

From 1973, 320 Squadron deployed to the Dutch Antilles to assume the duties of the S-2N Trackers of 1 Squadron. Three Neptunes formed the Det/Curaçao at NAS Hato. Each aircraft was replaced every three months to allow maintenance at NAS Valkenburg.

In 1975 the Dutch Minister of Defence, Ir. H. Vredeling, announced the phasing-out of the Neptune and disbandment of 320 Squadron on 1 January 1976. The day after the announcement, five Neptunes made a formation fly-by over the Binnenhof in The Hague, the Dutch seat of government. The press reported the episode as a protest against the disbandment, but squadron's commander insisted it had been merely a formation flying practice and a welcome gesture to the new Navy commander.

Curaçao detachment

The MLD has maintained an MPA detachment at Hato since 1952. 320 Squadron operated there from 1974 until 1981 with a permanent detachment of three SP-2H Neptunes. From 1981 to 2000, the RNLAF operated two F27 Maritime Friendships from Curaçao on coastguard, SAR, fisheries, environmental and anti-drug smuggling work. In July 2000, the RNLAF withdrew the two F27s to Holland, where they are now stored at Eindhoven AB pending disposal. Three P-3C Orions are now operated from Hato (two were assigned from 1992); one is permanently based there, while the other two are on TDY detachments.

The importance of the counter-drug tasking has increased in recent years. The US has established a forward operating location at Hato to enable US aircraft, such as E-2C Hawkeyes, E-3 AWACS, KC-135s and detachments of F-16s, to undertake counter-drugs sorties. The Joint Inter Agency Task Force – East is a US International Drug Enforcement organisation and is responsible for counter-drug operations in the Caribbean.

In addition to US and Dutch aircraft, French and British aircraft regularly patrol the area. A major success was achieved in June 2000 when, in a joint operation involving US, French and Dutch forces, over 4,410 lb (2000 kg) of cocaine were intercepted. Later in the year, in another combined operation, this time with the US Coast Guard, 11,000 lb (5000 kg) of marijuana was seized on a ship sailing from Colombia. A total of 29,691 lb (13468 kg) of drugs was seized by the Dutch Navy and Coast Guard in 2000.

An MLD Orion shadows a roll-on/roll-off truck ferry on the Harwich, England – Holland route. As well as undertaking ASW patrols over the North Sea, the MLD's Netherlands-based P-3C fleet has a secondary coastguard tasking.

He was later severely rebuked for his actions.

However, the squadron survived the announcement and soldiered on. In March 1980 the Det/Curaçao came to an end when No. 336 Squadron of the Royal Netherlands Air Force, with two F27M aircraft, took over the duties of No. 320 Squadron. The last Neptune on station did not return to Valkenburg; it remained at NAS Hato as a museum piece but was later scrapped and served as a training airframe by the fire service.

HrMs *Karel Doorman* was withdrawn from service in the late 1960s. In order to maintain the MLD's ASW capability a new maritime patrol aircraft was required. The Navy tried to persuade the government to buy the Lockheed P-3 Orion, but political considerations led to a decision to obtain nine French Breguet Br 1150 Atlantics (Dutch designation SP-13A) for assignment to No. 321 Squadron.

Briefly, during the 1970s, the Dutch government considered abandoning the Navy's maritime patrol role altogether, but before long the search was on for replacements for the ageing SP-2H Neptunes. A shortlist was defined, comprising the BAe Nimrod, Breguet Atlantic NG and Lockheed P-3C Orion. The Atlantic NG was quickly dropped due to its unpopularity in the MLD – two aircraft had been lost at sea as a result of control problems – and the Nimrod was passed over because of its high operating costs and noise levels.

Thus, in September 1981, the first Dutch crews started P-3 training at NAS Jacksonville

with VP-30. In November of that year, the first Orion was delivered to the Dutch liaison team. After completing their training, a mixed force of Dutch and US Navy crews flew the first four Orions to Valkenburg on 21 July 1982. The first to arrive, tail number 300, was welcomed in Dutch airspace by an SP-13A Atlantic (258) and the last operational SP-2H Neptune (204). The following day, Neptune 204 was flown to the UK for static display in a museum. Originally scheduled to be withdrawn from use in 1976, the last Neptunes served until the first Orions were delivered in July 1982.

Despite being only recently operational, the Dutch P-3s participated in various international exercises and achieved a top score during Exercise Ocean Safari in June 1983. In 1984 an important milestone was reached when the first contact was made with a Soviet submarine, during a detachment to NAS Keflavik, Iceland. The final Orion was delivered on 14 September 1984 and, soon after, the last SP-13A of No. 321 Squadron was withdrawn from use following an incident in which three crew members died when they ditched their aircraft after experiencing steering problems.

This left the MLD with 13 P-3C Orions as its only fixed-wing aircraft. No. 321 Squadron converted to the P-3 shortly afterwards, each squadron drawing aircraft from an Orion 'pool' as required.

On 29 May 1985 the 10,000th MLD P-3 flight hour was reached; 6,677 landings had been made (many by VSQ2, the training squadron). Aircraft 311 had performed the most landings – a total of 667. In March 1987 the 20,000th flight hour was flown, and on 29 December 1999 a No. 320 Squadron crew, in Orion 312/V, flew the 100,000th flight hour in a Dutch P-3.

Eric Katerberg

Above: To mark 75 years of Dutch naval aviation in 1992, a number of aircraft received commemorative markings, including this MLD P-3C.

Right: In 2001 320 Squadron has 11 operational crews, each comprising 10 airmen and women, plus five non-flying personnel. As part of a personnel exchange programme with the US Navy, five USN personnel are detached to the squadron, including two pilots, one inflight technician and one crew chief; similarly five MLD personnel are detached to NAS Jacksonville. Five crews are also on detachment abroad; three are based at Curaçao, one is at Keflavik, Iceland, and two are based at NAS Sigonella, Sicily. The Keflavik detachment, which began on 18 October 1985, sees a Dutch P-3C and crew based on Iceland all year round, usually working in conjunction with USN P-3 units on six-month TDYs.

Myasishchev M-50 'Bounder'

The M-50A is seen prior to its first flight at State Aircraft Factory No. 23's airfield. This image is one of a number of black and white photographs that Myasishchev's team doctored by adding colours for a presentation to Soviet military and political leaders in Moscow.

While Soviet fighter designs had kept pace with their Western counterparts throughout the 1950s, the existence of the B-58 Hustler and subsequent B-70 supersonic bomber projects caused great alarm to both political and military leaders. It was seen as imperative that the Soviet Union produced its own strategic supersonic bomber, and the monumental task of designing such an aircraft was given to OKB-23 – under the leadership of Vladimir Myasishchev.

On 9 July 1961 an unorthodox-looking aircraft of impressive size roared at high speed over the Tushino airfield in Moscow. Approaching the stands of spectators, it pulled into a steep climb before disappearing into a layer of cloud.

Its sheer size, futuristic shape and ear-splitting performance prompted a crescendo of delighted comment among spectators. The shutters of the cameras operated by representatives of western news agencies and air attachés started working in overdrive. The next day photographs of the aircraft appeared in newspapers around the world, with its very existence seeming to herald a new era in the development of Soviet bomber aircraft. However, what none of the spectators appreciated, with the exception of a handful of high-ranking Soviet officials sitting in the main stand, was that they had just witnessed the last flight of the Myasishchev M-50 and the end of supersonic long-range bomber development in the USSR for a number years.

The M-50's untimely demise was a direct consequence of the evolving Soviet political and military thinking in the early 1960s. Soviet leader N.K. Khrushchev and senior military officials within the Kremlin had fallen under the spell of 'missile euphoria' to a greater extent than even the US and UK governments. It had become widely accepted that manned aircraft were becoming obsolete and that ground and ship- or submarine-based missiles would resolve any large scale conflict that erupted.

In a misguided decision, Khrushchev, The Council of Ministers and the Central Committee of the Communist Party issued a joint decree terminating all new work on aircraft of this class. A number of aviation Experimental Design Bureaus (OKBs) were disbanded or reorientated to missile design projects, the first victims being Myasishchev (OKB-23) and Lavochkin (OKB-301). Hundreds of drawing-board designs were consigned to the waste-paper bin and several valuable prototypes succumbed to the scrapman's torch.

In 1961 OKB-23 was absorbed into the missile design bureau OKB-52, headed by V.N. Chelomey, and Myasishchev himself was

Construction of the airframe's components was split between Myasishchev's OKB-23 and the State Aircraft Factory No. 23. The tail section (above) was constructed by the latter, with final assembly (left) conducted by the OKB (note the M-50 mock-up in the rear of the hangar). Despite the challenging timescale set for construction of the first M-50, the design team completed the task on time and the aircraft flew just a little over three years after the first metal was cut.

appointed director of the Central Aero-Hydro Dynamics Institute (TsAGI). The personal disappointment felt by Myasishchev was nothing compared to the drastic effect the cancellation of the M-50, and the projected M-52 and M-56 production versions, had on Soviet bomber development, which, when revived at the end of the 1960s, had lost technical parity with the West.

Work began, on what would eventually emerge as the M-50, in 1954. At that time, new fighter aircraft development had jumped well ahead of prospective bomber designs. In terms of speed and altitude performance fighters were a generation ahead and in the time one heavy bomber was designed and built, several new types of fighters were created.

Countering the B-58

By 1952 the Soviet Union was aware of US design plans for a new supersonic bomber (which eventually emerged as the B-58 Hustler), and a Soviet supersonic strategic bomber equivalent suddenly became a top priority. OKB-23 was heavily involved in the original design work, producing their own series of near- and supersonic bomber designs, allocated the designations VM-31, VM-32 and VM-34, in 1952-53.

In 1953 all design bureau involved in the initial concept work pooled their ideas under the SDB (SverkhDal'niy Bombardirovshchik; Super-Long-range Bomber) project. Myasishchev's work with the '31', '32' and '34' design concepts had provided valuable experience and understanding of the basic problems involved, and OKB-23 was rapidly established as the leading aircraft design bureau in the field.

One of the first problems encountered was the creation and installation of powerful turbojet engines to enable the aircraft to cruise at supersonic speeds, yet with specific fuel consumption rates to allow trans-continental range. Detailed studies were undertaken to achieve the optimum aerodynamic layout, combined with weight-saving measures and integration of new materials and equipment.

Among the first studies to be seriously considered was the compound aircraft, otherwise known as the 'Duck'. The aircraft consisted of two parts: a fuel-carrier and a piloted section carrying the warload. It was intended that the 'Duck' would take-off, gain altitude and progress towards the target at cruising speed in a joined state until the carrier had exhausted its fuel. The piloted section would then separate and continue the mission, leaving the expendable carrier section to fall to the ground. This arrangement was intended to provide a considerable increase in range.

This solution attracted approval of both aviation industry and political leaders and, on 30 July 1954, Enactment No. 1607-728 was issued

Myasishchev supersonic long-range bomber studies

Having been allocated such a technologically demanding task, Myasishchev OKB, with assistance from engine design bureau and the TsAGI, designed a whole range of possible configurations to meet the supersonic long-range bomber requirement. Many of the concepts were truly ingenious, but the advances needed in the fields of aerodynamics and engine development rapidly consigned them to the wastepaper basket.

One of the first design layouts to be seriously considered was the compound aircraft concept. Known as the 'Duck', it consisted of carrier and piloted sections, which were designed to separate en-route to the target, with the piloted front section completing the mission.

'Duck'-configuration composite aircraft

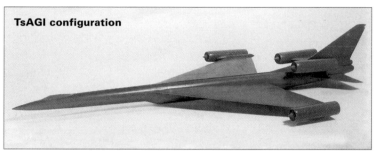

TsAGI configuration

The Central Aero-Hydro Dynamics Institute (TsAGI) was heavily involved in the M-50 design from the outset. It built and tested numerous wind tunnel models to assess various engine configurations and wing designs. This configuration was its initial favoured design.

The tanks/engines project comprised a jettisonable fuel tank suspended beneath each wing, each incorporating an NK-6 or VD-9 engine. However, the NK-6 was rejected due to an increase of 40 per cent in its design weight, and jettisoning the VD-9 was impossible due to a lack of thrust from the remaining engines.

Combined 'Tanks/Engines' project

External fuel tank design

Struggling to reach the range/payload demands resulted in large drop tanks featuring on many of the subsequent designs. The fact that fuel consumption increased and supersonic speed was not possible while the tank was being carried led to their eventual rejection.

By late 1955 the layout had been finalised as a four-engined design with a cropped-delta wing, leaving the placing of the engines as the next step. Four configurations were studied, including this under/overwing nacelle layout, which was eventually rejected in favour of the M-50's final design.

Under-/overwing engines design

The front main landing gear could be extended, giving the aircraft a characteristic nose-high attitude. The sole purpose of this feature was to increase the aircraft's angle-of-attack during the take-off run, thus significantly increasing lift and decreasing the length of the take-off. Between the two open main undercarriage doors are the open weapons bay doors. The M-50's principal weapon would have been a single free-fall nuclear device, intended for western military installations or cities if the Cold War ever turned hot. A powerful tracked towing vehicle was needed to manoeuvre the aircraft on the ground, as seen in this OKB photograph, which again has had colour artificially added for presentation purposes.

Above: Myasishchev engineers check the M-50's systems and avionics prior to the aircraft's transportation to Zhukovskiy for its maiden flight. The fairings running along the top and bottom of the fuselage held the control and communications cable runs.

Right: The LII's M-50 flight test crew, consisting of N.I. Goryaynov and A.S. Lipko, make the somewhat perilous descent from the cockpit at the triumphant conclusion of the M-50's maiden flight on 27 October 1959.

by the Council of Ministers requiring the development of the concept. On 10 August Order No. 488 of MAP (Ministry of Aviation Industry) assigned OKB-23 the task of designing and constructing an experimental supersonic separable bomber with either Dobrynin or Mikulin engines. By then allocated the profect designation '50', the aircraft was to be ready for state tests in the third quarter of 1958 and, in accordance with the Enactment, would have a range of 13000 km (8,078 miles) with a 5000-kg (11,023-lb) bomb load; maximum speed of 1800 km/h (1,118 mph); cruising speed of between 1500-1600 km/h (932-994 mph) and a cruising altitude of 14000-15000 m (45,932-49,213 ft).

Work on the selection and development of the turbojet engines was carried out by the TsIAM (Central Institute of Aviation Motorbuilding) in conjunction with Dobrynin at OKB-36 and Tumanski at OKB-300.

Design of the 'Duck'

The completion of the initial design studies in the spring of 1955 revealed a number of principal problems, which could not be resolved in the time period given for the design and construction of the aircraft. These included: insufficient study of the 'Duck' design (especially aerodynamic balancing and manoeuvrability for a wide range of speeds); inability to achieve performance/weight efficiency of the piloted section which performed the role of the horizontal tail in the combined aircraft; the complex nature of the flight tests (especially the separation of the modules) and problems with the creation of a launching device providing take-off speeds of up to 650 km/h (404 mph) at a take-off weight of 300,000 kg (661,390 lb).

In addition, wind-tunnel tests of the UA11 and UV11 models at TsAGI revealed that the aerodynamic qualities of the whole unit and of the

piloted section at supersonic speeds were much worse than that of a regular aircraft layout. Also, during take-off and landing, the large size of control surfaces needed to maintain aerodynamic balance resulted in a loss of aerodynamic quality and a reduction in weight efficiency. After comprehensive analysis of the data, the compound 'Duck'-type supersonic bomber was deemed unfeasible and further work ceased.

Looking to eradicate the insurmountable problems associated with the 'Duck' concept, TsAGI experts proposed a new design involving the towing of a piloted aircraft by a non-reusable large aircraft acting as a refueller.

Although seeming even more optimistic than the 'Duck' proposal, a similar system had in fact been tested in 1949-52. Known as the 'Burlaki' (barge hauler) system, the LII (Flight Test Institute) at Zhukovskiy had flown a heavy bomber towing in flight (with a cable and special tie) its own escort fighter with its engine shutdown, therefore increasing its radius of action. The TsAGI proposal was thoroughly studied by OKB-23 from the point of view of constructive configuration, control system and conditions of operation.

However, there were three areas of operation for which the problems could not be solved: take-off of the two aircraft attached together; creation of a system for co-ordinated remote control of the aircraft and assurance of the necessary manoeuvrability of the system. The 'towing' proposal was rejected in March 1955.

A more promising development of the compound idea involved a more conventional aircraft with large capacity fuel tanks suspended under the wings and engines suspended beneath the tanks, with further engines mounted on the rear fuselage. Take-off was performed using thrust from all engines and then, after exhaustion of the fuel from the suspended tanks,

they were jettisoned together with their engines. The thrust of the remaining engines was deemed sufficient for continuation of cruising flight. However, this variant had also drawbacks.

Firstly, the part of the flight with PTBs (suspended fuel tanks), of around 3000 km (1,864 miles), would be at subsonic speed and relatively low altitude, which was not only tactically undesirable, but resulted in increased fuel consumption. Secondly, the high take-off wing loading of 1100-1200 kg/m² (225-246 lb/sq ft) would result in a prohibitively long take-off run.

As results of all three of these studies became clear, the creation of compound aircraft and other exotic variants was seen as unrealistic, and all further studies were abandoned. Although the first design stage had produced negative results, the valuable experience gained yielded the necessary requirements for creation of a viable supersonic bomber design.

Configuration finalised

Analysis of the data helped create design parameters for a new configuration. The aircraft should have a wing of small relative thickness, minimal mid cross-section area of the fuselage and superstructures, all-moving vertical and horizontal control surfaces, automatic control, air navigation and bombing systems to reduce crew numbers, all-round reduction of airframe and equipment weight to attain a fuel mass efficiency not less than 70 per cent and use of in-flight refuelling.

To meet these requirements OKB-23 began design work on a configuration with four turbojets and suspended fuel tanks to meet a new Enactment issued on 19 July 1955 that demanded: maximum range with two refuellings of 14000-15000 km (8,700-9,320 miles); range without refuelling of 11000-12000 km (6,835-7,455 miles); maximum speed over the target of 1900-2000 km/h (1,180-1,240 mph); cruising speed of 1700-1800 km/h (1,055-1,120 mph); service ceiling of 15000-16000 m (49,215-52,490 ft); take-off run of 3000 m (9,845 ft) and bomb load of 5000 kg (11,025 lb). It was also to be able to be operated by a two-man crew.

The aircraft was to be powered by four Kuznetsov NK-6 turbofans or Dobrynin VD-9

The M-50 is seen here during early ground running tests with the original engine configuration of four VD-7A turbojets, prior to the availability of the afterburning VD-7M and eventually the Zubets M16-17. The latter engine would have been fitted to the pre-production prototype aircraft and production variants.

It had originally been intended to fit metal outrigger skids to the M-50, but these were changed, shortly before construction of the prototype began, to twin-wheeled units. These not only provided lateral stability on the ground, but prevented the inboard nacelles striking the runway in the event of a heavy landing.

M-50 development concepts

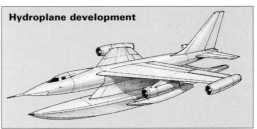

Ten-engined bomber variant

Hydroplane development

Tailless bomber project with drooping nose

turbojets. However, OKB and TsIAM studies showed that the application of turbofan engines (NK-6) would result in reduction of range in comparison with high-temperature turbojets. After completing the requirements for engine characteristics, the NK-6 was rejected in favour of the Zubets M16-17 turbojet.

The design and development timescale was extremely tough, with the Council of Ministers calling on the first aircraft to be ready for State tests in the first quarter of 1958.

Testing the configuration

OKB-23 immediately went to work, and several dozen configurations were reviewed. Aircraft of triangular and swept wing layout, with various arrangements of engines (on wing, underwing, on tailplane etc.), in addition to tailless aircraft were considered.

Large-scale aerodynamic studies on TsAGI and OKB models were carried out. As a result of these wind tunnel tests, the '50' design was finalised with a triangular wing and swept tailplane.

The most complex problem now facing the design team, headed by I. P. Tolstykh, was the placing of the engines. The frontal air resistance of the nacelles and their supports rendered significant influence on the aircraft's range. To find the optimum arrangement four concepts were studied:

1. Two engines on pylons under the wings and two on the wing tips.
2. The engines placed in pairs in two underwing nacelles (similar to the B-52's arrangement).
3. The engines placed in nacelles (one above the other below) on each wing

4. Two engines on pylons under the wings and two on pylons above the rear fuselage.

The frontal air resistance of options 1, 2 and 4 were near identical, however, the accommodation of engines in options 2 and 4 would not provide sufficient structural rigidity, resulting in a substantial increase in design weight, and were therefore rejected. In a technological sense option 3 was the most favourable as the inherent dynamic strength of the arrangement would give appreciable weight advantages more than compensating for a slight increase in drag. However, aerodynamic research data of such an arrangement was sparse and construction would have been more complex. Given the tight schedule, option 1 was chosen as the most favourable arrangement, although several technological hurdles, such as fastening engines to the wing tips and ensuring strength and rigidity of underwing pylons would have to be resolved.

To try to achieve the stringent range requirements, a number of measures for weight reduction were introduced including: wide application of large-size pressed panels on the airframe; reduction to two crew members through the application of automatic devices and use of fuel cells in the fuselage replacing the external fuel tank. At maximum take-off weight (253 tonnes/249 tons), the aircraft would be fitted with two additional self-orientating underwing four-wheel undercarriage units, to be jettisoned after a rocket-assisted take-off. These would supplement the two bicycle-type fuselage-mounted main landing gear and two outrigger units.

Despite these measures the maximum range

Supersonic civilian airliner

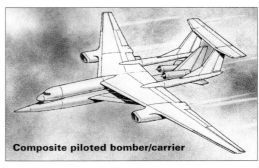

Composite piloted bomber/carrier

Increasing the range

Even as the M-50 prototype was nearing completion, Myasishchev's OKB was working on various ways of addressing the range shortfall. Incorporation of underwing fuel tanks (below left) was seriously considered, and may well have been incorporated into

the final design had production been approved.

A more adventurous solution was the M-50LL (below right), which incorporated a centre fuselage section of significantly greater diameter and an additional tailplane, as featured on the M-52. The integral fuel tanks would have much greater capacity and the weapons bay could also have been enlarged. Neither solution was, however, destined to leave the drawing board.

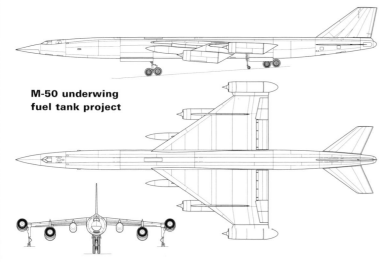

M-50 underwing fuel tank project

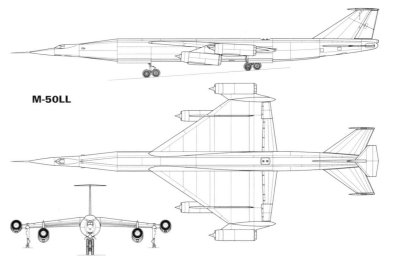

M-50LL

M-50A 'Bounder' – Technical description

Intended for long duration flights at supersonic speeds, the M-50A emerged as a four-engined, cantilever, all-metal monoplane incorporating a delta-shaped wing mounted high on the fuselage and retractable landing gear.

The M-50's slender **fuselage** was of semi-monocoque construction with circular cross-section. Four cut-off joints divided the fuselage into its five constituent parts comprising:

Forward fuselage and nose cone incorporating the radar and other auxiliary equipment. The bottom section comprised a radio-transparent radome.

Pressurised cockpit housed the pilot and navigator and the cockpit entrance hatches.

Fuel compartment divided into three sections by hermetically-sealed partitions. The first section contained the LAS-2 life-raft and its associated equipment that could be released through a ventral hatch. The other sections contained fuel tanks.

Central section (including the integral fuel tank of the wing centre-section) as the basic load-bearing unit with two reinforcing beams laid along its length. It incorporated the main landing gear, weapons bay and fuel tanks. In addition, the integral fuel tank/torsion box of the wing centre-section was integrated into the upper portion.

Rear section to which the all-moving horizontal and vertical tail was fastened and incorporated integral fuel tanks, fuel equipment, horizontal and vertical control surfaces' drives, control cable runs and the brake parachute compartment.

Of cropped delta-shape, the M-50's **wing** consisted of a central integral fuel tank/torsion box, mid-wing sections, outer-wing sections and detachable consoles. The centre-section had a front beam, four spars and load-bearing ribs. The inboard engine pylon was attached to a load-bearing structure on the join between the mid- and outer-wing sections. The outer-wing had its own torsion box and supported the nacelle of the external (outer) engine and the wing-mounted landing gear. On the top surface of each wing an aerodynamic fence was fitted to help prevent airflow over the wing migrate towards the wing tips at lower speeds. The trailing edge of the wing comprised retractable flaps (inboard) and ailerons (outboard).

The rigid **flight control system** actuated the vertical and horizontal tail surfaces, ailerons and trim tab mechanisms. Inputs to the control columns, pedals and steering wheels in the cockpit were carried through a system of rods. The translational motion was converted to rotational motion by ball converters. Roll dampers (fitted to the ailerons) and yaw dampers (fitted to the all-moving tail fin) were electrically driven.

An automatic longitudinal control device (APU), consisted of a spring loader and servo-mechanism, creating an artificial 'feel' for pitch control, using a combination of tailplane deflection and velocity of the slipstream.

In its final form the M-50's **powerplant** comprised four engines consisting of two 93.2-kN (20,944-lb) RKBM VD-7BA (*izdeliye* 15A) turbojets contained in nacelles fitted to the wing tips and two afterburning 156.9-kN (35,258-lb) VD-7MAs (*izdeliye* 17A) on pylons under the wing. During initial tests the aircraft was fitted with four VD-7BAs. The RD16-17 turbojet was the intended powerplant for the aircraft but was not available in time for flight tests. If the project had continued, production aircraft would have received four RD-16-17s each capable of producing 181.4 kN (40,764 lb) of thrust with afterburning.

Each engine was installed in an individual nacelle, with air intakes for cooling the engine and its systems fitted to the top and bottom of the nacelles. The engines could be controlled by either the pilot or co-pilot/navigator via an SDU-15 servo-electro-remote system.

Fuel capacity for the M-50 was provided by eight groups of tanks (on the first M-50A the 4th and part of the 5th groups of tanks were not installed). The fuel system and the tanks were pressurised using inert gas, and an emergency fuel jettison and fire safety system were installed.

Electrical supply was provided by an AC generator installed in the middle section of the fuselage above the reinforcing beams. The power was provided by compressed air, extracted from the engine compressors. Among the many electrical functions was the retraction/extension of the wing-mounted undercarriage.

The M-50's complex **hydraulics** consisted of six independent systems – two main (left and right) and four booster. The working pressure in the hydraulic system was 190 kg/cm² (2,700 lb/sq in).

The right main system controlled the steering of the front main landing gear, retraction/extension of the rear landing gear, emergency release of front landing gear, operation of the weapons bay doors, braking of the rear main wheels and operation of the front main gear braking skis.

The left main system controlled emergency steering of the front main gear,

Above: The nose section contained the navigation radar, and its associated equipment was housed in the bulged dorsal fairing seen here.

Above: For supersonic cruising performance, the M-50 was fitted with a thin and rigid delta-shaped wing, of which this is the outboard section.

Above: Typical of Soviet designs, the M-50's fuselage was immensely strong. Many miles of electrical cabling ran along the sides of the fuselage powering many of the aircraft's systems.

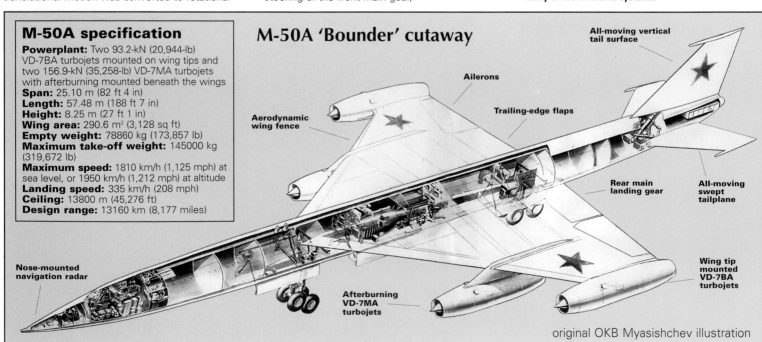

M-50A specification

Powerplant: Two 93.2-kN (20,944-lb) VD-7BA turbojets mounted on wing tips and two 156.9-kN (35,258-lb) VD-7MA turbojets with afterburning mounted beneath the wings
Span: 25.10 m (82 ft 4 in)
Length: 57.48 m (188 ft 7 in)
Height: 8.25 m (27 ft 1 in)
Wing area: 290.6 m² (3,128 sq ft)
Empty weight: 78860 kg (173,857 lb)
Maximum take-off weight: 145000 kg (319,672 lb)
Maximum speed: 1810 km/h (1,125 mph) at sea level, or 1950 km/h (1,212 mph) at altitude
Landing speed: 335 km/h (208 mph)
Ceiling: 13800 m (45,276 ft)
Design range: 13160 km (8,177 miles)

M-50A 'Bounder' cutaway

All-moving vertical tail surface
Ailerons
Trailing-edge flaps
Aerodynamic wing fence
Rear main landing gear
All-moving swept tailplane
Nose-mounted navigation radar
Afterburning VD-7MA turbojets
Wing tip mounted VD-7BA turbojets

original OKB Myasishchev illustration

M-50A fuselage cross-sections

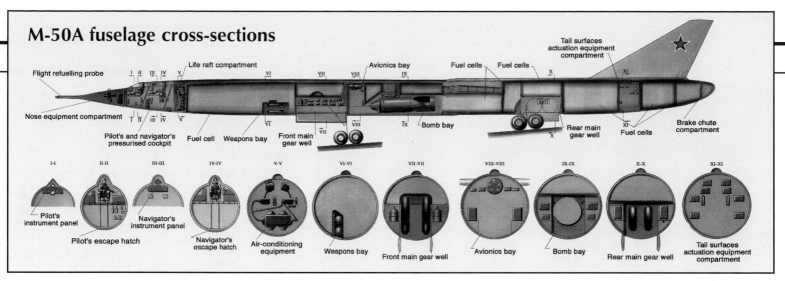

retraction/extension of the front landing gear, emergency release of the rear main landing gear, emergency operation of the weapons bay doors, and emergency braking of the rear main wheels.

The four booster systems operated the ailerons, flaps and horizontal and vertical tail surfaces.

For navigation and weapons delivery the M-50A was fitted with an integrated **navigation/bombing suite**, controlled by a central computer system, providing accurate day or night navigation and bomb-aiming assistance over land or water in any meteorological conditions, even in areas with high radar clutter. The system comprised: KS-6B course management system; TsSVS-25/2500 speed and altitude buses; TsGV-10 central hydraulic vertical line; NA automatic navigating device; 'Put' flight navigation instrument; BA automatic bombing device and 3SO star-solar orienting device. The primary sensors feeding the system consisted of: ARK-54B automatic radio compass; RV-U and RV-25 radio altimeters and SBP-50 radar with associated FARM-2 photo attachment. For assistance during the approach and landing phase, the 'Materik' radar landing

equipment was installed, featuring a KRP-F course-landing radio receiver, GRP-2 glide-path radio receiver and MRP-56P marker-landing radio receiver.

Radio equipment allowed communications with other aircraft, ground stations and other crew members. Long-distance communications were achieved using the 'Planeta' short-wave radio (which included a 'Kristall' transmitter), 'Rul-M' receiver and 'Kvarts' coding device. The RSIU-4V 'Dub' VHF radio, 'Geliy' radio transmitter and SPU-6 aircraft intercom system were also included. The M-50 was fitted with the 'Kremniy' identification system that included a 'Tsink-MM' IFF interrogator-responder.

Inside the pressurised **cockpit**, the pilot and navigator sat in tandem on downward-firing ejection seats. With a separate hatch beneath each crew member, ejection could be made in any sequence. For routine entry/egress from the cockpit the seats descended on rails to the entry hatches. Emergency escape hatches were also available in the roof of the cockpit. Due to the high-altitude nature of the operational role, the crew were also obliged to wear pressure suits in

case of cockpit de-pressurisation. In addition, an air conditioning system and heat/sound insulation ensured a comfortable working cockpit environment. The aircraft's **oxygen system** provided air for the cockpit and pressure suits. In addition two-piece KP-27M emergency oxygen systems were provided for the crew after ejection.

Fitted in a cylindrical container in the rear fuselage, the M-50's three-dome **brake parachute** had a total surface area of 216 m² (2,325 sq ft). The parachute was released at the moment of contact with the ground with the shutters of the container and locks controlled by compressed air.

The **weapons bay** was fully ventilated and encompassed the bomb suspension system, bomb release control system and weapons bay doors control system as well as the weapons themselves. The bombs were suspended on III, IV and VI Group cassette holders up to a maximum bomb load of 5000 kg (11,023 lb).

Auxiliary equipment for use by the air crew within the fuselage included: suitcases with in-flight food, first-aid kits, urinals, tools and axes.

M-50 landing gear system

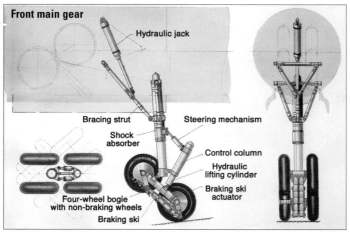

The M-50's landing gear was of bicycle type design. It consisted of two main fuselage-mounted undercarriage units with oil-air shock absorbers, and two underwing support units. The two main gear units retracted forward into fuselage undercarriage bays, with the underwing supports retracting backward into fairings just inboard of the external engine nacelles. The main gear wheels were 130 cm (51.2 in) in diameter and 38 cm (15 in) in width. The non-braking underwing support wheels measured 52 x 12.5 cm (20.5 x 4.9 in) and served to provide lateral stability on the ground and for protection of the internal engine nacelles from impact on landing.

The main gear units retracted/extended hydraulically, with the underwing wheels being driven electrically. For reduction of the take-off run the front main gear could extend, raising the nose to take-off attitude.

Standard brakes were fitted to the rear main wheels, although this could be supplemented by a braking ski fitted to the front gear which lowered onto the runway, sending up a shower of sparks. The only steerable unit for taxiing was the front main unit.

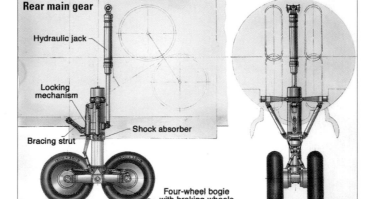

Above left: An original technical drawing from Myasishchev of the front main landing gear, showing the forward retraction mechanism and rotating bogie.

Above centre: The front main gear undergoing testing in a specially constructed pit. Note the braking ski is not yet fitted.

Above: This view of the M-50A clearly shows the positioning of the wing-support gear inboard of the wing tip engines.

Left: The non-rotating rear main gear bogie was fitted with standard braking wheels.

of some 9500 km (5,900 miles) without flight refuelling was still below the requirement, and additional measures, which could be implemented after the initial flight test stage, were examined. Although never implemented these included: dropping two engines and transitioning to subsonic speed 1000 km (621 miles) into the return leg from the target (increasing range by 900 km/560 miles); use of boundary layer control and increasing wing loading (additional 300 km/185 miles) and the 'flying chassis' consisting of a piloted launch vehicle with the M-50 mounted on top, with both aircraft return-

ing to make a conventional landing.

To increase the tactical capabilities of the aircraft a number of different launching devices for take-off from land or water were also studied including a wheeled launching carriage, a launching carriage on rails, a hydrocarriage and a zero-length launcher, which involved a rocket-assisted take-off with the M-50 mounted on a telescopic arm.

Of these, work on the seaplane version of Project '50' progressed furthest. Allocated the Project No. '70', preliminary hydrodynamic calculations, completed by the end of 1955,

showed great potential. An outline sketch of the design was sent to MAP for examination and, after discussion, the project was presented to the air force on 6 April 1956. The military staff were enthused by the concept and a mock-up model was built, which underwent successful trials at TsAGI's laboratory No. 12. Allowing expansion of areas of operation and quick dispersal in case of threat, the concept was highly appealing to the air force and OKB-23 was asked to execute a draft design for presentation in early 1957. However, the hard-pressed design team could not devote sufficient time to the project which was postponed and eventually abandoned.

Before any of the modifications to increase range had been implemented, the air force calculated that the M-50 would have a maximum range of 9600 km (5,965 miles) without refuelling and 12500 km (7,767 miles) with two refuellings, some 2000 km (1,243 miles) shorter

Delta wing
Apart from the XB-70, the M-50's wing is the largest ever built for supersonic flight, and was the aircraft's most outstanding aerodynamic feature. The wing was notably thin to aid supersonic performance with a thickness of 3.7 per cent at the root and 3.5 per cent at the tip. It was not quite a pure delta, as the wing incorporated a slight crank in the leading edge (57° 34' on the inner section, 54° 25' on the wing tip section).

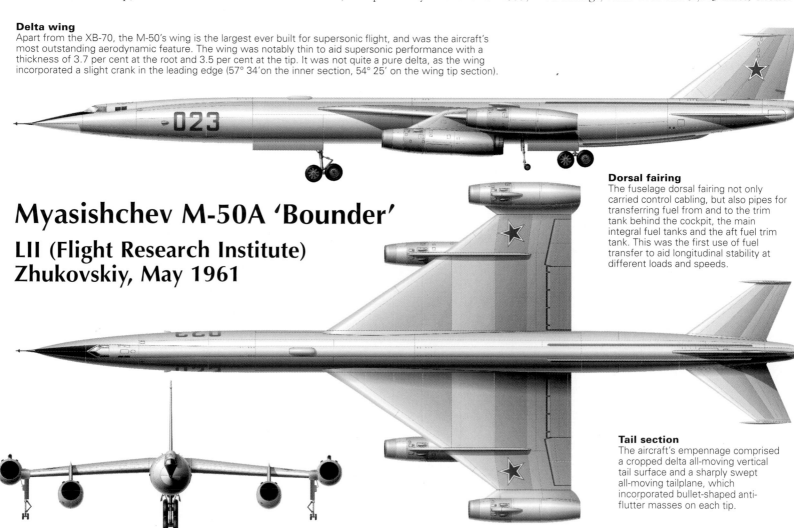

Myasishchev M-50A 'Bounder'
LII (Flight Research Institute) Zhukovskiy, May 1961

Dorsal fairing
The fuselage dorsal fairing not only carried control cabling, but also pipes for transferring fuel from and to the trim tank behind the cockpit, the main integral fuel tanks and the aft fuel trim tank. This was the first use of fuel transfer to aid longitudinal stability at different loads and speeds.

Tail section
The aircraft's empennage comprised a cropped delta all-moving vertical tail surface and a sharply swept all-moving tailplane, which incorporated bullet-shaped anti-flutter masses on each tip.

Above: Having been grounded for over six months, awaiting the delivery of the VD-7MA engines for the inboard nacelles, the M-50A resumed flying in the late spring of 1961. However, only eight more flights were ever made, of which at least four involved rehearsals for and participation in the Tushino Parade flypast. In an attempt to give western observers the impression there were a number of M-50s flying, the aircraft received the Bort number '05' (as seen on retouched photograph) for the rehearsals, before receiving additional fuselage markings and the number '12' for the flypast itself.

Right: The greater diameter of the engine exhaust on the inboard engines identifies them as afterburning VD-7MAs, dating the photograph as post-May 1961. The original VD-7As were retained in the wing tip positions.

than the range declared in the Enactment.

Other parameters also fell short of the requirements including the take-off run without assistance at maximum take-off weight which was calculated at 6000 m (19,685 ft) instead of 3000 m (9,843 ft). In addition, tactical requirements of the air force for defensive and bombing armament, and battle survivability issues had not yet been addressed. OKB-23 was asked to create a modified design, together with a mock-up of the plane for consideration by the air force.

Design refinements

Myasishchev's team returned weeks later with a number of design changes to the aircraft and its systems. The aircraft had been designed to use side-control sticks driving an electro-remote control system with triple redundancy (a primitive form of fly-by-wire), however, in accordance to TsAGI recommendations, the side sticks were replaced by conventional control columns. The landing gear was reworked with metal skis replacing the outrigger wheels and the two jettisonable 17-tonne rocket booster engines were replaced by four of 9-tonnes each installed in pairs under the fuselage. These, and other modifications, were incorporated into a mock-up which was presented for examination by air force and industry officials.

Myasishchev, well aware that the aircraft would still not meet the specification, made his case that creation of such an aircraft represented a huge technological leap forward, and had no precedent in the world of aviation.

Achievement of the desired range would only be possible with new technological developments in materials, construction techniques and engine efficiency. The air force accepted this opinion and, with an urgent need for the aircraft, suggested the building of a prototype for flight and system testing. A second experimental aircraft would then be built when the technological problems in achieving the desired performance had been solved.

Even before the construction of the plywood mock-up it was clear that the favoured M16-17 engines were delayed and would not be available for the initial flight tests. Their delivery was expected in the fourth quarter of 1957, while production tests of the aircraft were to start in the first quarter.

To overcome the problem, the OKB offered to build the first experimental prototype with available VD-7M engines. This powerplant would allow the study of problems associated with supersonic flight and achieve the basic performance with the exception of maximum range.

A restriction of the engines to a flight speed of 1500 km/h (932 mph) was placed on the advice of A.N. Tupolev (the engine was designed for the Tu-22 'Blinder'), although according to OKB-36 it could be used for speeds up to 2000 km/h (1,243 mph).

In April 1956 the M-50 was approved for design development work and, according to the MAP instructions, construction of the components was allocated. Already working on construction of Myasishchev's M-4 'Bison', State

Aircraft Factory No. 23 was allocated a proportion of the construction work, with the OKB completing the remainder as follows:

Plant No. 23	**OKB-23**
Forward fuselage	Experimental works
Crew cockpit	Fuel compartment
Rear fuselage	Middle fuselage section
Outer/rear wing section	Wing root
Tail unit	Main landing gear
Engine nacelles	Engine installation
Underwing wheel fairings	Fuel system installation
Bag fuel tanks	Aircraft controls
Castings	Navigation/radio systems
Engine pylons	Electrics
Heat treatment	Pilot's seats
Aircraft skin	Life-saving equipment
Internal equipment	Special equipment
Ground equipment	Sealing of fuel tanks
Ailerons	Test bench construction
Test bench design	Final assembly

A number of basic difficulties resulted in delays to the initial construction work, these including an absence of materials and semi-finished products along with the fact that Plant No. 23 was busy working on other projects.

The necessity for a number of design revisions came to light during testing of the aircraft's units and systems. An additional electro-hydraulic system was included, the underwing ski supports were replaced by wheeled units, a tail skid was installed and a new independent cooling system for the bomb bay and main landing gear compartments was integrated.

After elimination of minor defects, which were revealed at installation of equipment, the M-50 was rolled out at Myasishchev's factory airfield for the commencement of the first stage of ground tests in July 1958.

With the VD-7M turbojets not yet ready, 107.91-kN (24,250-lb) VD-7As were installed, and the aircraft received a new designation – M-50A.

Led by chief engineer S.P. Kazantsev the ground crew, working together with lab workers, commenced the ground tests in August. In

Preparing to begin a sortie in the spring of 1960 during the initial flight test programme, the M-50 has a temporary cross marking applied to the forward fuselage, possibly for photo-calibration. Note the air scoops on the upper surface of the VD-7A engines, which were absent on the later VD-7MAs.

addition to testing the aircraft's systems and equipment, the training of the flight crew, checks of the forward runway view from the cockpit and full adjustment of the steerable forward landing gear were conducted. In the autumn of 1958 taxiing tests were carried out and the remainder of the first stage of ground testing was completed.

The new flight control system was installed in October 1958, before the M-50A was declared ready for transportation to LII (Flight Research Institute) at Zhukovskiy. On 12 November the aircraft, with a number of units disassembled for the journey, was loaded onto a specially prepared barge, arriving safely the next day. The airframe was reassembled, and installation of the systems was complete by the end of December.

To aid the test programme, in March-April 1959 an M-4 'Bison' (No. 0201) was fitted with the SDU-15 remote control system for engine control. Testing of the SDU-15 was conducted in six flights, with a total duration of 16 hours 45 minutes, between 28 May and 20 June 1959.

The M-4 was later used to test a wide range of other M-50 equipment including the APU-50 automatic longitudinal control device, the pilot's SI-5 pressurised suit system and prolonged external compression of the aircraft at the maximum speed.

Preliminary training of the crew in the use of their pressurised suits was conducted at the NII (Scientific Test Institute) and, in early May 1959 the testing responsibility was passed to the dedicated LII crew for commencement of ground-running tests. These tests revealed a number of flaws which were eliminated before the second-stage – taxiing.

Five taxiing trials were performed between 28 May and 15 June 1959. Further problems were discovered and two teams of technicians began work on their elimination (one team in the field,

working on the aircraft, the other going back to the plant and working on redesign and construction of several new units).

During this delay the crew took the opportunity to use the flight simulator, developed by TsAGI, to check the control system and to acquaint the pilots with the aircraft's behaviour. Using electronics, the movement of the M-50 was projected on a screen installed in front of the cockpit. The crew later stated that the modelling was very effective, and the simulator can justly be regarded as a prototype of the modern flight simulator.

Four more taxi runs were performed in October 1959 for a final check of the recently introduced modifications to the aircraft and its systems. During the tests, speed was gradually increased and, on the final run, the aircraft actually became airborne for a few seconds, confirming the readiness of the M-50 for the commencement of flight tests.

"The test programme of the M-50A experimental supersonic bomber, with four VD-7A modified engines" was authorised on 22 October 1959 and was scheduled for a total of 35 test flights.

Flight test programme

On 27 October 1959, test pilots N.I. Goryaynov and A.S. Lipko made the M-50A's official first flight. The take-off was completed with the control dampers switched off. During the climb phase the pilots turned on the roll damper after 117 seconds of flight and the yaw damper after 165 seconds. After gaining altitude, the aircraft executed a 270° turn and passed over the runway, before completing two very wide circuits, returning after 35 minutes for a safe landing. A speed of 575 km/h (357 mph) and altitude of 1000 m (3,280 ft) were reached during the flight. The landing gear was extended for the entire flight as a safety measure.

Myasishchev M-52 missile carrier

In an attempt to rectify the ongoing problems of achieving the range criteria with the basic M-50, the design was developed into the M-52 missile carrier which would form the basis of the M-52K strategic strike complex. This consisted of the M-52, two Kh-22 (AS-4 'Kitchen') stand-off air-to-surface missiles and the K-22 guidance system. The aircraft was to be powered by four RD16-17 engines (as intended for the M-50) and included a number of internal and structural changes, which included repositioning of the outer nacelles, shortening of the wings and fuselage and rearrangement of the cockpit crew positions from tandem to side-by-side.

An unrefuelled range of 5300 km (3,293 mile) or 8100 km (5,033 miles) with air-to-air refuelling, even taking into account the 500-km (311-miles) range of the Kh-22 missile, was not really sufficient for the aircraft to be considered an intercontinental bomber, and was a major reason for its eventual abandonment. This did not happen before a single prototype had been completed which, by mid-1961, was in an airworthy state but for the installation of the engines. The aircraft was destined never to fly and was replaced by the M-56 canard-configured design, which itself would never reach fruition.

Outboard engines
To improve aerodynamic qualities, the outboard RD16-17 engines were housed in nacelles incorporated into a redesigned wing tip section.

Tail turret
A remotely-controlled twin-cannon tail turret armament with radar-ranging was planned for the M-52K complex.

Tail section
The tail section was extensively redesigned, incorporating an additional smaller tailplane positioned at the top of the tail fin.

A second flight was was made on 31 October, also with the landing gear extended, testing major aircraft and flight control systems.

After work was completed on the undercarriage retraction gear, two more flights, were completed on 23 and 28 November, during which the retraction/extension of the gear was extensively tested, the flap system worked and the aircraft's acceleration and deceleration and behaviour in banks and turns was examined. The 40-minute flight on 28 November saw the aircraft reach a speed of 1010 km/h (628 mph) and a height of 5000 m (16,404 ft).

The take-off weight was brought up to 118 tonnes (116 tons) and the stability and control characteristics were surveyed at a range of speeds and altitudes. The landing flaps were also tested at angles of up to 18°. The crew began evaluation of stability/controllability of the aircraft in the surveyed range of speeds and altitudes, and also of its behaviour on take-off and landing.

As the programme entered the 1960s, the OKB's primary goal was the completion of the initial flight tests with the VD-7A engines.

This had successfully been achieved by 5 October 1960, by which time the M-50A had completed 11 test flights lasting 8 hours and 33 minutes. On 16 September a speed of 1090 km/h (677 mph) was achieved. According to the flight data this corresponded to Mach 1.01 although subsequent calculations by LII engineers quoted Mach 0.99 as the speed reached. It is impossible to define the true value of the Mach number achieved due to a lack of sophisticated equipment on the aircraft. However, the physical phenomena experienced by pilots in this flight, such as disappearance of jolting and lagging of the engine noise give credence to the belief that the aircraft had reached supersonic speed.

The only major setback encountered during initial flight tests was a major accident, which occurred during ground running on 12 May 1960. This work was being completed to check performance of the engines after several minor adjustments. Flight Engineer A.I. Shchelokov, OKB-36 engineer E.G. Alkhimenkov and Flight Electrician B.A. Golikov were conducting the test. After engines No. 1 and No. 2 were tested

Despite flying only 19 times and never progressing beyond the prototype stage, the M-50 contributed greatly to Soviet knowledge of supersonic large bomber design, construction and operation. Much of the data collected later proved valuable in designing what eventually became the Tu-160 'Blackjack'. The M-50 is currently on display at the Air Force Museum, Monino.

and shut down, engine No. 3 was started and brought close to nominal power. At this juncture, the braking shoes (chocks) under the front main wheels were crushed by the force, and the sudden movement forward pushed the braking shoes under the main rear wheels apart. As the M-50A lurched forward, DC and AC feeding cables from ground electrical power sources broke. The aircraft's internal electricity supply was provided by the No. 3 engine which had been shut down, therefore the flight engineer had no way of reducing RPM or shutting the No. 1 engine down.

The slowly accelerating M-50 collided with a 3ME 'Bison' bomber on the apron, striking its cockpit with the right wing and causing severe damage. Luckily the impact dislodged the protective cover from the M-50's cockpit, which was sucked into the running engine, causing it to surge and stop. The aircraft, however, continued to move under inertia and struck the corner of a hangar with the left wing before coming to a halt. Unfortunately, mechanic-radio operator A.F. Kruchinkin, in a vain attempt to stop the runaway aircraft, was run over by one of the underwing supports and subsequently died of his injuries. In addition, engineers V.V. Koliupanov and E.G. Alkhimenkov, in the cockpit of the 'Bison', also suffered injuries.

Both aircraft were badly damaged. The M-50's leading edges of the wing were crumpled, pylons damaged, engine gondolas destroyed, the frame of the entrance hatch torn away and the skin damaged in several places. The damage to the 'Bison' was severe enough that the decision was made to scrap the aircraft. However, the M-50 was repaired within two months.

In order to continue testing in a wider range of speeds, the decision had been made to install two VD-7MA afterburning turbojets. The flight test programme was halted at the end of the initial test phase pending the delivery of the delayed VD-7MA engines from OKB-36. In the

During both rehearsals for and participation in the Tushino flypast, the M-50A was closely accompanied by a pair of early production MiG-21F-13 ('Fishbed-Cs'), each carrying a pair of K-13 (R-3 'Atoll') air-to-air missiles on underwing pylons.

meantime, system adjustments and preparations for the second stage were implemented, these being completed in late March 1961.

During the lull, fatigue tests were performed on several of the most critical units including hoists (lifts), steering drives, hydraulic pumps, turbo-refrigerators etc.

At last, in April 1961, the two inner VD-7A engines were replaced with 157-kN (35,274-lb) VD-7MAs. According to calculations the M-50A's maximum speed would now be Mach 1.35.

'Bounder' finale

Testing with the new engines was, however, limited to a few short test flights and preparation flights for the 1961 flypast at Tushino. In this configuration, the M-50A made only eight flights (including the parade), and the true performance of the aircraft was never tested as afterburner was only ever used on take-off.

In total, the M-50 made 19 flights before the programme was closed. OKB-23 was shut down, and the government lost all interest in the M-50 and M-52 development aircraft, the latter of which was then complete and, apart from the installation of the engines, ready for flight.

With the remainder of the flight tests cancelled, the opportunity for aviation OKBs, TsAGI, LII and other scientific research institutes to gather valuable data on aerodynamics, stability, flight dynamics etc. of supersonic bomber designs was lost. After spending some years in a neglected state at Zhukovskiy, the M-50A was placed in the Air Force Museum at Monino, where it still resides today.

Boris Puntus and Konstantin Udalov
Translated by Rauf Eylanbekov

Index

Picture acknowledgments

Front cover: Dassault, Ted Carlson/Fotodynamics, US Air Force. **4:** BAE Systems, Lockheed Martin, Boeing. **5:** via Tom Kaminski, Boeing. **6:** USAF, Peter R. Foster, Bell. **7:** Shlomo Aloni (two), Carlo Kuit/Paul Kievit. **8:** NASA (two), USAF. **9:** Carlo Kuit/Paul Kievit (five). **10:** via Tom Kaminski, Peter R. Foster (two). **11:** Peter R. Foster (two), Carlo Kuit/Paul Kievit. **12:** Lockheed Martin, NH Industries. **13:** Carlo Kuit/Paul Kievit, Peter R. Foster (two). **14:** Peter R. Foster, via Tom Kaminski. **15:** USAF (two), via Tom Kaminski, US Navy. **16-29:** US Air Force, US Navy, US Marine Corps, US DoD, US Central Command, UK MoD. **30:** Shlomo Aloni (six). **31:** Shlomo Aloni, USAF (three). **32-43:** Ted Carlson/Fotodynamics. **44-49:** Heinz Berger and Erich Strobl/Skyhawk. **50-51:** François Robineau/Dassault-Aviaplans (DA) via Henri-Pierre Grolleau (HPG), Henri-Pierre Grolleau (two). **52:** Frédéric Lert, François Robineau/DA. **53:** Henri-Pierre Grolleau, François Robineau/DA, ECPA. **54:** François Robineau/DA via HPG. **55:** David Donald (three), J.P. Thiery/Dassault via HPG, François Robineau/DA via HPG. **56:** Frédéric Lert (two), François Robineau/DA via HPG. **57:** François Robineau/DA via HPG, J.P. Thiery/Dassault via HPG. **58:** Frédéric Lert, Henri-Pierre Grolleau (two). **59:** François Robineau/DA via HPG, Henri-Pierre Grolleau (two). **60:** Dassault, David Donald, Dassault. **61:** François Robineau/DA via HPG, Thales via HPG. **62:** Katsuhiko Tokunaga/Dassault, Thales via HPG, HPG. **63:** Dassault via HPG, François Robineau/DA via HPG. **64:** MBDA/DGA, GIAT via HPG, Henri-Pierre Grolleau, MBDA. **65:** Henri-Pierre Grolleau (three), François Robineau/DA via HPG, Dassault Istres via HPG. **66:** François Robineau/DA via HPG, Henri-Pierre Grolleau (two). **67:** François Robineau/DA via HPG, Henri-Pierre Grolleau (four). **71:** David Donald, Henri-Pierre Grolleau (four). **72:** Frédéric Lert, François Robineau/DA via HPG. **73:** Henri-Pierre Grolleau (six), Dassault, Antoine Gonin via HPG. **74:** Frédéric Lert (two), François Robineau/DA via HPG. **75:** Henri-Pierre Grolleau, Martin-Baker, CEL Biscarosse via Martin-Baker. **76:** François Robineau/DA, Henri-Pierre Grolleau (two). **77:** François Robineau/DA via HPG (two). **78:** Henri-Pierre Grolleau, François Robineau/DA via HPG (two). **79:** Burdin/Nicolas Masini via HPG, François Robineau/DA via HPG. **80:** Jelle Sjoerdsma, François Robineau/DA via HPG. **81:** François Robineau/DA via HPG (two), Thales via HPG. **83:** François Robineau/DA, François Robineau/DA via EADS. **83:** Henri-Pierre Grolleau (two). **84:** François Robineau/DA via HPG, Henri-Pierre Grolleau, François Robineau/DA via Matra. **85:** Frédéric Lert (two), Henri-Pierre Grolleau. **86:** Greg L. Davis. **87:** USAF (five), Greg L. Davis (two), Ted Carlson/Fotodynamics. **89:** Ted Carlson/Fotodynamics (four), Greg L. Davis (three), USAF (three). **91:** Greg L. Davis, Ted Carlson/Fotodynamics (four), USAF (four). **92:** Ted Carlson/Fotodynamics (three). **93:** Ted Carlson/Fotodynamics (seven), Greg L. Davis (three), USAF. **94-97:** Luigino Caliaro. **98:** Piotr Butowski (two), Beriev. **99:** Beriev, Zinchuk. **100:** Piotr Butowski. **101:** Beriev (three), Hugo Mambour. **102:** Zinchuk (three), Piotr Butowski, Hugo Mambour, Artemyev Archive, Beriev. **104:** Beriev (two), Hugo Mambour. **105:** Beriev (four). **106:** Aerospace, Beriev (two), Artemyev Archive. **107:** Beriev (four). **108:** Zinchuk (two), Flygvapnet, Beriev. **109:** Zinchuk (two), Beriev. **110:** Beriev (four). **111:** Beriev (two), Piotr Butowski, Gordon Upton. **112:** USAF via Larry Davis (LD) (two), via Les King via LD, Major Tom Fields via LD. **113:** USAF via LD (two), Jack Morris via LD, Duke McEntee via LD. **114:** Dave Menard via LD (two), Al Adcock via LD, LD Collection. **115:** LD Collection (three), USAF via LD. **116:** USAF via LD (three), C.A. Shaw via LD, LD Collection. **117:** USAF via LD (three), LD Collection. **118:** USAF via LD (two), Steve Alexander via LD. **119:** Major Tom Fields via LD, USAF via LD, Tom Hansen via LD. **120:** Dave Menard via LD, USAF via LD, Duke McEntee via LD. **121:** Duke McEntee via LD, LD Collection, Robert Mikesh via LD. **122:** USAF via LD, Nick Williams via LD. **123:** USAF via LD, Cullen Cline via LD, William Platt via LD (two). **124:** J. Ward Boyce via LD (two), USAF via LD, LD Collection (two). **125:** USAF via LD, LD Collection, Bob McGarry via LD. **126-127:** via Stan Piet. **128:** Consolidated (three). **129:** NASM via Stan Piet, Aerospace, via Martin Bowman. **130:** IWM (two), via Martin Bowman. **131:** NASM via Stan Piet (two), Aerospace. **132:** via Martin Bowman (two), via Philip Jarrett, IWM. **133:** IWM, British Airways, via Martin Bowman, via Harry Gann. **134:** via Stan Piet, via Harry Gann, IWM. **135:** Aerospace, USAF, USAAF via Harry Gann. **136:** via Martin Bowman, Aerospace, Douglas via Harry Gann. **137:** via Martin Bowman, via Harry Gann, Convair, USAF. **143:** USAF. **144:** USAF (two), IWM. **145:** Aerospace (three), USAF (two), via Martin Bowman (two). **146:** Convair (two), Aerospace. **147:** Convair, Aerospace. **148:** Aerospace (two), US National Archives via Martin Bowman, Consolidated (two). **149:** via Martin Bowman, via Stan Piet, Consolidated (two). **150:** via Martin Bowman. **151:** via Martin Bowman, USAF via Harry Gann. **152:** USAF (two), Australian War Memorial. **153:** USAF (two), US Navy via IWM, Aerospace. **154:** via Martin Bowman (two). **155:** Aerospace, via Martin Bowman (two). **156:** IWM, via Harry Gann, Aerospace, via Martin Bowman. **157:** IWM, via Harry Gann, Aerospace. **158:** via Martin Bowman (two), USAF, NASM. **159:** Aerospace (two), via Martin Bowman. **160:** IWM, A&AEE, MAP. **161:** John Meyer via Stan Piet (two), Aerospace, via Martin Bowman. **162:** USMC via Harry Gann, US Navy, via Martin Bowman (two). **163:** US Navy via Stan Piet (two). **164:** IWM, Institute of Maritime History/Royal Netherlands Navy (IMHRNN) via Eric Katerberg, Audiovisuele Dienst Koninklijke via Eric Katerberg. **165:** IWM, Aerospace, IMHRNN via Eric Katerberg. **166-168:** IMHRNN via Eric Katerberg. **169:** IMHRNN via Eric Katerberg (two), Henk Eldering via Eric Katerberg. **170:** Henk Eldering via Eric Katerberg, Eric Katerberg. **171:** Audiovisuele Dienst Koninklijke via Eric Katerberg, Eric Katerberg (two). **172-177:** Avico Press. **178:** Piotr Butowski, Avico Press. **179:** Avico Press (two), Aerospace. **180:** Aerospace (two). **181:** Aerospace, Piotr Butowski.